Brownian Motion Calculus

For other titles in the Wiley Finance series
please see www.wiley.com/finance

Brownian Motion Calculus

Ubbo F Wiersema

John Wiley & Sons, Ltd

Copyright © 2008   John Wiley & Sons Ltd, The Atrium, Southern Gate, Chichester,
West Sussex PO19 8SQ, England

Telephone   (+44) 1243 779777

Email (for orders and customer service enquiries): cs-books@wiley.co.uk
Visit our Home Page on www.wiley.com

Other Wiley Editorial Offices

John Wiley & Sons Inc., 111 River Street, Hoboken, NJ 07030, USA

Jossey-Bass, 989 Market Street, San Francisco, CA 94103-1741, USA

Wiley-VCH Verlag GmbH, Boschstr. 12, D-69469 Weinheim, Germany

John Wiley & Sons Australia Ltd, 42 McDougall Street, Milton, Queensland 4064, Australia

John Wiley & Sons (Asia) Pte Ltd, 2 Clementi Loop #02-01, Jin Xing Distripark, Singapore 129809

John Wiley & Sons Canada Ltd, 6045 Freemont Blvd, Mississauga, ONT, L5R 4J3, Canada

Wiley also publishes its books in a variety of electronic formats. Some content that appears in print
may not be available in electronic books.

Library of Congress Cataloging-in-Publication Data

Wiersema, Ubbo F.
  Brownian motion calculus / Ubbo F Wiersema.
    p. cm. – (Wiley finance series)
  Includes bibliographical references and index.
  ISBN 978-0-470-02170-5 (pbk. : alk. paper)
  1. Finance–Mathematical models.   2. Brownian motion process.   I. Title.
  HG106.W54 2008
  332.64'2701519233–dc22
                                                    2008007641

British Library Cataloguing in Publication Data

A catalogue record for this book is available from the British Library

ISBN 978-0-470-02170-5 (PB)

Typeset in 11/13pt Times by Aptara Inc., New Delhi, India
Printed and bound in Great Britain by Antony Rowe Ltd, Chippenham, Wiltshire

Models are, for the most part, caricatures of reality,
but if they are good, like good caricatures,
they portray, though perhaps in a disturbed manner,
some features of the real world.

Marc Kač

voor

Margreet

Contents

ANNEXES

Preface

This is a text which presents the basics of stochastic calculus in an elementary fashion with plenty of practice. It is aimed at the typical student who wants to learn quickly about the use of these tools in financial engineering, particularly option valuation, and who in the first instance can accept (and usually prefers) certain propositions without the full mathematical proofs and technical conditions. Elementary ordinary calculus has been successfully taught along these lines for decades. Concurrent numerical experimentation, using Excel/VBA and Mathematica, forms an integral part of the learning. Useful side readings are given with each topic. Annexes provide background and elaborate some more technical aspects. The technical prerequisites are elementary probability theory and basic ordinary calculus.

OUTLINE

The sequence of chapters in this text is best explained by working backwards from the ultimate use of Brownian motion calculus, namely the valuation of an option. An option is a financial contract that produces a random payoff, which is non-negative, at some contractually specified date. Because of the absence of downside risk, options are widely used for risk management and investment. The key task is to determine what it should cost to buy an option prior to the payoff date. What makes the payoff uncertain is the value of the so-called underlying asset of the option on the date of payoff. In the standard case the underlying asset is a share of stock. If T denotes the payoff date, and $S(T)$ the value of the stock at T, then a standard European call option allows its holder to acquire the stock, of value $S(T)$, for a fixed payment of K that is specified in the contract. The European call has payoff $\max[S(T) - K, 0]$ which is positive if $S(T) > K$ and zero otherwise. So the behaviour of the stock price needs to be modelled. In the standard case it is assumed to be driven by a random process called Brownian motion, denoted B. The basic model for the stock price is as follows. Over a small time step dt there is a random increment $dB(t)$ and this affects the rate of return on the stock price by a scaling factor σ. In addition there is a regular growth component $\mu\, dt$. If at time t the stock has the known value $S(t)$, then the resulting change in stock price dS is specified by

$$\frac{dS(t)}{S(t)} = \mu\, dt + \sigma\, dB(t)$$

The change in stock price over a future period of length T, from its initial price $S(0)$, is then the sum (integral) of the changes over all time steps dt

$$\int_{t=0}^{T} dS(t) = S(T) - S(0) = \int_{t=0}^{T} \mu S(t)\, dt + \int_{t=0}^{T} \sigma S(t)\, dB(t)$$

This sets the agenda for what needs to be worked out mathematically. First the Brownian motion process needs to be specified. This is the subject of Chapter 1. Then an integral of the form $\int_{t=0}^{T} \sigma S(t)\, dB(t)$ needs to be defined; that requires a new concept of integration which is introduced in Chapter 3; the other term, $\int_{t=0}^{T} \mu S(t)\, dt$, can be handled by ordinary integration. Then the value of $S(T)$ needs to be obtained from the above equation. That requires stochastic calculus rules which are set out in Chapter 4, and methods for solving stochastic differential equations which are described in Chapter 5. Once all that is in place, a method for the valuation of an option needs to be devised. Two methods are presented. One is based on the concept of a martingale which is introduced in Chapter 2. Chapter 7 elaborates on the methodology for the change of probability that is used in one of the option valuation methods. The final chapter discusses how computations can be made more convenient by the suitable choice of the so-called numeraire.

The focus of this text is on Brownian motion in one dimension, and the time horizon is always taken as finite. Other underlying random processes, such as jump processes and Lévy processes, are outside the scope of this text.

The references have been selected with great care, to suit a variety of backgrounds, desire and capacity for rigour, and interest in application.

They should serve as a companion. After an initial acquaintance with the material of this text, an excellent way to gain a further understanding is to explore how specific topics are explained in the references. In view of the perceived audience, several well-known texts that are mathematically more demanding have not been included. In the interest of readability, this text uses the Blackwood Bold font for probability operations; a probability distribution function is denoted as $\mathbb{P}$, an expectation as $\mathbb{E}$, a variance as $\mathbb{V}$ar, a standard deviation as $\mathbb{S}$tdev, a covariance as $\mathbb{C}$ov, and a correlation as $\mathbb{C}$orr.

1
Brownian Motion

The exposition of Brownian motion is in two parts. Chapter 1 introduces the properties of Brownian motion as a random process, that is, the true technical features of Brownian motion which gave rise to the theory of stochastic integration and stochastic calculus. Annex A presents a number of useful computations with Brownian motion which require no more than its probability distribution, and can be analysed by standard elementary probability techniques.

1.1 ORIGINS

In the summer of 1827 Robert Brown, a Scottish medic turned botanist, microscopically observed minute pollen of plants suspended in a fluid and noticed increments[1] that were highly irregular. It was found that finer particles moved more rapidly, and that the motion is stimulated by heat and by a decrease in the viscosity of the liquid. His investigations were published as *A Brief Account of Microscopical Observations Made in the Months of June, July and August 1827*. Later that century it was postulated that the irregular motion is caused by a very large number of collisions between the pollen and the molecules of the liquid (which are microscopically small relative to the pollen). The hits are assumed to occur very frequently in any small interval of time, independently of each other; the effect of a particular hit is thought to be small compared to the total effect. Around 1900 Louis Bachelier, a doctoral student in mathematics at the Sorbonne, was studying the behaviour of stock prices on the Bourse in Paris and observed highly irregular increments. He developed the first mathematical specification of the increment reported by Brown, and used it as a model for the increment of stock prices. In the 1920s Norbert Wiener, a mathematical physicist at MIT, developed the fully rigorous probabilistic framework for this model. This kind of increment is now called a Brownian motion or a Wiener process. The position of the process is commonly denoted

---

[1] This is meant in the mathematical sense, in that it can be positive or negative.

by B or W. Brownian motion is widely used to model randomness in economics and in the physical sciences. It is central in modelling financial options.

1.2 BROWNIAN MOTION SPECIFICATION

The physical experiments suggested that:

- the increment is continuous
- the increments of a particle over disjoint time intervals are independent of one another
- each increment is assumed to be caused by independent bombardments of a large number of molecules; by the Central Limit Theorem of probability theory the sum of a large number of independent identically distributed random variables is approximately normal, so each increment is assumed to have a normal probability distribution
- the mean increment is zero as there is no preferred direction
- as the position of a particle spreads out with time, it is assumed that the variance of the increment is proportional to the length of time the Brownian motion has been observed.

Mathematically, the random process called Brownian motion, and denoted here as $B(t)$, is defined for times $t \geq 0$ as follows. With time on the horizontal axis, and $B(t)$ on the vertical axis, at each time t, $B(t)$ is the position, in one dimension, of a physical particle. It is a random variable. The collection of these random variables indexed by the continuous time parameter t is a *random process* with the following properties:

(a) The increment is continuous; when recording starts, time and position are set at zero, $B(0) = 0$
(b) Increments over non-overlapping time intervals are independent random variables
(c) The increment over any time interval of length u, from any time t to time $(t + u)$, has a normal probability distribution with mean zero and variance equal to the length of this time interval.

As the probability density of a normally distributed random variable with mean μ and variance σ^2 is given by

$$\frac{1}{\sigma\sqrt{2\pi}}\exp\left[-\frac{1}{2}\left(\frac{x-\mu}{\sigma}\right)^2\right]$$

the probability *density* of the position of a Brownian motion at the end of time period $[0, t]$ is obtained by substituting $\mu = 0$ and $\sigma = \sqrt{t}$, giving

$$\frac{1}{\sqrt{t}\sqrt{2\pi}} \exp\left[-\frac{1}{2}\left(\frac{x}{\sqrt{t}}\right)^2\right]$$

where x denotes the value of random variable $B(t)$. The probability *distribution* of the increment $B(t + u) - B(t)$ is

$$\mathbb{P}[B(t+u) - B(t) \leq a] = \int_{x=-\infty}^{a} \frac{1}{\sqrt{u}\sqrt{2\pi}} \exp\left[-\frac{1}{2}\left(\frac{x}{\sqrt{u}}\right)^2\right] dx$$

Note that the starting time of the interval does not figure in the expression for the probability distribution of the increment. The probability distribution depends only on the time spacing; it is the same for all time intervals that have the same length. As the standard deviation at time t is $\sqrt{t}$, the longer the process has been running, the more spread out is the density, as illustrated in Figure 1.1.

As a reminder of the randomness, one could include the state of nature, denoted ω, in the notation of Brownian motion, which would then be $B(t, \omega)$, but this is not commonly done. For each fixed time t^*, $B(t^*, \omega)$ is a function of ω, and thus a random *variable*. For a particular ω^* over the time period $[0, t]$, $B(t, \omega^*)$ is a function of t which is known as a *sample path* or trajectory. In the technical literature this is often denoted as $t \longmapsto B(t)$. On the left is an element from the domain, on the right the corresponding function value in the range. This is as in ordinary calculus where an expression like $f(x) = x^2$ is nowadays often written as $x \longmapsto x^2$.

As the probability distribution of $B(t)$ is normal with standard deviation $\sqrt{\Delta t}$, it is the same as that of $\sqrt{\Delta t}\, Z$, where Z is a standard normal random variable. When evaluating the probability of an expression involving $B(t)$, it can be convenient to write $B(t)$ as $\sqrt{\Delta t}Z$.

The Brownian motion distribution is also written with the cumulative standard normal notation $N(\text{mean, variance})$ as $B(t + u) - B(t) \sim N(0, u)$, or for any two times $t_2 > t_1$ as $B(t_2) - B(t_1) \sim N(0, t_2 - t_1)$. As $\mathbb{V}\text{ar}[B(t)] = \mathbb{E}[B(t)^2] - \{\mathbb{E}[B(t)]\}^2 = t$, and $\mathbb{E}[B(t)] = 0$, the second moment of Brownian motion is $\mathbb{E}[B(t)^2] = t$. Over a time step Δt, where $\Delta B(t) \stackrel{\text{def}}{=} B(t + \Delta t) - B(t)$, $\mathbb{E}\{[\Delta B(t)]^2\} = \Delta t$. A normally distributed random variable is also known as a Gaussian random variable, after the German mathematician Gauss.

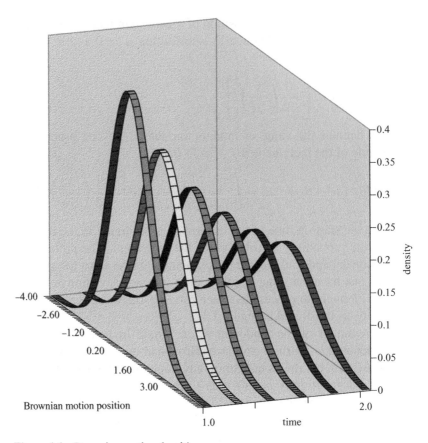

Figure 1.1   Brownian motion densities

1.3   USE OF BROWNIAN MOTION IN STOCK PRICE DYNAMICS

Brownian motion arises in the modelling of the evolution of a stock price (often called the stock price dynamics) in the following way. Let Δt be a time interval, $S(t)$ and $S(t + \Delta t)$ the stock prices at current time t and future time $(t + \Delta t)$, and $\Delta B(t)$ the Brownian motion increment over Δt. A widely adopted model for the stock price dynamics, in a discrete time setting, is

$$\frac{S(t + \Delta t) - S(t)}{S(t)} = \mu \, \Delta t + \sigma \, \Delta B(t)$$

where μ and σ are constants. This is a stochastic *difference* equation which says that the change in stock price, relative to its current value at time t, $[S(t + \Delta t) - S(t)]/S(t)$, grows at a non-random rate of μ per unit of time, and that there is also a random change which is proportional to the increment of a Brownian motion over Δt, with proportionality parameter σ. It models the *rate of return* on the stock, and evolved from the first model for stock price dynamics postulated by Bachelier in 1900, which had the change in the stock price itself proportional to a Brownian motion increment, as

$$\Delta S(t) = \sigma \, \Delta B(t)$$

As Brownian motion can assume negative values it implied that there is a probability for the stock price to become negative. However, the limited liability of shareholders rules this out. When little time has elapsed, the standard deviation of the probability density of Brownian motion, $\sqrt{t}$, is small, and the probability of going negative is very small. But as time progresses the standard deviation increases, the density spreads out, and that probability is no longer negligible. Half a century later, when research in stock price modelling began to take momentum, it was judged that it is not the level of the stock price that matters to investors, but the rate of return on a given investment in stocks.

In a continuous time setting the above discrete time model becomes the stochastic *differential* equation

$$\frac{dS(t)}{S(t)} = \mu \, dt + \sigma \, dB(t)$$

or equivalently $dS(t) = \mu \, S(t) \, dt + \sigma S(t) \, dB(t)$, which is discussed in Chapter 5. It is shown there that the stock price process $S(t)$ which satisfies this stochastic differential equation is

$$S(t) = S(0) \exp[(\mu - \tfrac{1}{2}\sigma^2)t + \sigma B(t)]$$

which cannot become negative. Writing this as

$$S(t) = S(0) \exp(\mu t) \exp[\sigma B(t) - \tfrac{1}{2}\sigma^2 t]$$

gives a decomposition into the non-random term $S(0) \exp(\mu t)$ and the random term $\exp[\sigma B(t) - \tfrac{1}{2}\sigma^2 t]$. The term $S(0) \exp(\mu t)$ is $S(0)$ growing at the continuously compounded constant rate of μ per unit of time, like a savings account. The random term has an expected value of 1. Thus the expected value of the stock price at time t, given $S(0)$, equals $S(0) \exp(\mu t)$. The random process $\exp[\sigma B(t) - \frac{1}{2}\sigma^2 t]$ is an example of a martingale, a concept which is the subject of Chapter 2.

1.4 CONSTRUCTION OF BROWNIAN MOTION FROM A SYMMETRIC RANDOM WALK

Up to here the reader may feel comfortable with most of the mathematical specification of Brownian motion, but wonder why the variance is proportional to time. That will now be clarified by constructing Brownian motion as the so-called *limit in distribution* of a symmetric random walk, illustrated by computer simulation. Take the time period $[0, T]$ and partition it into n intervals of equal length $\Delta t \overset{\text{def}}{=} T/n$. These intervals have endpoints $t_k \overset{\text{def}}{=} k \Delta t, k = 0, \ldots, n$. Now consider a particle which moves along in time as follows. It starts at time 0 with value 0, and moves up or down at each discrete time point with equal probability. The magnitude of the increment is specified as $\sqrt{\Delta t}$. The reason for this choice will be made clear shortly. It is assumed that successive increments are *independent* of one another. This process is known as a symmetric (because of the equal probabilities) random walk. At time-point 1 it is either at level $\sqrt{\Delta t}$ or at level $-\sqrt{\Delta t}$. If at time-point 1 it is at $\sqrt{\Delta t}$, then at time-point 2 it is either at level $\sqrt{\Delta t} + \sqrt{\Delta t} = 2\sqrt{\Delta t}$ or at level $\sqrt{\Delta t} - \sqrt{\Delta t} = 0$. Similarly, if at time-point 1 it is at level $-\sqrt{\Delta t}$, then at time-point 2 it is either at level 0 or at level $-2\sqrt{\Delta t}$, and so on. Connecting these positions by straight lines gives a continuous path. The position at any time between the discrete time points is obtained by linear interpolation between the two adjacent discrete time positions. The complete picture of all possible discrete time positions is given by the nodes in a so-called binomial tree, illustrated in Figure 1.2 for $n = 6$. At time-point n, the node which is at the end of a path that has j up-movements is labelled (n, j), which is very convenient for doing tree arithmetic.

When there are n intervals, there are $(n + 1)$ terminal nodes at time T, labelled $(n, 0)$ to (n, n), and a total of 2^n different paths to these terminal nodes. The number of paths ending at node (n, j) is given by a Pascal triangle. This has the same shape as the binomial tree. The upper and lower edge each have one path at each node. The number of paths going to any intermediate node is the sum of the number of paths going to the preceding nodes. This is shown in Figure 1.3. These numbers are the binomial coefficients from elementary probability theory.

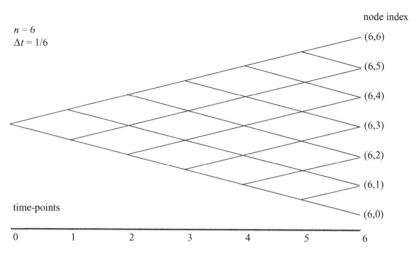

Figure 1.2   Symmetric binomial tree

Each path has a probability $(\frac{1}{2})^n$ of being realized. The total probability of terminating at a particular node equals the number of different paths to that node, times $(\frac{1}{2})^n$. For $n = 6$ these are shown on the Pascal triangle in Figure 1.2. It is a classical result in probability theory that as n goes to infinity, the terminal probability distribution of the symmetric

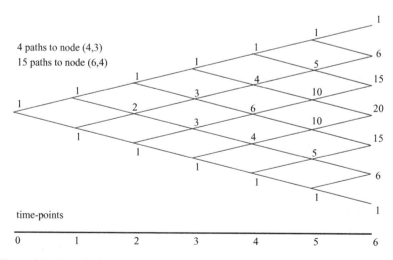

Figure 1.3   Pascal triangle

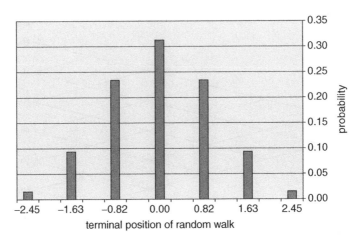

Figure 1.4 Terminal probabilities

random walk tends to that of a normal distribution. The picture of the terminal probabilities for the case $n = 6$ is shown in Figure 1.4.

Let the increment of the position of the random walk from time-point t_k to t_{k+1} be denoted by discrete two-valued random variable X_k. This has an expected value of

$$\mathbb{E}[X_k] = \tfrac{1}{2}\sqrt{\Delta t} + \tfrac{1}{2}(-\sqrt{\Delta t}) = 0$$

and variance

$$\mathbb{V}\text{ar}[X_k] = \mathbb{E}[X_k^2] - \{\mathbb{E}[X_k]\}^2$$
$$= \mathbb{E}[X_k^2] = \tfrac{1}{2}(\sqrt{\Delta t})^2 + \tfrac{1}{2}(-\sqrt{\Delta t})^2 = \Delta t$$

The position of the particle at terminal time T is the sum of n independent identically distributed random variables X_k, $S_n \overset{\text{def}}{=} X_1 + X_2 + \cdots + X_n$. The expected terminal position of the path is

$$\mathbb{E}[S_n] = \mathbb{E}[X_1 + X_2 + \cdots + X_n]$$
$$= \mathbb{E}[X_1] + \mathbb{E}[X_2] + \cdots + \mathbb{E}[X_n] = n0 = 0$$

Its variance is

$$\mathbb{V}\text{ar}[S_n] = \mathbb{V}\text{ar}\left[\sum_{k=1}^{n} X_k\right]$$

As the X_k are independent this can be written as the sum of the variances $\sum_{k=1}^{n} \mathbb{V}\text{ar}[X_k]$, and as the X_k are identically distributed they have the same variance Δt, so

$$\mathbb{V}\text{ar}[S_n] = n\Delta t = n(T/n) = T$$

For larger n, the random walk varies more frequently, but the magnitude of the increment $\sqrt{\Delta t} = \sqrt{T/n}$ gets smaller and smaller. The graph of the probability distribution of

$$Z_n \stackrel{\text{def}}{=} \frac{S_n - \mathbb{E}[S_n]}{\sqrt{\mathbb{V}\text{ar}[S_n]}} = \frac{S_n}{\sqrt{T}}$$

looks more and more like that of the standard normal probability distribution.

Limiting Distribution   The probability distribution of S_n is determined uniquely by its moment generating function.[2] This is $\mathbb{E}[\exp(\theta S_n)]$, which is a function of θ, and will be denoted $m(\theta)$.

$$m(\theta) \stackrel{\text{def}}{=} \mathbb{E}[\exp(\theta\{X_1 + \cdots + X_k + \cdots + X_n\})]$$
$$= \mathbb{E}[\exp(\theta X_1) \cdots \exp(\theta X_k) \cdots \exp(\theta X_n)]$$

As the random variables $X_1, \ldots, X_n$ are independent, the random variables $\exp(\theta X_1), \ldots, \exp(\theta X_n)$ are also independent, so the expected value of the product can be written as the product of the expected values of the individual terms

$$m(\theta) = \prod_{k=1}^{n} \mathbb{E}[\exp(\theta X_k)]$$

As the X_k are identically distributed, all $\mathbb{E}[\exp(\theta X_k)]$ are the same, so

$$m(\theta) = \{\mathbb{E}[\exp(\theta X_k)]\}^n$$

As X_k is a discrete random variable which can take the values $\sqrt{\Delta t}$ and $-\sqrt{\Delta t}$, each with probability $\frac{1}{2}$, it follows that $\mathbb{E}[\exp(\theta X_k)] = \frac{1}{2}\exp(\theta\sqrt{\Delta t}) + \frac{1}{2}\exp(-\theta\sqrt{\Delta t})$. For small Δt, using the power series expansion of exp and neglecting terms of order higher than Δt, this can be approximated by

$$\tfrac{1}{2}(1 + \theta\sqrt{\Delta t} + \tfrac{1}{2}\theta^2 \Delta t) + \tfrac{1}{2}(1 - \theta\sqrt{\Delta t} + \tfrac{1}{2}\theta^2 \Delta t) = 1 + \tfrac{1}{2}\theta^2 \Delta t$$

so

$$m(\theta) \approx (1 + \tfrac{1}{2}\theta^2 \Delta t)^n$$

---

[2] See Annex A, *Computations with Brownian motion.*

As $n \to \infty$, the probability distribution of S_n converges to the one determined by the limit of the moment generating function. To determine the limit of m as $n \to \infty$, it is convenient to change to ln.

$$\ln[m(\theta)] \approx n \ln(1 + \tfrac{1}{2}\theta^2 \Delta t)$$

Using the property, $\ln(1 + y) \approx y$ for small y, gives

$$\ln[m(\theta)] \approx n \tfrac{1}{2}\theta^2 \Delta t$$

and as $\Delta t = T/n$

$$m(\theta) \approx \exp(\tfrac{1}{2}\theta^2 T)$$

This is the moment generating function of a random variable, Z say, which has a normal distribution with mean 0 and variance T, as can be readily checked by using the well-known formula for $\mathbb{E}[\exp(\theta Z)]$. Thus in the continuous-time limit of the discrete-time framework, the probability density of the terminal position is

$$\frac{1}{\sqrt{T}\sqrt{2\pi}} \exp\left[-\frac{1}{2}\left(\frac{x}{\sqrt{T}}\right)^2\right]$$

which is the same as that of a Brownian motion that has run an amount of time T. The probability distribution of $S_n = \sqrt{T}Z_n$ is then normal with mean 0 and variance T.

The full proof of the convergence of the symmetric random walk to Brownian motion requires more than what was shown. Donsker's theorem from advanced probability theory is required, but that is outside the scope of this text; it is covered in *Korn/Korn* Excursion 7, and in *Capasso/Bakstein* Appendix B. The construction of Brownian motion as the limit of a symmetric random walk has the merit of being intuitive. See also *Kuo* Section 1.2, and *Shreve II* Chapter 3. There are several other constructions of Brownian motion in the literature, and they are mathematically demanding; see, for example, *Kuo* Chapter 3. The most accessible is *Lèvy's interpolation method*, which is described in *Kuo* Section 3.4.

Size of Increment   Why the size of the random walk increment was specified as $\sqrt{\Delta t}$ will now be explained. Let the increment over time-step Δt be denoted y. So $X_k = y$ or $-y$, each with probability $\frac{1}{2}$, and

$$\mathbb{E}[X_k] = \tfrac{1}{2}y + \tfrac{1}{2}(-y) = 0$$

$$\mathbb{V}\mathrm{ar}[X_k] = \mathbb{E}[X_k^2] - \{\mathbb{E}[X_k]\}^2 = \tfrac{1}{2}y^2 + \tfrac{1}{2}(-y)^2 - 0^2 = y^2$$

Then $\mathbb{V}\mathrm{ar}[S_n] = n\mathbb{V}\mathrm{ar}[X_k]$ as the successive X_k are independent

$$\mathbb{V}\mathrm{ar}[S_n] = ny^2 = \frac{T}{\Delta t}y^2 = T\frac{y^2}{\Delta t}$$

Now let both $\Delta t \to 0$ and $y \to 0$, in such a way that $\mathbb{V}\mathrm{ar}[S_n]$ stays finite. This is achieved by choosing $\frac{y^2}{\Delta t} = c$, a positive constant, so $\mathbb{V}ar[S_n] = Tc$. As time units are arbitrary, there is no advantage in using a c value other than 1.

So if one observes Brown's experiment at equal time intervals, and models this as a symmetric random walk with increment y, then the continuous-time limit is what is called Brownian.

This motivates why Brownian motion has a variance equal to the elapsed time. Many books introduce the variance property of Brownian motion without any motivation.

Simulation of Symmetric Random Walk   To simulate the symmetric random walk, generate a succession of n random variables X with the above specified two-point probabilities and multiply these by $\pm\sqrt{\Delta t}$. The initial position of the walk is set at zero. Three random walks over the time period $[0, 1]$ are shown in Figure 1.5.

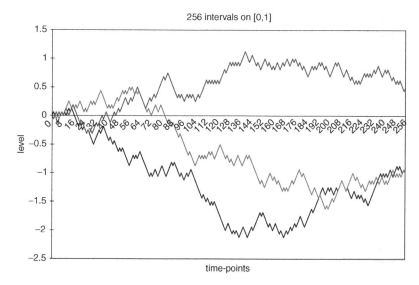

Figure 1.5   Simulated symmetric random walks

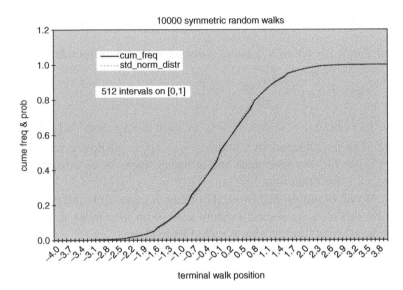

Figure 1.6   Simulated versus exact

For a batch of 100 simulated symmetric walks of 512 steps, the cumulative frequency of the terminal positions is shown in Figure 1.6, together with the limiting standard normal probability distribution.

The larger the number of simulations, the closer the cumulative frequency resembles the limiting distribution. For 10 000 simulated walks the difference is not graphically distinguishable. The simulation statistics for the position at time 1 are shown in Figure 1.7.

1.5   COVARIANCE OF BROWNIAN MOTION

A *Gaussian process* is a collection of normal random variables such that any finite number of them have a multivariate normal distribution. Thus Brownian motion increments are a Gaussian process. Consider the covariance between Brownian motion positions at any times s and t, where $s < t$. This is the expected value of the product of the deviations

	sample	exact
mean	0.000495	0
variance	1.024860	1

Figure 1.7   Simulation statistics

of these random variables from their respective means

$$\mathbb{C}\text{ov}[B(s), B(t)] = \mathbb{E}[\{B(s) - \mathbb{E}[B(s)]\}\{B(t) - \mathbb{E}[B(t)]\}]$$

As $\mathbb{E}[B(s)]$ and $\mathbb{E}[B(t)]$ are zero, $\mathbb{C}\text{ov}[B(s), B(t)] = \mathbb{E}[B(s)B(t)]$. Note that the corresponding time intervals $[0, s]$ and $[0, t]$ are overlapping. Express $B(t)$ as the sum of independent random variables $B(s)$ and the subsequent increment $\{B(t) - B(s)\}$, $B(t) = B(s) + \{B(t) - B(s)\}$. Then

$$\begin{aligned}\mathbb{E}[B(s)B(t)] &= \mathbb{E}[B(s)^2 + B(s)\{B(t) - B(s)\}] \\ &= \mathbb{E}[B(s)^2] + \mathbb{E}[B(s)\{B(t) - B(s)\}]\end{aligned}$$

Due to independence, the second term can be written as the product of $\mathbb{E}$s, and

$$\begin{aligned}\mathbb{E}[B(s)B(t)] &= \mathbb{E}[B(s)^2] + \mathbb{E}[B(s)]\mathbb{E}[B(t) - B(s)] \\ &= s + 0\,0 = s\end{aligned}$$

If the time notation was $t < s$ then $\mathbb{E}[B(s)B(t)] = t$. Generally for any times s and t

$$\mathbb{E}[B(s)B(t)] = \min(s, t)$$

For increments during any two non-overlapping time intervals $[t_1, t_2]$ and $[t_3, t_4]$, $\Delta B(t_1)$ is independent of $\Delta B(t_3)$, so the expected value of the product of the Brownian increments over these non-overlapping time intervals (Figure 1.8) equals the product of the expected values

$$\begin{aligned}\mathbb{E}[\{B(t_2) - B(t_1)\}\{B(t_4) - B(t_3)\}] \\ = \mathbb{E}[B(t_2) - B(t_1)]\mathbb{E}[B(t_4) - B(t_3)] = 0\,0 = 0\end{aligned}$$

whereas $\mathbb{E}[B(t_1)B(t_3)] = t_1 \neq \mathbb{E}[B(t_1)]\mathbb{E}[B(t_3)]$.

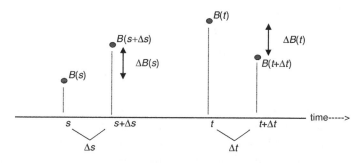

Figure 1.8 Non-overlapping time intervals

1.6   CORRELATED BROWNIAN MOTIONS

Let B and B^* be two *independent* Brownian motions. Let $-1 \leq \rho \leq 1$ be a given number. For $0 \leq t \leq T$ define a new process

$$Z(t) \stackrel{\text{def}}{=} \rho B(t) + \sqrt{1 - \rho^2} B^*(t)$$

At each t, this is a linear combination of independent normals, so $Z(t)$ is normally distributed. It will first be shown that Z is a Brownian motion by verifying its expected value and variance at time t, and the variance over an arbitrary time interval. It will then be shown that Z and B are correlated.

The expected value of $Z(t)$ is

$$\begin{aligned}
\mathbb{E}[Z(t)] &= \mathbb{E}[\rho B(t) + \sqrt{1 - \rho^2}\, B^*(t)] \\
&= \rho \mathbb{E}[B(t)] + \sqrt{1 - \rho^2}\, \mathbb{E}[B^*(t)] \\
&= \rho 0 + \sqrt{1 - \rho^2}\, 0 = 0
\end{aligned}$$

The variance of $Z(t)$ is

$$\begin{aligned}
\mathbb{V}\mathrm{ar}[Z(t)] &= \mathbb{V}\mathrm{ar}[\rho B(t) + \sqrt{1 - \rho^2}\, B^*(t)] \\
&= \mathbb{V}\mathrm{ar}[\rho B(t)] + \mathbb{V}\mathrm{ar}[\sqrt{1 - \rho^2}\, B^*(t)]
\end{aligned}$$

as the random variables $\rho B(t)$ and $\sqrt{1 - \rho^2}\, B^*(t)$ are independent. This can be written as

$$\rho^2 \mathbb{V}\mathrm{ar}[B(t)] + \left(\sqrt{1 - \rho^2}\right)^2 \mathbb{V}\mathrm{ar}[B^*(t)] = \rho^2 t + \left(1 - \rho^2\right) t = t$$

Now consider the increment

$$\begin{aligned}
Z(t + u) - Z(t) &= [\rho B(t + u) + \sqrt{1 - \rho^2}\, B^*(t + u)] \\
&\quad - [\rho B(t) + \sqrt{1 - \rho^2}\, B^*(t)] \\
&= \rho[B(t + u) - B(t)] \\
&\quad + \sqrt{1 - \rho^2}\, [B^*(t + u) - B^*(t)]
\end{aligned}$$

$B(t + u) - B(t)$ is the random increment of Brownian motion B over time interval u and $B^*(t + u) - B^*(t)$ is the random increment of Brownian motion B^* over time interval u. These two random quantities are independent, also if multiplied by constants, so the $\mathbb{V}$ar of the sum is the sum of Var

$$
\begin{aligned}
\text{Var}[Z(t+u) - Z(t)] &= \text{Var}\{\rho[B(t+u) - B(t)] \\
&\quad + \sqrt{1-\rho^2}\,[B^*(t+u) - B^*(t)]\} \\
&= \text{Var}\{\rho[B(t+u) - B(t)]\} \\
&\quad + \text{Var}\{\sqrt{1-\rho^2}\,[B^*(t+u) - B^*(t)]\} \\
&= \rho^2 u + \left(\sqrt{1-\rho^2}\right)^2 u = u
\end{aligned}
$$

This variance does not depend on the starting time t of the interval u, and equals the length of the interval. Hence Z has the properties of a Brownian motion. Note that since $B(t+u)$ and $B(t)$ are not independent

$$
\begin{aligned}
\text{Var}[B(t+u) - B(t)] &\neq \text{Var}[B(t+u)] + \text{Var}[B(t)] \\
&= t+u+t = 2t+u
\end{aligned}
$$

but

$$
\begin{aligned}
\text{Var}[B(t+u) - B(t)] &= \text{Var}[B(t+u)] + \text{Var}[B(t)] \\
&\quad - 2\text{Cov}[B(t+u), B(t)] \\
&= (t+u) + t - 2\min(t+u, t) \\
&= (t+u) + t - 2t = u
\end{aligned}
$$

Now analyze the correlation between the processes Z and B at time t. This is defined as the covariance between $Z(t)$ and $B(t)$ scaled by the product of the standard deviations of $Z(t)$ and $B(t)$:

$$
\text{Corr}[Z(t), B(t)] = \frac{\text{Cov}[Z(t), B(t)]}{\sqrt{\text{Var}[Z(t)]}\sqrt{\text{Var}[B(t)]}}
$$

The numerator evaluates to

$$
\begin{aligned}
\text{Cov}[Z(t), B(t)] &= \text{Cov}[\rho B(t) + \sqrt{1-\rho^2}\,B^*(t), B(t)] \\
&= \text{Cov}[\rho B(t), B(t)] + \text{Cov}[\sqrt{1-\rho^2}\,B^*(t), B(t)] \\
&\quad \text{due to independence} \\
&= \rho\,\text{Cov}[B(t), B(t)] + \sqrt{1-\rho^2}\,\text{Cov}[B^*(t), B(t)] \\
&= \rho\,\text{Var}[B(t), B(t)] + \sqrt{1-\rho^2}\,0 \\
&= \rho t
\end{aligned}
$$

Using the known standard deviations in the denominator gives

$$\mathbb{Corr}[Z(t), B(t)] = \frac{\rho t}{\sqrt{t}\sqrt{t}} = \rho$$

Brownian motions B and Z have correlation ρ at all times t. Thus if two correlated Brownian motions are needed, the first one can be B and the second one Z, constructed as above. Brownian motion B^* only serves as an intermediary in this construction.

1.7 SUCCESSIVE BROWNIAN MOTION INCREMENTS

The increments over non-overlapping time intervals are independent random variables. They all have a normal distribution, but because the time intervals are not necessarily of equal lengths, their variances differ. The joint probability distribution of the positions at times t_1 and t_2 is

$$\mathbb{P}[B(t_1) \leq a_1, B(t_2) \leq a_2]$$
$$= \int_{x_1=-\infty}^{a_1} \int_{x_2=-\infty}^{a_2} \frac{1}{\sqrt{t_1}\sqrt{2\pi}} \exp\left[-\frac{1}{2}\left(\frac{x_1-0}{\sqrt{t_1}}\right)^2\right]$$
$$\times \frac{1}{\sqrt{t_2-t_1}\sqrt{2\pi}} \exp\left[-\frac{1}{2}\left(\frac{x_2-x_1}{\sqrt{t_2-t_1}}\right)^2\right] dx_1\, dx_2$$

This expression is intuitive. The first increment is from position 0 to x_1, an increment of $(x_1 - 0)$ over time interval $(t_1 - 0)$. The second increment starts at x_1 and ends at x_2, an increment of $(x_2 - x_1)$ over time interval $(t_2 - t_1)$. Because of independence, the integrand in the above expression is the product of conditional probability densities. This generalizes to any number of intervals. Note the difference between the increment of the motion and the position of the motion. The increment over any time interval $[t_{k-1}, t_k]$ has a normal distribution with mean zero and variance equal to the interval length, $(t_k - t_{k-1})$. This distribution is not dependent on how the motion got to the starting position at time t_{k-1}. For a known position $B(t_{k-1}) = x$, the position of the motion at time t_k, $B(t_k)$, has a normal density with mean x and variance as above. While this distribution is not dependent on how the motion got to the starting position, it is dependent on the position of the starting point via its mean.

1.7.1 Numerical Illustration

A further understanding of the theoretical expressions is obtained by carrying out numerical computations. This was done in the mathematical software Mathematica. The probability density function of a increment was specified as

[f[u_,w_]: = (1/(Sqrt[u] * Sqrt[2 * Pi])) * Exp[−0.5 * (w/Sqrt[u]) ∧ 2]

A time interval of arbitrary length uNow = 2.3472 was specified. The expectation of the increment over this time interval, starting at time 1, was then specified as

NIntegrate[(x2−x1) * f[1, x1] * f[uNow, x2−x1],
{x1, −10, 10},{x2, −10, 10}]

Note that the joint density is multiplied by (x2−x1). The normal densities were integrated from −10 to 10, as this contains nearly all the probability mass under the two-dimensional density surface. The result was 0, in accordance with the theory. The variance of the increment over time interval uNow was computed as the expected value of the second moment

NIntegrate[((x2−x1)∧ 2) * f[1, x1] * f[uNow, x2−x1],
{x1, −10, 10},{x2, −10, 10}]

Note that the joint density is multiplied by (x2−x1) ∧ 2. The result was 2.3472, exactly equal to the length of the time interval, in accordance with the theory.

Example 1.7.1   This example (based on *Klebaner* example 3.1) gives the computation of $\mathbb{P}[B(1) \leq 0, B(2) \leq 0]$. It is the probability that both the position at time 1 and the position at time 2 are not positive. The position at all other times does not matter. This was specified in Mathematica as

NIntegrate[f[1, x1] * f[1, x2−x1],{x1, −10, 0},{x2, −10, 0}]

To visualize the joint density of the increment (Figure 1.9), a plot was specified as

Plot3D[f[1, x1] * f[1, x2−x1],{x1, −4, 4},{x2, −4, 4}]

The section of the probability density surface pertaining to this example is plotted in Figure 1.10.

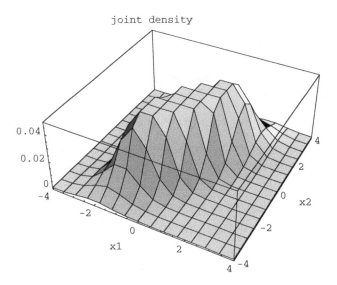

Figure 1.9   Joint density

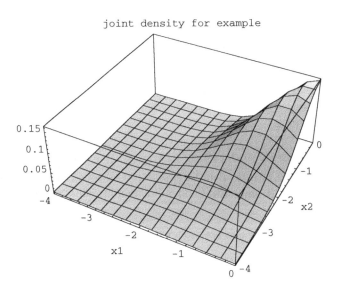

Figure 1.10   Joint density for example

The result of the numerical integration was 0.375, which agrees with Klebaner's answer derived analytically. It is the volume under the joint density surface shown below, for $x_1 \leq 0$ and $x_2 \leq 0$. $\mathbb{P}[B(1) \leq 0] = 0.5$ and $\mathbb{P}[B(2) \leq 0] = 0.5$. Multiplying these probabilities gives 0.25, but that is not the required probability because random variables $B(1)$ and $B(2)$ are not independent.

1.8 FEATURES OF A BROWNIAN MOTION PATH

The properties shown thus far are simply manipulations of a normal random variable, and anyone with a knowledge of elementary probability should feel comfortable. But now a highly unusual property comes on the scene. In what follows, the time interval is again $0 \leq t \leq T$, partitioned as before.

1.8.1 Simulation of Brownian Motion Paths

The path of a Brownian motion can be simulated by generating at each time-point in the partition a normally distributed random variable with mean zero and standard deviation $\sqrt{\Delta t}$. The time grid is discrete but the values of the position of the Brownian motion are now on a continuous scale. Sample paths are shown in Figure 1.11.

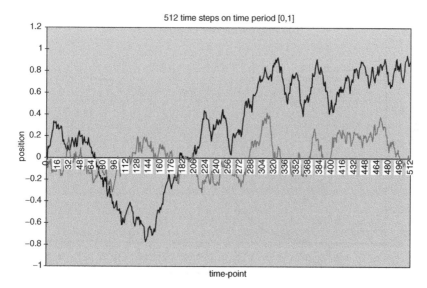

Figure 1.11   Simulated Brownian motion paths

	sample	exact
mean	0.037785	0
variance	1.023773	1

Figure 1.12 Brownian motion path simulation statistics

A batch of 1000 simulations of a standard Brownian motion over time period [0, 1] gave the statistics shown in Figure 1.12 for the position at time 1. The cumulative frequency of the sample path position at time 1 is very close to the exact probability distribution, as shown in Figure 1.13. For visual convenience cume_freq is plotted as continuous.

1.8.2 Slope of Path

For the symmetric random walk, the magnitude of the slope of the path is

$$\frac{|S_{k+1} - S_k|}{\Delta t} = \frac{\sqrt{\Delta t}}{\Delta t} = \frac{1}{\sqrt{\Delta t}}$$

This becomes infinite as $\Delta t \to 0$. As the symmetric random walk converges to Brownian motion, this puts in question the differentiability

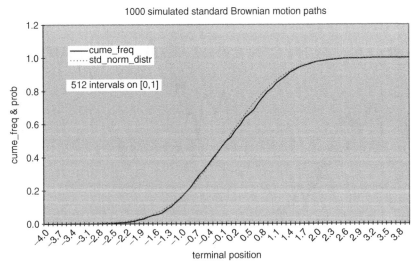

Figure 1.13 Simulated frequency versus exact Brownian motion distribution

of a Brownian motion path. It has already been seen that a simulated Brownian motion path fluctuates very wildly due to the independence of the increments over successive small time intervals. This will now be discussed further.

1.8.3 Non-Differentiability of Brownian Motion Path

First, non-differentiability is illustrated in the absence of randomness. In ordinary calculus, consider a continuous function f and the expression $[f(x + h) - f(x)]/h$. Let h approach 0 from above and take the limit $\lim_{h \downarrow 0}\{[f(x + h) - f(x)]/h\}$. Similarly take the limit when h approaches 0 from below, $\lim_{h \uparrow 0}\{[f(x + h) - f(x)]/h\}$. If both limits exist, and if they are equal, then function f is said to be differentiable at x. This limit is called the derivative (or slope) at x, denoted $f'(x)$.

Example 1.8.1

$$f(x) \overset{\text{def}}{=} x^2$$
$$\frac{f(x + h) - f(x)}{h} = \frac{(x + h)^2 - x^2}{h} = \frac{x^2 + 2xh + h^2 - x^2}{h} = \frac{2xh + h^2}{h}$$

Numerator and denominator can be divided by h, since h is not equal to zero but approaches zero, giving $(2x + h)$, and

$$\lim_{h \downarrow 0} (2x + h) = 2x \qquad \lim_{h \uparrow 0} (2x + h) = 2x$$

Both limits exist and are equal. The function is differentiable for all x, $f'(x) = 2x$.

Example 1.8.2 (see Figure 1.14)

$$f(x) \overset{\text{def}}{=} |x|$$

For $x > 0$, $f(x) = x$ and if h is also > 0 then $f(x + h) = x + h$

$$\lim_{h \downarrow 0} \frac{f(x + h) - f(x)}{h} = \frac{x + h - x}{h} = 1$$

For $x < 0$, $f(x) = -x$, and if h is also < 0, then $f(x + h) = -(x + h)$

$$\lim_{h \uparrow 0} \frac{f(x + h) - f(x)}{h} = \frac{-(x + h) - (-x)}{h} = -1$$

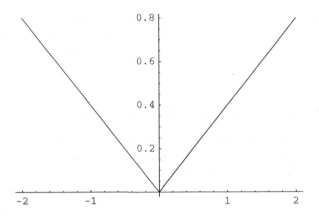

Figure 1.14   Function modulus x

Here both limits exist but they are not equal, so $f'(x)$ does not exist. This function is not differentiable at $x = 0$. There is not one single slope at $x = 0$.

Example 1.8.3 (see Figure 1.15)

$$f(x) = c_1 \mid x - x_1 \mid + c_2 \mid x - x_2 \mid + c_3 \mid x - x_3 \mid$$

This function is not differentiable at x_1, x_2, x_3, a *finite* number of points.

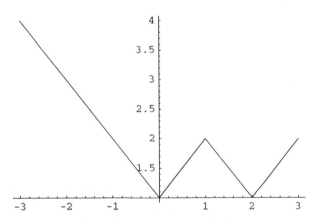

Figure 1.15   Linear combination of functions modulus x

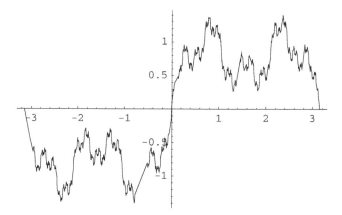

Figure 1.16 Approximation of non-differentiable function

Example 1.8.4

$$f(x) = \sum_{i=0}^{\infty} \frac{\sin(3^i x)}{2^i}$$

It can be shown that this function is non-differentiable at *any* point x. This, of course, cannot be shown for $i = \infty$, so the variability is illustrated for $\sum_{i=0}^{10}$ in Figure 1.16.

Brownian Motion   Now use the same framework for analyzing differentiability of a Brownian motion path. Consider a time interval of length $\Delta t = 1/n$ starting at t. The rate of change over time interval $[t, t + \Delta t]$ is

$$X_n \stackrel{\text{def}}{=} \frac{B(t + \Delta t) - B(t)}{\Delta t} = \frac{B(t + 1/n) - B(t)}{1/n}$$

which can be rewritten as $X_n = n[B(t + 1/n) - B(t)]$. So X_n is a normally distributed random variable with parameters

$$\mathbb{E}[X_n] = n^2 \left[B\left(t + \frac{1}{n} \right) - B(t) \right] = n0 = 0$$

$$\mathbb{V}ar[X_n] = n^2 \mathbb{V}ar \left[B\left(t + \frac{1}{n} \right) - B(t) \right]$$

$$= n^2 \frac{1}{n} = n$$

$$\mathbb{S}tdev[X_n] = \sqrt{n}$$

X_n has the same probability distribution as $\sqrt{n}\, Z$, where Z is standard normal. Differentiability is about what happens to X_n as $\Delta t \to 0$, that is, as $n \to \infty$. Take any positive number K and write X_n as $\sqrt{n}\, Z$. Then

$$\mathbb{P}[|X_n| > K] = \mathbb{P}[|\sqrt{n}Z| > K] = \mathbb{P}[\sqrt{n}|Z| > K] = \mathbb{P}\left[|Z| > \frac{K}{\sqrt{n}}\right]$$

As $n \to \infty$, $K/\sqrt{n} \to 0$ so

$$\mathbb{P}[|X_n| > K] = \mathbb{P}\left[|Z| > \frac{K}{\sqrt{n}}\right] \to \mathbb{P}[|Z| > 0]$$

which equals 1. As K can be chosen arbitrarily large, the rate of change at time t is not finite, and the Brownian motion path is not differentiable at t. Since t is an arbitrary time, the *Brownian motion path is nowhere differentiable*. It is impossible to say at any time t in which direction the path is heading.

The above method is based on the expositions in *Epps* and *Klebaner*. This is more intuitive than the 'standard proof' of which a version is given in *Capasso/Bakstein*.

1.8.4   Measuring Variability

The variability of Brownian motion will now be quantified. From t_k to t_{k+1} the absolute Brownian motion increment is $|B(t_{k+1}) - B(t_k)|$. The sum over the entire Brownian motion path is $\sum_{k=0}^{n-1} |B(t_{k+1}) - B(t_k)|$. This is a random variable which is known as the *first variation* of Brownian motion. It measures the length of the Brownian motion path, and thus its variability. Another measure is the sum of the square increments, $\sum_{k=0}^{n-1} [B(t_{k+1}) - B(t_k)]^2$. This random second-order quantity is known as the *quadratic* variation (or second variation). Now consider successive refinements of the partition. This keeps the original time-points and creates additional ones. Since for each partition the corresponding variation is a random variable, a sequence of random variables is produced. The question is then whether this sequence converges to a limit in some sense. There are several types of convergence of sequences of random variables that can be considered.[3] As the time intervals in the composition of the variation get smaller and smaller, one may be inclined to think that the variation will tend to zero. But it turns out that regardless of the size of an interval, the increment over

---

[3] See Annex E, *Convergence Concepts*.

steps	dt	first_var	quadr_var	third_var
2000	0.00050000	16.01369606	0.2016280759	0.0031830910
4000	0.00025000	19.39443203	0.1480559146	0.0014367543
8000	0.00012500	25.84539243	0.1298319380	0.0008117586
16000	0.00006250	32.61941799	0.1055395009	0.0004334750
32000	0.00003125	40.56883140	0.0795839944	0.0001946600
64000	0.00001563	43.36481866	0.0448674991	0.0000574874
128000	0.00000781	44.12445062	0.0231364852	0.0000149981
256000	0.00000391	44.31454677	0.0116583498	0.0000037899
512000	0.00000195	44.36273548	0.0058405102	0.0000009500
1024000	0.00000098	44.37481932	0.0029216742	0.0000002377
	limit	about 44.3	0	0

Figure 1.17   Variation of function which has a continuous derivative

that interval can still be infinite. It is shown in Annex C that as n tends to infinity, the first variation is not finite, and the quadratic variation is positive. This has fundamental consequences for the way in which a stochastic integral may be constructed, as will be explained in Chapter 3. In contrast to Brownian motion, a function in ordinary calculus which has a derivative that is continuous, has positive first variation and zero quadratic variation. This is shown in *Shreve II*. To support the derivation in Annex C, variability can be verified numerically. This is the object of Exercise [1.9.12] of which the results are shown in Figure 1.17 and 1.18.

Time period [0,1]				
steps	dt	first_var	quadr_var	third_var
2000	0.00050000	36.33550078	1.0448863386	0.0388983241
4000	0.00025000	50.47005112	1.0002651290	0.0253513781
8000	0.00012500	71.85800329	1.0190467736	0.0184259646
16000	0.00006250	101.65329098	1.0155967391	0.0129358213
32000	0.00003125	142.19694118	0.9987482348	0.0089475369
64000	0.00001563	202.67088291	1.0085537303	0.0063915246
128000	0.00000781	285.91679729	1.0043769437	0.0045014163
256000	0.00000391	403.18920472	0.9969064552	0.0031386827
512000	0.00000195	571.17487195	1.0005573262	0.0022306000
1024000	0.00000098	807.41653827	1.0006685086	0.0015800861
	limit	not finite	time period	0

Figure 1.18   Variation of Brownian motion

1.9 EXERCISES

The numerical exercises can be carried out in Excel/VBA, Mathematica, MatLab, or any other mathematical software or programming language.

[1.9.1] *Scaled Brownian motion* Consider the process $X(t) \overset{\text{def}}{=} \sqrt{\gamma} B(t/\gamma)$ where B denotes standard Brownian motion, and γ is an arbitrary positive constant. This process is known as scaled Brownian motion. The time scale of the Brownian motion is reduced by a factor γ, and the magnitude of the Brownian motion is multiplied by a factor $\sqrt{\gamma}$. This can be interpreted as taking snapshots of the position of a Brownian motion with a shutter speed that is γ times as fast as that used for recording a standard Brownian motion, and magnifying the results by a factor $\sqrt{\gamma}$.

(a) Derive the expected value of $X(t)$
(b) Derive the variance of $X(t)$
(c) Derive the probability distribution of $X(t)$
(d) Derive the probability density of $X(t)$
(e) Derive $\mathbb{V}\text{ar}[X(t + u) - X(t)]$, where u is an arbitrary positive constant
(f) Argue whether $X(t)$ is a Brownian motion

Note: By employing the properties of the distribution of Brownian motion this exercise can be done without elaborate integrations.

[1.9.2] *Seemingly Brownian motion* Consider the process $X(t) \overset{\text{def}}{=} \sqrt{t} Z$, where $Z \sim N(0, 1)$.

(a) Derive the expected value of $X(t)$
(b) Derive the variance of $X(t)$
(c) Derive the probability distribution of $X(t)$
(d) Derive the probability density of $X(t)$
(e) Derive $\mathbb{V}\text{ar}[X(t + u) - X(t)]$ where u is an arbitrary positive constant
(f) Argue whether $X(t)$ is a Brownian motion

[1.9.3] *Combination of Brownian motions* The random process $Z(t)$ is defined as $Z(t) \overset{\text{def}}{=} \alpha B(t) - \sqrt{\beta} B^*(t)$, where B and B^* are independent standard Brownian motions, and α and β are arbitrary positive constants. Determine the relationship between α and β for which $Z(t)$ is a Brownian motion.

[1.9.4] *Correlation* Derive the correlation coefficient between $B(t)$ and $B(t + u)$.

[1.9.5] *Successive Brownian motions* Consider a standard Brownian motion which runs from time $t = 0$ to time $t = 4$.

(a) Give the expression for the probability that its path position is positive at time 4. Give the numerical value of this probability

(b) For the Brownian motion described above, give the expression for the joint probability that its path position is positive at time 1 as well as positive at time 4. No numerical answer is requested.

(c) Give the expression for the expected value at time 4 of the position of the path described in (a). No numerical answer is requested.

[1.9.6] *Brownian motion through gates* Consider a Brownian motion path that passes through two gates situated at times t_1 and t_2.

(a) Derive the expected value of $B(t_1)$ of all paths that pass through gate 1.

(b) Derive the expected value of $B(t_2)$ of all paths that pass through gate 1 and gate 2.

(c) Derive an expression for the expected value of the increment over time interval $[t_1, t_2]$ for paths that pass through both gates.

(d) Design a simulation program for Brownian motion through gates, and verify the answers to (a), (b), and (c) by simulation.

[1.9.7] *Simulation of symmetric random walk*

(a) Construct the simulation of three symmetric random walks for $t \in [0, 1]$ on a spreadsheet.

(b) Design a program for simulating the terminal position of thousands of symmetric random walks. Compare the mean and the variance of this sample with the theoretical values.

(c) Derive the probability distribution of the terminal position. Construct a frequency distribution of the terminal positions of the paths in (b) and compare this with the probability distribution.

[1.9.8] *Simulation of Brownian motion*

(a) Construct the simulation of three Brownian motion paths for $t \in [0, 1]$ on a spreadsheet.
(b) Construct a simulation of two Brownian motion paths that have a user specified correlation for $t \in [0, 1]$ on a spreadsheet, and display them in a chart.

[1.9.9] *Brownian bridge* Random process X is specified on $t \in [0, 1]$ as $X(t) \stackrel{\text{def}}{=} B(t) - t B(1)$. This process is known as a Brownian bridge.

(a) Verify that the terminal position of X equals the initial position.
(b) Derive the covariance between $X(t)$ and $X(t + u)$.
(c) Construct the simulation of two paths of X on a spreadsheet.

[1.9.10] *First passage of a barrier* Annex A gives the expression for the probability distribution and the probability density of the time of first passage, T_L. Design a simulation program for this, and simulate $\mathbb{E}[T_L]$.

[1.9.11] *Reflected Brownian motion* Construct a simulation of a reflected Brownian motion on a spreadsheet, and show this in a chart together with the path of the corresponding Brownian motion.

[1.9.12] *Brownian motion variation*

(a) Design a program to compute the *first variation*, *quadratic variation*, and *third variation* of the differentiable ordinary function in Figure 1.16 over $x \in [0, 1]$, initially partitioned into $n = 2000$ steps, with successive doubling to 1024000 steps
(b) Copy the program of (a) and save it under another name. Adapt it to simulate the *first variation*, *quadratic variation*, and *third variation* of Brownian motion

1.10   SUMMARY

Brownian motion is the most widely used process for modelling randomness in the world of finance. This chapter gave the mathematical specification, motivated by a symmetric random walk. While this looks innocent enough as first sight, it turns out that Brownian motion has highly unusual properties. The independence of subsequent increments produces a path that does not have the smoothness of functions in ordinary calculus, and is not differentiable at any point. This feature is difficult to comprehend coming from an ordinary calculus culture. It leads to the definition of the stochastic integral in Chapter 3 and its corresponding calculus in Chapter 4.

More on Robert Brown is in the *Dictionary of Scientific Biography*, Vol. II, pp. 516–522. An overview of the life and work of Bachelier can be found in the conference proceedings *Mathematical Finance Bachelier Congress 2000*, and on the Internet, for example in *Wikipedia*. Also on the Internet is the original thesis of Bachelier, and a file named *Bachelier 100 Years*.

2

Martingales

The theory of financial options makes extensive use of expected values of random processes which are computed in the knowledge of some history. These are so-called conditional expectations. Martingale is the name for a random process whose conditional expectation has a particularly useful property. The exposition here is mainly in discrete time. Chapter 6 explains how martingales are used in the valuation of options.

2.1 SIMPLE EXAMPLE

To introduce the notion of a conditional expectation, consider a discrete sequence of times at which a stock trades. Assume the simple model in which at each trading time the stock price S can move as follows:

	by factor	*probability*
up	u	p
down	d	$1 - p$

Let S_n be the stock price at the close of trading time n. At the close of time $(n + 1)$, the price will be uS_n with probability p, or dS_n with probability $(1 - p)$, see Figure 2.1.

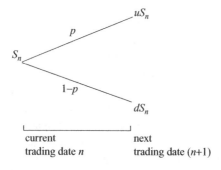

Figure 2.1   Stock price tree – one period

The expected value of S_{n+1}, given that the price at time n was S_n, is $p(uS_n) + (1 - p)(dS_n)$. This expectation is taken at time n. It is a conditional expectation, the condition being the known value of S_n, and is written as $\mathbb{E}[S_{n+1}|S_n]$. In this notation the condition is written behind the vertical bar $|$. Generally, an expectation of a random process in finance is taken as conditional upon the history of the process. This history is formally called a filtration and is described further in the next section.

To illustrate this numerically, suppose that $u = 1.25$, $d = 0.8$, and $p = 0.4444$. If S_n was recorded as 100, then S_{n+1} in the upstate equals $(1.25)(100) = 125$ and in the down state $(0.8)(100) = 80$. So

$$\mathbb{E}[S_{n+1}|S_n] = (0.4444)(125) + (1 - 0.4444)(80) = 100.$$

For these numbers the expected value equals the currently known value, and the stock price process S is said to be a martingale. This means that in expected terms there is no gain. But if $p \neq 0.4444$ then S is no longer a martingale. To be a martingale, p must be such that $\mathbb{E}[S_{n+1}|S_n] = S_n$ for given u and d, that is, p must satisfy $p(uS_n) + (1 - p)(dS_n) = S_n$. Dividing both sides by S_n and rearranging gives that p must equal $p = (1 - d)/(u - d)$. Indeed in the numerical example, $p = (1 - 0.8)/(1.25 - 0.8) = 0.4444$. If the magnitude of the up-movement were larger, say $u = 0.3$, and d is unchanged at 0.8, then S is a martingale if $p = (1 - 0.8)/(1.3 - 0.8) = 0.4$. This is lower than the previous 0.4444 because a larger up-movement has to happen less often in order to give the average of 100.

2.2 FILTRATION

Let $X_1, X_2, \ldots, X_n, \ldots$ be a sequence of random variables which measure a particular random phenomena at successive points in time. This is a random process. The first value that becomes known is that of X_1. The information revealed by the realization of X_1 is commonly denoted with a script font as $\Im_1$. Subsequently the value of X_2 becomes known. The information accumulated thus far by X_1 and X_2 is denoted $\Im_2$. And so on. It is assumed that all information is kept, so there is no less information as time goes on. This is written as $\Im_1 \subseteq \Im_2 \subseteq \cdots \subseteq \Im_n$ where the symbol is $\subseteq$ used to compare two sets; $\Im_1 \subseteq \Im_2$ means that all information in set $\Im_1$ is contained in set $\Im_2$; it looks like the symbol $\leq$ used to compare two numbers. The increasing sequence of information revealed by this random process is commonly called a filtration in the technical literature. Other names are: information set, information structure, history.

2.3   CONDITIONAL EXPECTATION

The concept of conditional expectation is discussed further using the example of the stock price tree for two periods, shown in Figure 2.2.

At time 1, the history of the process at node $(1, 1)$ is $\Im(1, 1) = u$, and at node $(1, 0)$ is $\Im(1, 0) = d$. The expected value of the stock price at time 2, S_2, taken at node $(1, 1)$ in the knowledge of $\Im(1, 1)$, is

$$\mathbb{E}[S_2|\Im(1, 1)] = p[u(uS)] + (1 - p)[d(uS)]$$

This is a conditional expectation, the condition being that the stock price is at node $(1, 1)$. Similarly, taken at node $(1, 0)$,

$$\mathbb{E}[S_2|\Im(1, 0)] = p[u(dS)] + (1 - p)[d(dS)]$$

So the conditional expectation of S_2, $\mathbb{E}[S_2|\text{history}]$, has two possible values. As node $(1, 1)$ is reached with probability p, and node $(1, 0)$ with probability $(1 - p)$, the conditional expectation of S_2 is $\mathbb{E}[S_2|\Im(1, 1)]$ with probability p and $\mathbb{E}[S_2|\Im(1, 0)]$ with probability $(1 - p)$. Thus the probability distribution of $\mathbb{E}[S_2|\text{history}]$ is fully

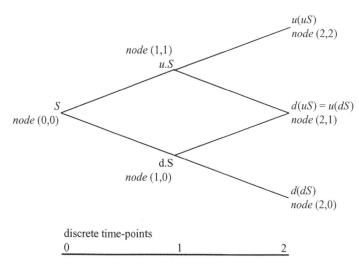

Figure 2.2   Stock price tree – two periods

specified by two values and their probabilities p and $(1 - p)$. The *conditional expectation is a random variable*, unlike an ordinary (unconditional) expectation which is a single number. The history up to the point where the expectation is taken determines the possible future values to be incorporated in the conditional expectation, and this history is random. Node $(2, 2)$ corresponds to two up-movements. There is one path to $(2, 2)$ which is reached with probability p^2. Node $(2, 1)$ corresponds to one up-movement and one down-movement. This has probability $p(1 - p)$. There are two mutually exclusive paths to $(2, 1)$, so the probability of reaching $(2, 1)$ is $p(1 - p) + (1 - p)p$. Node $(2, 0)$ corresponds to two down-movements. There is one path to $(2, 0)$ which is reached with probability $(1 - p)^2$. As a conditional expectation is a random variable, its expectation can be taken. This requires knowledge of the probabilities with which the conditions are realized. The probability of an up-movement is p, that is the probability of getting $\Im(1, 1)$. Similarly, $(1 - p)$ is the probability of getting $\Im(1, 0)$. Taking the expected value of the conditional expectations $\mathbb{E}[S_2|\Im(1, 1)]$ and $\mathbb{E}[S_2|\Im(1, 0)]$ using these probabilities performs the *unconditioning*, and gives the ordinary (unconditional) expected value of S_2 as

$$
\begin{aligned}
&p\mathbb{E}[S_2|\Im(1, 1)] + (1 - p)\mathbb{E}[S_2|\Im(1, 0)] \\
&= p\{p[u(uS)] + (1 - p)[d(uS)]\} \\
&\quad + (1 - p)\{p[u(dS)] + (1 - p)[d(dS)]\} \\
&= p^2 u^2 S + p(1 - p)duS + (1 - p)pudS + (1 - p)d^2 S \\
&= p^2 u^2 S + 2p(1 - p)duS + (1 - p)d^2 S
\end{aligned}
$$

This can be readily verified by taking the (unconditional) ordinary expected value of S_2. The possible values of S_2 are $d^2 S$, $udS = duS$, $u^2 S$, and the corresponding probabilities are p^2, $p(1 - p)$, $(1 - p)^2$. Thus

$$
\mathbb{E}[S_2] = p^2(u^2 S) + 2p(1 - p)(udS) + (1 - p)^2(d^2 S)
$$

as above.

2.3.1 General Properties

The unconditional expected value of random variable X exists if $\mathbb{E}[|X|]$ is finite; then X is said to be integrable. The notation $|X|$ stands for the sum of the positive and the negative part of X; both parts must have a finite integral. Let Y denote the conditional expectation of an integrable random variable X, $Y \stackrel{\text{def}}{=} \mathbb{E}[X|\Im]$. Then $\mathbb{E}[Y]$ is a single number which equals the unconditional expected value of X

$$\boxed{\mathbb{E}\{\mathbb{E}[X|\Im]\} = \mathbb{E}[X]}$$

So in reverse, $\mathbb{E}[X]$ can be expressed as $\mathbb{E}\{\mathbb{E}[X|\Im]\}$; an ordinary *unconditional expectation is converted into a conditional expectation*. The latter is used when the expected value of the product of two random variables has to be determined, and where the value of one of the variables is included in a history $\Im$. Then $\mathbb{E}[XY]$ is written as $\mathbb{E}\{\mathbb{E}[XY|\Im]\}$, and if Y is known given $\Im$, it can be taken outside $\mathbb{E}$ leaving $\mathbb{E}\{Y\mathbb{E}[X|\Im]\}$. Typically $\mathbb{E}[X|\Im]$ is known and can be substituted. This is known as *computing an expectation by conditioning*. It will be used in chapter 3.

Now consider the expected value of a random process X, conditioned on the history up to time s, $\mathbb{E}[X(t)|\Im(s)\ s < t]$. If this random variable is conditioned on the history up to a time r earlier than time s, the result is the same as conditioning on the earlier time r only.

$$\mathbb{E}[\mathbb{E}\{X(t)|\Im(s)\}|\Im(r)] = \mathbb{E}[X(t)|\Im(r)] \text{ where } r < s < t$$

This is known as the *tower property*. An equivalent way of writing it is

$$\boxed{\mathbb{E}[\mathbb{E}(X|\Im_2)|\Im_1] = \mathbb{E}(X|\Im_1)}$$

where $\Im_1$ and $\Im_2$ are two information sets, and there is less information in set $\Im_1$ than in set $\Im_2$; in technical notation $\Im_1 \subseteq \Im_2$.

A random variable X, whose value is known given history $\Im$, is said to be *measurable* with respect to $\Im$. Then the expectation of X conditional upon $\Im$ equals the value of X, $\mathbb{E}[X|\Im] = X$. In reverse, if in some application it is known that $\mathbb{E}[X|\Im] = X$, then it can be concluded that X is independent of $\Im$. If Z is another random variable, then a measurable X in the unconditional expectation $\mathbb{E}[ZX]$ can be replaced by $\mathbb{E}[X|\Im]$ so that $\mathbb{E}[ZX] = \mathbb{E}[Z\mathbb{E}[X|\Im]]$. In the literature, conditional expectation $\mathbb{E}[X(t)|\Im(s)]$ is also written in shorthand as $\mathbb{E}_s[X(t)]$, but not in this text.

A conditional expectation is linear. For any two random variables X and Y and constants α and β

$$\boxed{\mathbb{E}[\alpha X + \beta Y|\Im] = \alpha \mathbb{E}[X|\Im] + \beta \mathbb{E}[Y|\Im]}$$

These properties hold in a discrete- and in a continuous-time framework. Derivations can be found in the references.

2.4   MARTINGALE DESCRIPTION

Consider the random process X in discrete-time or continuous-time. Suppose the value of $X(s)$ at time s has just become known. If the conditional expectation $\mathbb{E}[X(t)|\Im(s)\ t > s]$ equals $X(s)$ then random process X is called a martingale. Writing $X(s) = \mathbb{E}[X(s)|\Im(s)]$ and moving this term to the left-hand side, an alternative formulation is $\mathbb{E}[X(t) - X(s)|\Im(s)\ t > s] = 0$. For a martingale the expected value of the increment over any future period is zero. If, on the other hand, $X(s)$ is greater than the expectation of the future, then the random process is called a supermartingale. Then the present is a 'super' situation so to speak relative to the future. A supermartingale 'decreases on average'. The reverse ($<$) is called a submartingale. A random process may be a martingale under one probability distribution but not under another. To verify whether a random process is a martingale, it must be established that $X(t)$ is integrable, and the filtration and the probability distribution of X must be known.

2.4.1   Martingale Construction by Conditioning

A martingale can be constructed from a random variable by conditioning the random variable on a filtration. That turns the random variable into a random process. Consider a discrete-time framework and a random variable Z. Let the information available at time n be denoted $\Im_n$. Define a new random variable X_n as the conditional expectation of Z, given the information contained in $\Im_n$

$$X_n \stackrel{\text{def}}{=} \mathbb{E}[Z|\Im_n] \quad \text{for} \quad n = 0, 1, 2, \ldots$$

This is a collection of random variables indexed by a time point, and is thus a random process. Fix two times, the present time n and the future time $n + 1$, and consider at time n the conditional expectation of X at future time $n + 1$, given the information up to the present time n, $\mathbb{E}[X_{n+1}|\Im_n]$. The expression for X_{n+1} is $\mathbb{E}[Z|\Im_{n+1}]$. Substituting this gives

$$\mathbb{E}[X_{n+1}|\Im_n] = \mathbb{E}[\mathbb{E}[Z|\Im_{n+1}]|\Im_n].$$

Using $\Im_n \subseteq \Im_{n+1}$, and the tower rule for expectations, the above equals $\mathbb{E}[Z|\Im_n]$, which is X_n by definition. It has thus been shown that the X process is a discrete-time martingale. This also holds in continuous time.

2.5   MARTINGALE ANALYSIS STEPS

Verifying whether a process is a martingale is best guided by the following. The first step is to translate what needs to be verified in mathematical terms. It is useful to draw a simple diagram of the time notation, showing the present time-point and the future time-point. In continuous time the notation t_1 and t_2 is self-explanatory; the notations s and t are also useful as they are alphabetical, but t and s are counter-intuitive, although often seen. An expression which involves the future random value $B(t)$, given the history of the process up to the present time s, should be decomposed into the known non-random quantity $B(s)$ and the random increment $[B(t) - B(s)]$. Once $B(t)$ has been decomposed as $B(t) = B(s) + [B(t) - B(s)]$, the increment $[B(t) - B(s)]$ must be treated as a single random variable. When taking terms outside the expectation operator $\mathbb{E}$, it should be justified why these terms can be taken out. There is frequent use for the expression $\mathbb{E}[e^Z] = e^{\mathbb{E}(Z) + \frac{1}{2}\mathrm{Var}(Z)}$ but be aware that this only holds when Z has a normal distribution. So when using this formula one needs to justify that random variable Z is indeed normally distributed. Also recall that the expected value of the product of two random variables X and Y can only be written as the product of the respective expected values, if X and Y are known to be independent, $\mathbb{E}[XY] = \mathbb{E}[X]\mathbb{E}[Y]$. In particular, $\mathbb{E}[B(s)B(t)]$ cannot be written as $\mathbb{E}[B(s)]\mathbb{E}[B(t)]$. As the analysis progresses from line to line, always repeat the condition, '$|\Im(s)$' or similar, because without that, an expectation is not a conditional expectation. Progress in simple steps. Do not jump to expressions that are not derived. If part of the analysis entails rather long expressions, give them a name and analyze them separately. Then put all the pieces together. That reduces the chance of mistakes. Finally, write down a conclusion, in relation to what was to be shown.

2.6   EXAMPLES OF MARTINGALE ANALYSIS

First some discrete time processes.

2.6.1   Sum of Independent Trials

Let $X_1, X_2, \ldots, X_n, \ldots$ be a sequence of independent identically distributed random variables with mean zero. The random variable X_i could be the numerical outcome of the ith step in a random walk where the movement is ± 1 with equal probability. The position after n steps is then $S_n \stackrel{\text{def}}{=} X_1 + X_2 + \cdots + X_n$. The filtration which contains the results of the first n increments is denoted $\Im_n$. To see whether this process is a martingale, evaluate $\mathbb{E}[S_{n+1}|\Im_n]$. The key is to express S_{n+1} as the known position S_n after the first n steps, plus the unknown outcome X_{n+1} of the $(n+1)^{th}$ step, $S_{n+1} = S_n + X_{n+1}$. This gives

$$\mathbb{E}[S_{n+1}|\Im_n] = \mathbb{E}[S_n + X_{n+1}|\Im_n] = \mathbb{E}[S_n|\Im_n] + \mathbb{E}[X_{n+1}|\Im_n]$$

Since $\Im_n$ is given, S_n is known, so $\mathbb{E}[S_n|\Im_n] = S_n$. In the second term, X_{n+1} is independent of the outcomes that have been produced by the first n trials, so $\mathbb{E}[X_{n+1}|\Im_n]$ can be written as its unconditional expectation, $\mathbb{E}[X_{n+1}|\Im_n] = \mathbb{E}[X_{n+1}] = 0$. Thus $\mathbb{E}[S_{n+1}|\Im_n] = S_n$ and the process S is a discrete martingale. It says that 'on average' there is neither increase nor decrease.

2.6.2   Square of Sum of Independent Trials

Is S_n^2 a martingale? It has to be verified whether

$$\mathbb{E}[S_{n+1}^2|\Im_n] = S_n^2$$

Using the decomposition of $S_{n+1} = S_n + X_{n+1}$, the left-hand side becomes

$$\mathbb{E}[S_n^2 + 2S_n X_{n+1} + X_{n+1}^2|\Im_n]$$
$$= \mathbb{E}[S_n^2|\Im_n] + \mathbb{E}[2S_n X_{n+1}|\Im_n] + \mathbb{E}[X_{n+1}^2|\Im_n]$$

As S_n is a known value when $\Im_n$ is given, this can be written as

$$S_n^2 + 2S_n\mathbb{E}[X_{n+1}|\Im_n] + \mathbb{E}[X_{n+1}^2|\Im_n]$$

Since X_{n+1} is independent of the past, $\mathbb{E}[X_{n+1}^2|\Im_n]$ equals the unconditional expectation $\mathbb{E}[X_{n+1}^2]$ which is $1^2\frac{1}{2} + (-1)^2\frac{1}{2} = 1$. With $\mathbb{E}[X_{n+1}|\Im_n] = \mathbb{E}[X_{n+1}] = 1(\frac{1}{2}) + (-1)\frac{1}{2} = 0$ this gives $\mathbb{E}[S_{n+1}^2|\Im_n] = S_n^2 + 1$. So S_n^2 is not a martingale because of the presence of the term $+1$. But subtracting n gives the discrete process $S_n^2 - n$, which *is* a martingale, as can be shown as follows.

For S_{n+1} there are two possible values, $S_n + 1$ with probability $\frac{1}{2}$, and $S_n - 1$ with probability $\frac{1}{2}$. Thus

$$\mathbb{E}[S_{n+1}^2 - (n+1)|\Im_n]$$
$$= \{(S_n + 1)^2 - (n+1)\}\tfrac{1}{2} + \{(S_n - 1)^2 - (n+1)\}\tfrac{1}{2}$$
$$= \{S_n^2 + 2S_n + 1 - n - 1\}\tfrac{1}{2} + \{S_n^2 - 2S_n + 1 - n - 1\}\tfrac{1}{2}$$
$$= S_n^2 - n$$

2.6.3   Product of Independent Identical Trials

As above, let $X_1, X_2, \ldots, X_n, \ldots$ be a sequence of independent identically distributed random variables but now with mean 1. Define the product $M_n \stackrel{\text{def}}{=} X_1 X_2 \ldots X_n$. Evaluate $\mathbb{E}[M_{n+1}|\Im_n]$. Write $M_{n+1} = M_n X_{n+1}$, the by now familiar decomposition into known and unknown. Then

$$\mathbb{E}[M_{n+1}|\Im_n] = \mathbb{E}[M_n X_{n+1}|\Im_n] = M_n \mathbb{E}[X_{n+1}|\Im_n]$$

as M_n is known when $\Im_n$ is given. With $\mathbb{E}[X_{n+1}|\Im_n] = \mathbb{E}[X_{n+1}] = 1$, the final result is $\mathbb{E}[M_{n+1}|\Im_n] = M_n$ so the process M is a discrete time martingale.

Now some continuous time processes.

2.6.4   Random Process $B(t)$

The expected value of a Brownian motion position at future time t, taken at present time s, given the entire history of the Brownian motion process, is $\mathbb{E}[B(t)|\Im(s)\ s < t]$. Decompose $B(t)$ into the known value $B(s)$ and the random increment $\{B(t) - B(s)\}$. That gives

$$\mathbb{E}[B(t)|\Im(s)] = \mathbb{E}[B(s) + \{B(t) - B(s)\}|\Im(s)]$$
$$= \mathbb{E}[B(s)|\Im(s)] + \mathbb{E}[B(t) - B(s)|\Im(s)]$$

In the first term, the value of $B(s)$, given the history to time s, is not a random variable, but the known value $B(s)$, so $\mathbb{E}[B(s)|\Im(s)] = B(s)$. In the second term, the increment from s to t, $B(t) - B(s)$, is independent of the value of the Brownian motion at time s or earlier, because Brownian motion is by definition a process of independent increments. Thus $\mathbb{E}[B(t) - B(s)|\Im(s)]$ is equal to its unconditional expectation $\mathbb{E}[B(t) - B(s)] = 0$. Hence $\mathbb{E}[B(t)|\Im(s)] = B(s)$, which shows that *Brownian motion is a martingale.*

2.6.5 Random Process $\exp[B(t) - \frac{1}{2}t]$

To evaluate whether $\mathbb{E}[\exp[B(t) - \frac{1}{2}t]|\Im(s)\ s < t] = \exp[B(s) - \frac{1}{2}s]$ substitute $B(t) = B(s) + \{B(t) - B(s)\}$ and $t = s + (t - s)$. This gives

$$\mathbb{E}[\exp[B(s) + \{B(t) - B(s)\} - \frac{1}{2}s - \frac{1}{2}(t - s)]|\Im(s)]$$

$$= \exp[B(s) - \frac{1}{2}s]\mathbb{E}[\exp[B(t) - B(s) - \frac{1}{2}(t - s)]|\Im(s)]$$

The exponent $[B(t) - B(s) - \frac{1}{2}(t - s)]$ is a normally distributed random variable because Brownian motion is normally distributed, and a normal random variable minus a constant is also normal. Write this exponent as $Y(t) \overset{\text{def}}{=} B(t) - B(s) - \frac{1}{2}(t - s)$. Use the rule for the expected values of an exponential with a normal as exponent

$$\mathbb{E}\{\exp[Y(t)]\} = \exp\{\mathbb{E}[Y(t)] + \frac{1}{2}\mathbb{V}\mathrm{ar}[Y(t)]\}$$
$$\mathbb{E}[Y(t)] = \mathbb{E}[B(t) - B(s) - \frac{1}{2}(t - s)] = -\frac{1}{2}(t - s)$$
$$\mathbb{V}\mathrm{ar}[Y(t)] = \mathbb{V}\mathrm{ar}[B(t) - B(s) - \frac{1}{2}(t - s)]$$
$$= \mathbb{V}\mathrm{ar}[B(t) - B(s)]$$
$$= (t - s)$$

So

$$\mathbb{E}\{\exp[Y(t)]\} = \exp[-\frac{1}{2}(t - s) + \frac{1}{2}(t - s)] = \exp(0) = 1$$

Thus

$$\mathbb{E}\{\exp[B(t) - \frac{1}{2}t|\Im(s)]\} = \exp[B(s) - \frac{1}{2}s]1 = \exp[B(s) - \frac{1}{2}s]$$

Hence $\exp[B(t) - \frac{1}{2}t]$ is a martingale.

2.6.6 Frequently Used Expressions

When t_k is the present time and t the future time

$$\mathbb{E}[B(t)|\Im(t_k)] = B(t_k)$$

because Brownian motion is a martingale.

A Brownian motion increment starting at t_k is independent of the history up to t_k

$$\mathbb{E}[B(t_{k+1}) - B(t_k)|\Im(t_k)] = \mathbb{E}[B(t_{k+1}) - B(t_k)]$$

by the definition of Brownian motion.

2.7   PROCESS OF INDEPENDENT INCREMENTS

It will now be shown that if a random process X has the property that the increment over an arbitrary time interval $[s, t]$, $[X(t) - X(s)]$, is independent of the information up to time s, $\Im(s)$, then the increments of X over non-overlapping time intervals are independent. This property is used in Section 4.5.

Consider the increments $[X(t_2) - X(t_1)]$ and $[X(t_3) - X(t_2)]$ over the successive intervals $[t_1, t_2]$ and $[t_2, t_3]$. These two random variables are independent if their joint moment generating function (mgf) can be written as the product of the individual mgfs. So it must be shown that

$$\mathbb{E}[e^{\theta_1[X(t_2)-X(t_1)]+\theta_2[X(t_3)-X(t_2)]}]$$
$$= \mathbb{E}[e^{\theta_1[X(t_2)-X(t_1)]}]\mathbb{E}[e^{\theta_2[X(t_3)-X(t_2)]}]$$

This can then be extended to any collection of successive increments.

Write the left-hand side as the expected value of a conditional expectation, where the conditioning is on the information up to the one-but-last time t_2, $\Im(t_2)$.

$$\mathbb{E}\{\mathbb{E}[e^{\theta_1[X(t_2)-X(t_1)]+\theta_2[X(t_3)-X(t_2)]}|\Im(t_2)]\}$$
$$= \mathbb{E}\{\mathbb{E}[e^{\theta_1[X(t_2)-X(t_1)]}e^{\theta_2[X(t_3)-X(t_2)]}|\Im(t_2)]\}$$

Given $\Im(t_2)$, $e^{\theta_1[X(t_2)-X(t_1)]}$ is known and can be taken outside the second $\mathbb{E}$, leaving

$$\mathbb{E}\{e^{\theta_1[X(t_2)-X(t_1)]}\mathbb{E}[e^{\theta_2[X(t_3)-X(t_2)]}|\Im(t_2)]\} \qquad (*)$$

As $[X(t_3) - X(t_2)]$ is assumed to be independent of $\Im(t_2)$, then so is the function $e^{\theta_2[X(t_3)-X(t_2)]}$ of this random variable, and

$$\mathbb{E}[e^{\theta_2[X(t_3)-X(t_2)]}|\Im(t_2)] = \mathbb{E}[e^{\theta_2[X(t_3)-X(t_2)]}]$$

Thus $(*)$ becomes

$$\mathbb{E}\{e^{\theta_1[X(t_2)-X(t_1)]}\mathbb{E}[e^{\theta_2[X(t_3)-X(t_2)]}]\}$$

Inside $\{\cdots\}$, the term $\mathbb{E}[e^{\theta_2[X(t_3)-X(t_2)]}]$ is an unconditional expectation; it is a single number that can be taken outside, giving

$$\mathbb{E}\,e^{\theta_1[X(t_2)-X(t_1)]}]\mathbb{E}\,e^{\theta_2[X(t_3)-X(t_2)]}]$$

as was to be shown.

2.8 EXERCISES

[2.8.1] Show that the symmetric random walk is a discrete martingale.

[2.8.2] $S_n^* \stackrel{\text{def}}{=} S_n/(pu + qd)^n$ where S_n is the stock price process discussed in Section 2.1, and $(1 - p)$ has been denoted q to simplify the notation. S_n^* is a process that is defined as the process S_n divided by $(pu + qd)$ to the power n. Verify whether it is a martingale.

[2.8.3] Verify whether $B(t) + 4t$ is a martingale.

[2.8.4] Verify whether $B(t)^2$ is a martingale. Also derive its probability density.

[2.8.5] Verify whether $B(t)^2 - t$ is a martingale. Also derive its probability density.

[2.8.6] Verify whether $Z \stackrel{\text{def}}{=} \exp[-\varphi B(t) - \frac{1}{2}\varphi^2 t]$ is a martingale; φ is a given constant. Also derive its probability density, its mean and its variance. Construct a graph of its density for parameter values $t = 1.5$, $\varphi = 0.5$.

[2.8.7] If M is a martingale, show that $\mathbb{E}\{M(u) - M(s)\}^2|\Im(s)] = \mathbb{E}[M(u)^2 - M(s)^2|\Im(s)]$.

[2.8.8] Write a simulation program for $Z \stackrel{\text{def}}{=} \exp[-\varphi B(t) - \frac{1}{2}\varphi^2 t]$ on the time interval $[0, 1]$ for a user specified constant φ. Show three paths on a spreadsheet.

2.9 SUMMARY

This chapter gave a brief introduction to the very basics of martingales. A conditional expectation is a random variable, not a fixed number. An ordinary unconditional expectation can be expressed as an iterated conditional expectation. The conditional expectation of a random process which is a martingale equals any earlier known value of the process. Thus if somehow the expectation is known, but the earlier value is not, then that earlier value can be found if the process is a martingale. That is how the martingale property will be used in reverse in the valuation of options in Chapter 6. A random variable can be transformed into a martingale by conditioning.

Excellent elementary coverage of discrete probability, conditional expectations, and discrete martingales, in the context of finance, is given in *Roman*. For continuous time, a good start is *Lin*. Other sources are *Kuo*, *Shreve II*, *Mikosch*, *Brzeźniak/Zastawniak*. The integrability condition $\mathbb{E}[|X|] < \infty$ is discussed in books on the foundations of the expected value concept. Particularly recommended are *Shreve II* Sections 1.3 and 1.5, and *Epps*, pp. 34–36 and 56–57. The concept of a conditional expectation in its full generality is complicated, but the comforting news is that it can be used without knowing the intricacies of its specification or the details on $\mathbb{E}[|X|]$.

Itō Stochastic Integral

Chapters 3 to 5 present the concepts of stochastic calculus created by the Japanese mathematician Kiyosi Itō[1] in the 1940s and 1950s. It is the mathematics that is used for modelling financial options, and has wide application in other fields. This chapter sets out the concept of the so-called stochastic integral in which the integrator is Brownian motion and the integrand is a random process dependent on Brownian motion; Chapter 4 introduces stochastic calculus rules for manipulating and evaluating such integrals; and Chapter 5 introduces the dynamics of random processes which are driven by Brownian motion. A tribute to Kiyosi Itō is presented as Section 3.11.

3.1 HOW A STOCHASTIC INTEGRAL ARISES

Consider the time period $[0, T]$ partitioned into n intervals of equal length $\Delta t \overset{\text{def}}{=} T/n$ with endpoints $t_k \overset{\text{def}}{=} k\Delta t, k = 0, ..., n$. These are the times at which the market is open and trade in shares can take place. An investor buys $q(t_0)$ shares at time 0 at a price of $S(t_0)$ each. At time t_1, market trading establishes a new share price $S(t_1)$. Once this price has been revealed the investor can change the quantity of shares held, from $q(t_0)$ to $q(t_1)$. The same at the subsequent times t_2 through t_{n-1}. At time t_n the entire portfolio is liquidated as soon as share price $S(t_n)$ becomes known. A portfolio of $q(t_k)$ shares is held from just after the share price $S(t_k)$ has been revealed, to when the share price $S(t_{k+1})$ at time t_{k+1} becomes known. This portfolio is worth $q(t_k)S(t_k)$ at time t_k and $q(t_k)S(t_{k+1})$ at time t_{k+1}. So from time t_k to time t_{k+1}, with the change in share price $S(t_{k+1}) - S(t_k)$ denoted by ΔS_k, the portfolio value changes by $q(t_k)\Delta S_k$. The gain[2] over all trades can then be expressed as $I_n \overset{\text{def}}{=} \sum_{k=0}^{n-1} q(t_k)\Delta S_k$. This expression is an example of a so-called *discrete stochastic integral*. At each trading time, the quantity of shares to be held until the next trading opportunity arises, is decided by the

---

[1] The standard Hepburn romanization (Itō) is used here. Alternatives are Itô and Ito, as explained in Wikipedia.
[2] To be understood in the mathematical sense, in that it can be negative as well as positive.

investor. Some investors will buy shares and others will sell as a trade requires two parties. The factors which an investor takes into account in determining the shareholdings are unknown, so q is a discrete-time random process. Thus the stochastic integral is constructed from the two random processes q and S. Note that in the expression for I_n, the upper limit in the summation is $k = n - 1$, not $k = n$, as the share price at t_n is applicable to the shareholding set previously at t_{n-1}. Using the stock price dynamics of Section 1.3, $\Delta S(t_k) = \mu S(t_k) \Delta t + \sigma S(t_k) \Delta B_k$, gives for a particular n,

$$I_n = \sum_{k=0}^{n-1} q(t_k)[\mu S(t_k) \Delta t + \sigma S(t_k) \Delta B(t_k)]$$

$$= \mu \sum_{k=0}^{n-1} q(t_k) S(t_k) \Delta t + \sigma \sum_{k=0}^{n-1} q(t_k) S(t_k) \Delta B(t_k)$$

Increasing n makes the time step Δt smaller and discrete-time trading approaches continuous-time trading. The question now is how the gain of continuous trading can be expressed. In other words, does the corresponding sequence of random variables I_n converge to a limit in some sense.

The first term is a summation with respect to time step Δt. For a particular market realization ω^*, $q(t_k, \omega^*)S(t_k)$ is a known step function (here ω is added to emphasize the random nature of q) which becomes known progressively with $\Im(t_k)$. It looks like a Riemann sum, and indeed, for each path ω^*, as $\Delta t \to 0$, the expression $\sum_{k=0}^{n-1} q(t_k, \omega^*)S(t_k) \Delta t$ converges to the ordinary integral $\int_{t=0}^{T} q(t, \omega^*)S(t) \, dt$; this is known as *pathwise integration*.[3]

The second term is a summation with respect to Brownian motion increment $[B(t_{k+1}) - B(t_k)]$. A first thought is to restate it as a Riemann sum by using the mean-value theorem from calculus: expressing $[B(t_{k+1}) - B(t_k)]$ as the derivative of B at some point in the time interval, times the length of the interval. But B is not differentiable so this cannot be done. Another thought is to interpret $\sum_{k=0}^{n-1} q(t_k)S(t_k)[B(t_{k+1}) - B(t_k)]$ as a Riemann–Stieltjes sum[4] $\sum_{k=0}^{n-1} f(t_k)[g(t_{k+1}) - g(t_k)]$, which converges as $(t_{k+1} - t_k) \to 0$. Using f at the left endpoint t_k makes it match the expression for the discrete stochastic integral. So in order to see if the Riemann–Stieltjes integral

---

[3] See Annex B.2.2. Also *Capasso/Bakstein* Annex A.5.
[4] The Riemann–Stieltjes integral is reviewed in Annex B, *Ordinary Integration*.

can be used in this stochastic setting, it has to be checked whether $B(t)$ is of bounded variation on $0 \le t \le T$. It is shown in Annex C that this is *not* the case. Therefore the Riemann–Stieltjes integral cannot be used for stochastic integration of a general random integrand with respect to Brownian motion.[5] There is a need for a new concept. It turns out that as $n \to \infty$, the sequence of stochastic integrals I_n has a limit in the mean-square sense, and that limit is denoted $\int_{t=0}^{T} q(t)S(t) \, dB(t)$.

This construction will now be developed for the integration with respect to Brownian motion of a random integrand f which satisfies the properties specified below. The discrete stochastic integral for this is written as

$$I_n(f) \overset{\text{def}}{=} \sum_{k=0}^{n-1} f(t_k, \omega) \, \Delta B(t_k),$$

sometimes as I_n for greater readability.

3.2   STOCHASTIC INTEGRAL FOR NON-RANDOM STEP-FUNCTIONS

In ordinary calculus, the construction of a Riemann integral starts from a partition of the domain of the variable into intervals. That also applies here. As before, the time period $[0, T]$ is partitioned into n intervals of equal length $\Delta t = T/n$, with endpoints $t_k \overset{\text{def}}{=} k\Delta t$, $k = 0, \ldots, n$. The simplest integrand f is a step-function whose values are non-random, so the only random quantity is the Brownian motion integrator. This will be considered first. In the next section the construction is repeated for a step-function with a random level. Finally this is extended to an integrand which is a general random process.

In the simplest case, suppose that at time 0, f is fixed for all intervals; for $t_k \le t < t_{k+1}$ at level $f(t_k)$. As a financial model this is not realistic because in the setting of Section 3.1 an investor would not need to fix all shareholdings at time 0. But this first construction is merely meant to set out the methodology.

Step-function f (Figure 3.1) can be written with indicator notation as

$$f = f(t_0) \, 1_{[t_0, t_1)} + \cdots + f(t_k) \, 1_{[t_k, t_{k+1})} + \cdots + f(t_{n-1}) \, 1_{[t_{n-1}, t_n)}$$

---

[5] For non-random integrands f of bounded variation, use can be made of the Riemann–Stieltjes integral. See *Kuo* section 2.3. Lesser known conditions for the existence of the Riemann–Stieltjes integal are discussed in *Mikosch* section 2.1.2.

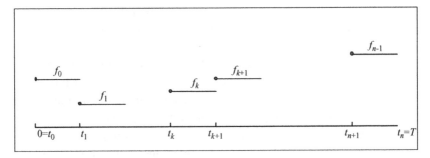

Figure 3.1   Non-random step-function f

where $1_{[t_k,t_{k+1})}$ denotes the indicator function which has value 1 on the interval shown in the subscript, and 0 elsewhere. *Define* the *discrete stochastic integral $I_n(f)$* as

$$I_n(f) \overset{\text{def}}{=} \sum_{k=0}^{n-1} f(t_k)[B(t_{k+1}) - B(t_k)]$$

The sum $I_n(f)$ is a random variable. Some of its properties are now discussed.

(i) For any constants α and β, and functions f and g which are defined on the same partition, $I_n(\alpha f + \beta g) = \alpha I_n(f) + \beta I_n(g)$. The integral is a *linear* function. This can be readily seen from

$$I_n(\alpha f + \beta g) = \sum_{k=0}^{n-1}[\alpha f(t_k) + \beta g(t_k)][B(t_{k+1}) - B(t_k)]$$

$$= \alpha \sum_{k=0}^{n-1} f(t_k)[B(t_{k+1}) - B(t_k)]$$

$$+ \beta \sum_{k=0}^{n-1} g(t_k)[B(t_{k+1}) - B(t_k)]$$

$$= \alpha I_n(f) + I_n(g)$$

If the partitions of f and g do not coincide, use the union of their partitions.

(ii) $I_n(f)$ is a linear combination of independent normal random variables $[B(t_{k+1}) - B(t_k)]$ with constant coefficients $f(t_k)$, and is thus a *normal random variable*. Its mean equals the sum of the means of the terms, which are all zero, therefore $\mathbb{E}[I_n(f)] = 0$. Its variance is the sum of the variances of the terms, since the random terms are independent, so

$$\mathbb{V}ar[I_n(f)] = \mathbb{E}[I_n(f)^2] = \sum_{k=0}^{n-1} f(t_k)^2 \, \Delta t$$

(iii) $I_n(f)$ is a discrete-time *martingale*. This can be shown as follows. If the value of the integral is known on $[0, t_n]$ then $B(t_n)$ is the last known position. The history is $\Im(t_n)$, and

$$\mathbb{E}\left\{ \sum_{k=0}^{n-1} f(t_k)[B(t_{k+1}) - B(t_k)] | \Im(t_n) \right\} = \sum_{k=0}^{n-1} f(t_k)[B(t_{k+1}) - B(t_k)]$$

Then, using only discrete time-points

$$\mathbb{E}\left\{ \sum_{k=0}^{n} f(t_k)[B(t_{k+1}) - B(t_k)] | \Im(t_n) \right\}$$

$$= \mathbb{E}\left\{ \sum_{k=0}^{n-1} f(t_k)[B(t_{k+1}) - B(t_k)] | \Im(t_n) \right\}$$

$$+ \mathbb{E}\{ f(t_n)[B(t_{n+1}) - B(t_n)] | \Im(t_n) \}$$

In the second term, $f(t_n)$ is known when $\Im(t_n)$ is given, so this term can be written as

$$f(t_n)\mathbb{E}\{B(t_{n+1}) - B(t_n) | \Im(t_n)\} = f(t_n)0 = 0$$

That leaves

$$\mathbb{E}\left\{ \sum_{k=0}^{n-1} f(t_k)[B(t_{k+1}) - B(t_k)] | \Im(t_n) \right\} = \sum_{k=0}^{n-1} f(t_k)[B(t_{k+1}) - B(t_k)],$$

which confirms that $I_n(f)$ is a martingale.

3.3   STOCHASTIC INTEGRAL FOR NON-ANTICIPATING RANDOM STEP-FUNCTIONS

Now f will be at random level $f(t_k, \omega)$ for $t_k \le t < t_{k+1}$. This function must have the following properties:

- When the history of the Brownian motion process becomes known progressively at each time t_k, it must be possible to determine $f(t_k)$ from this history alone. The current terminology is that $f(t_k)$ must be *adapted* to the filtration $\Im(t_k)$. The value of f at t_k should not depend on values of B beyond t_k. That is, it must not be possible for f to anticipate the future beyond t_k. The original terminology, coined by Itō, is that f must be a *non-anticipating* function with respect to the Brownian motion process, and that terminology is used in this text. Section 3.1 showed that this condition is typically satisfied in financial modelling, because $f(t_k)$ is the position an investor takes in the market, and $\Im(t_k)$ describes the flow of information available to an investor.

- $\int_{t=0}^{T} \mathbb{E}[f(t, \omega)^2]\, dt$ must be finite for finite T, the so-called *square integrability* condition. The reason for this will become clear shortly.

Assume now that each $f(t_k, \omega)$ is *continuous,* and *non-anticipating,* and that $\mathbb{E}[f(t_k, \omega)^2]$ is finite. The discrete stochastic integral is

$$I_n(f) \stackrel{\text{def}}{=} \sum_{k=0}^{n-1} f(t_k, \omega)[B(t_{k+1}) - B(t_k)]$$

Its properties are as follows:

(i) *Linearity,* $I_n(\alpha f + \beta g) = \alpha I_n(f) + \beta I_n(g)$.
(ii) The *distribution* of $I_n(f)$ is now a random mix of Brownian motion distributions, thus not normal. As each term in the expression for $I_n(f)$ is the product of two random quantities, the computation of $\mathbb{E}[I_n(f)]$ and $\mathbb{V}ar[I_n(f)]$ is more involved and requires the conditioning that was explained in Chapter 2.

$$\mathbb{E}[I_n(f)] = \mathbb{E}\left\{\sum_{k=0}^{n-1} f(t_k, \omega)[B(t_{k+1}) - B(t_k)]\right\}$$

$$= \sum_{k=0}^{n-1} \mathbb{E}\{f(t_k, \omega)[B(t_{k+1}) - B(t_k)]\}$$

Write $\mathbb{E}\{f_k(\omega)[B(t_{k+1}) - B(t_k)]\}$ as $\mathbb{E}$ of the conditional expectation $\mathbb{E}\{f_k(\omega)[B(t_{k+1}) - B(t_k)]|\Im(t_k)\}$. The increment $[B(t_{k+1}) - B(t_k)]$ is independent of $f(t_k, \omega)$. With $\Im(t_k)$ given, $f(t_k)$ is a known number, not a random variable, so it can be taken outside $\mathbb{E}$ giving $f(t_k)\mathbb{E}[B(t_{k+1}) - B(t_k)|\Im(t_k)] = f(t_k)0 = 0$. This holds progressively for each term in the summation, so $\mathbb{E}[I_n(f)] = 0$, the same as in the integral of a non-random step-function.

The expression for the variance then reduces to

$$
\mathbb{E}[I_n(f)^2]
$$

$$
= \mathbb{E}\left\{ \sum_{k=0}^{n-1} f(t_k, \omega)[B(t_{k+1}) - B(t_k)] \right.
$$

$$
\left. \times \sum_{m=0}^{n-1} f(t_m, \omega)[B(t_{m+1}) - B(t_m)] \right\}
$$

$$
= \sum_{k=0}^{n-1} \mathbb{E}\{ f(t_k, \omega)^2 [B(t_{k+1}) - B(t_k)]^2 \}
$$

$$
+ \sum_{k<m}^{n-1} \mathbb{E}\{ f(t_k, \omega) f(t_m, \omega)[B(t_{k+1}) - B(t_k)][B(t_{m+1}) - B(t_m)] \}
$$

$$
+ \sum_{m<k}^{n-1} \mathbb{E}\{ f(t_m, \omega) f(t_k, \omega)[B(t_{m+1}) - B(t_m)][B(t_{k+1}) - B(t_k)] \}
$$

In the summation for $k < m$, $f(t_k, \omega)$ and $f(t_m, \omega)$ and $[B(t_{k+1}) - B(t_k)]$ are independent of $[B(t_{m+1}) - B(t_m)]$ which is generated later. Thus $\mathbb{E}\{ f(t_k, \omega) f(t_m, \omega)[B(t_{k+1}) - B(t_k)][B(t_{m+1}) - B(t_m)] \}$ can be written as the product of expectations

$$
\mathbb{E}\{ f(t_k, \omega) f(t_m, \omega)[B(t_{k+1}) - B(t_k)] \} \mathbb{E}[B(t_{m+1}) - B(t_m)]
$$

As the last expectation is zero, the $k < m$ summation is zero. Exactly the same reasoning applies to the summation for $m < k$, and for this reason the order of the terms has been written with m first. To evaluate the terms in the first summation, conditioning is introduced as in the evaluation of the mean, by writing it as $\mathbb{E}$ of the conditional expectation, conditioned on $\Im(t_k)$.

$$
\mathbb{E}[\mathbb{E}\{ f(t_k, \omega)^2 [B(t_{k+1}) - B(t_k)]^2 | \Im(t_k) \}]
$$

$$
= \mathbb{E}[f(t_k, \omega)^2 \mathbb{E}\{ [B(t_{k+1}) - B(t_k)]^2 | \Im(t_k) \}]
$$

$$
= \mathbb{E}[f(t_k, \omega)^2] \Delta t
$$

So

$$
\mathbb{V}\mathrm{ar}[I_n(f)] = \mathbb{E}[I_n(f)^2] = \sum_{k=0}^{n-1} \mathbb{E}[f(t_k, \omega)^2] \Delta t
$$

This is the same type of expression as before, but as f is now random its expected value is used.

By exactly the same method, the expectation of the product expression $I_n(f)I_n(g)$ can be evaluated; just change the notation of the second f to g. The result is

$$\mathbb{E}[I_n(f)I_n(g)] = \sum_{k=0}^{n-1} \mathbb{E}[f(t_k)g(t_k)]\,\Delta t$$

(iii) $I_n(f)$ is a discrete *martingale*. The derivation is similar to the case of a non-random step-function.

3.4   EXTENSION TO NON-ANTICIPATING GENERAL RANDOM INTEGRANDS

The Itō stochastic integral for a non-anticipating general random process will now be constructed by approximation. A result from probability theory says that a general random integrand f can be approximated on domain $[0, T]$ with any desired degree of accuracy by a random step-function on n intervals of length $\Delta t = T/n$ by taking n large enough.[6] For a particular n, the discrete stochastic integral of the corresponding random step-function is as shown above. Successive doubling of the partition produces a *sequence* of random variables $I_n, I_{2n}, I_{4n}, \ldots$. The question is now whether this sequence converges to a limit in some sense. Such a limit will generally be a random variable as the sequence consists of random variables. There are several types of convergence of sequences of random variables that can be considered. In the knowledge that Brownian motion has a finite second moment, consider *convergence in mean-square*.[7] For this type of convergence, the difference between the values of any I_n and I_{2n} is considered. This difference is squared, $[I_n - I_{2n}]^2$, and the expected value of this random variable is taken, $\mathbb{E}\{[I_n - I_{2n}]^2\}$. For increasing values of n, this gives a sequence of numbers $\mathbb{E}\{[I_n - I_{2n}]^2\}$. If this sequence tends to zero as $n \to \infty$, then the sequence of random variables I_n has a limit, in the mean-square sense, which is a random variable, and is here denoted as I. *The Itō stochastic integral is defined as the limit in mean-square of the sequence of discrete stochastic integrals I_n of non-anticipating random step-functions which approximate the non-anticipating general random integrand.* This

---

[6] See for example *Klebaner* p. 31.
[7] See Annex E, *Convergence Concepts*.

construction uses the concept of a norm as discussed in Annex D. Consider the random step-function of Section 3.3:

$$f = f(t_0, \omega) \, 1_{[t_0,t_1)} + \cdots + f(t_k, \omega) \, 1_{[t_k,t_{k+1})} + \cdots$$
$$+ f(t_{n-1}, \omega) \, 1_{[t_{n-1},t_n)}$$

The squared norm of a random process $f(n)$ is

$$||f^{(n)}||^2 = \int_{t=0}^{T} \mathbb{E}[f^{(n)^2}] \, dt$$

where

$$f^{(n)^2} = f(t_0, \omega)^2 \, 1_{[t_0,t_1)} + \cdots + f(t_k, \omega)^2 \, 1_{[t_k,t_{k+1})} + \cdots$$
$$+ f(t_{n-1}, \omega)^2 \, 1_{[t_{n-1},t_n)}$$

is without cross products as they are on non-overlapping intervals. So

$$||f^{(n)}||^2 = \int_{t=0}^{T} \mathbb{E}\left[\sum_{k=0}^{n-1} f(t_k, \omega)^2 \, 1_{[t_k,t_{k+1})}\right] dt = \sum_{k=0}^{n-1} \mathbb{E}[f(t_k, \omega)^2] \, \Delta t$$

The squared norm of the corresponding random variable $I_n(f^{(n)}) \stackrel{\text{def}}{=} \sum_{k=0}^{n-1} f(t_k, \omega) \, \Delta B(t_k)$ is

$$||I(f^{(n)})||^2 = \mathbb{E}[I(f^{(n)})^2]$$

This was evaluated in Section 3.3 as $\sum_{k=0}^{n-1} \mathbb{E}[f(t_k, \omega)^2] \, \Delta t$. Thus in this discrete setting

$$\boxed{||I(f^{(n)})|| = ||f^{(n)}||}$$

The norm of the stochastic integral and the norm of the random integrand f have equal value. Note that this equality stems from different definitions. The squared norm of I is defined as its second moment. The squared norm of f is defined as an average of the second moment of f. The equality of norms is known technically as an *isometry*, where 'iso' comes from the Greek isos and means equal or identical, and 'metric' refers to distance. It is the mathematical name for any two quantities whose norms are equal, not necessarily those in a probabilistic setting. The isometry between f and $I_n(f)$ is the key to constructing the Itō stochastic integral for a general random non-anticipating square integrable integrand f. This goes as follows.

Step 1   *Approximate general random f by random step-function $f^{(n)}$*
$f^{(n)}$ is used to approximate a general integrand f arbitrarily close in
the norm of f, $||f^{(n)} - f|| \to 0$ as $n \to \infty$. It can be shown that this is
equivalent to $||f^{(n)} - f^{(2n)}|| \to 0$ as $n \to \infty$, where $f^{(2n)}$ is the random
step-function for the finer partition with $2n$ intervals.

Step 2   *Write the norm for the difference of two integrals*
The difference between the stochastic integrals of step-functions $f^{(n)}$
and $f^{(2n)}$ as measured by the norm of I is $||I[f^{(n)}] - I[f^{(2n)}]||$. The
difference of the stochastic integrals $I[f^{(n)}]$ and $I[f^{(2n)}]$ equals the
stochastic integral with respect to the difference of the integrands,
$||I[f^{(n)}] - I[f^{(2n)}]|| = ||I[f^{(n)} - f^{(2n)}]||$.

Step 3   *Apply the isometry*
According to the isometry, $||I[f^{(n)} - f^{(2n)}]||$ equals $||f^{(n)} - f^{(2n)}||$.
Since $||f^{(n)} - f^{(2n)}|| \to 0$ as $n \to \infty$, it follows that $||I[f^{(n)} - f^{(2n)}]|| \to 0$ as $n \to \infty$. Thus $I[f^{(n)}]$ is a sequence of random vari-
ables that has a limit as $n \to \infty$. This *limit is called the Itō stochastic
integral* and denoted $I(f) = \int_{t=0}^{T} f(t, \omega) \, dB(t)$.

The approximation of f and the corresponding discrete integral are now illustrated
for the time period $[0, 4]$, first partitioned into $n = 2$ intervals, then refined to $n = 4$
intervals.

Norm of f   The discrete approximations of f based on left endpoint values, denoted
$f^{(2)}$ and $f^{(4)}$, are specified below and sketched in Figures 3.2–3.4.

$$f^{(2)} = f(0) \, 1_{[0,2)} + f(2) \, 1_{[2,4)}$$
$$f^{(4)} = f(0) \, 1_{[0,1)} + f(1) \, 1_{[1,2)} + f(2) \, 1_{[2,3)} + f(3) \, 1_{[3,4)}$$
$$f^{(2)} - f^{(4)} = [f(0) - f(1)] \, 1_{[1,2)} + [f(2) - f(3)] \, 1_{[3,4)}$$
$$[f^{(2)} - f^{(4)}]^2 = [f(0) - f(1)]^2 \, 1_{[1,2)} + [f(2) - f(3)]^2 \, 1_{[3,4)}$$

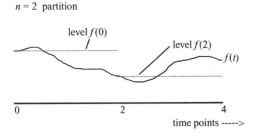

$n = 2$ partition

Figure 3.2   Approximation of f on two intervals

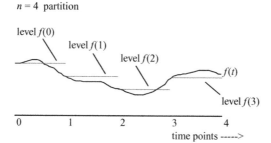

Figure 3.3   Approximation of f on four intervals

as all cross products are zero since all pairs of intervals are non-overlapping

$$\mathbb{E}\{[f^{(2)} - f^{(4)}]^2\} = \mathbb{E}\{[f(0) - f(1)]^2 \, 1_{[1,2)} + [f(2) - f(3)]^2 \, 1_{[3,4)}\}$$

Then

$$\|f^{(2)} - f^{(4)}\|^2 = \int_{t=0}^{4} [\mathbb{E}\{[f(0) - f(1)]^2 \, 1_{[1,2)}\} + \mathbb{E}\{[f(2) - f(3)]^2 \, 1_{[3,4)}\}] \, dt$$

$$= \int_{t=1}^{2} \mathbb{E}\{[f(0) - f(1)]^2\} \, dt + \int_{t=3}^{4} \mathbb{E}\{[f(2) - f(3)]^2\} \, dt$$

In the case where f is Brownian motion

$$\|f^{(2)} - f^{(4)}\|^2 = \int_{t=1}^{2} \mathbb{E}\{[B(0) - B(1)]^2\} \, dt + \int_{t=3}^{4} \mathbb{E}\{[B(2) - B(3)]^2\} \, dt$$

Substituting $\mathbb{E}\{[B(0) - B(1)]^2\} = 1$ and $\mathbb{E}\{[B(2) - B(3)]^2\} = 1$ gives

$$\|f^{(2)} - f^{(4)}\|^2 = \int_{t=1}^{2} 1 \, dt + \int_{t=3}^{4} 1 \, dt = 1(2 - 1) + 1(4 - 3) = 2$$

Using the norm of f, the difference between these two step-functions is

$$\|f^{(2)} - f^{(4)}\| = \sqrt{2}$$

Norm of I   The corresponding stochastic integrals can be compared similarly. This is illustrated for the case where f is Brownian motion. Figure 3.5 shows $I[f^{(2)}] - I[f^{(4)}]$ and Figure 3.6 shows $I[f^{(2)} - f^{(4)}]$.

Subinterval	$f^{(2)}$	$f^{(4)}$	$f^{(2)} - f^{(4)}$	$[f^{(4)} - f^{(2)}]^2$
$[0, 1)$	$f(0)$	$f(0)$	$f(0) - f(0) = 0$	0
$[1, 2)$	$f(0)$	$f(1)$	$f(0) - f(1)$	$[f(0) - f(1)]^2$
$[2, 3)$	$f(2)$	$f(2)$	$f(2) - f(2) = 0$	0
$[3, 4)$	$f(2)$	$f(3)$	$f(2) - f(3)$	$[f(2) - f(3)]^2$

Figure 3.4   Difference between f approximations

Subinterval	$I[f^{(2)}]$	$I[f^{(4)}]$	$I[f^{(2)}] - I[f^{(4)}]$
[0, 1)	$B(0).[B(1) - B(0)]$	$B(0).[B(1) - B(0)]$	0
[1, 2)	$B(0).[B(2) - B(1)]$	$B(1).[B(2) - B(1)]$	$[B(0) - B(1)].[B(2) - B(1)]$
[2, 3)	$B(2).[B(3) - B(2)]$	$B(2).[B(3) - B(2)]$	0
[3, 4)	$B(2).[B(4) - B(3)]$	$B(3).[B(4) - B(3)]$	$[B(2) - B(3)].[B(4) - B(3)]$

Figure 3.5   Difference between discrete integrals

The values in the right-hand columns of Figures 3.5 and 3.6 are the same, which confirms that

$$I[f^{(2)}] - I[f^{(4)}] = I[f^{(2)} - f^{(4)}]$$
$$I[f^{(2)} - f^{(4)}] = [B(0) - B(1)][B(2) - B(1)] \, 1_{[1,2)}$$
$$+ [B(2) - B(3)][B(4) - B(3)] \, 1_{[3,4)}$$
$$I[f^{(2)} - f^{(4)}]^2 = [B(0) - B(1)]^2[B(2) - B(1)]^2 \, 1_{[1,2)}$$
$$+ [B(2) - B(3)]^2[B(4) - B(3)]^2 \, 1_{[3,4)}$$

Then

$$||I[f^{(2)} - f^{(4)}]||^2 = \mathbb{E}[(I[f^{(2)} - f^{(4)}])^2]$$
$$= \mathbb{E}\{[B(0) - B(1)]^2[B(2) - B(1)]^2 \, 1_{[1,2)}\}$$
$$+ \mathbb{E}\{[B(2) - B(3)]^2[B(4) - B(3)]^2 \, 1_{[3,4)}\}$$
$$= \mathbb{E}\{[B(0) - B(1)]^2\}\mathbb{E}\{[B(2) - B(1)]^2\}$$
$$+ \mathbb{E}\{[B(2) - B(3)]^2\}\mathbb{E}\{[B(4) - B(3)]^2\}$$
$$= (1)(1) + (1)(1) = 2$$
$$||I[f^{(2)} - f^{(4)}]|| = \sqrt{2} = ||f^{(2)} - f^{(4)}||$$

In summary, because the approximation of f by step-function $f^{(n)}$ is arbitrarily close, as measured by the norm of $f - f^{(n)}$, the stochastic integral of f can also be approximated arbitrarily close by the discrete stochastic integral of $f^{(n)}$ as measured by the norm of the integral. Clearly, for the norm of f to exist, f must be such that $\int_{t=0}^{T} \mathbb{E}[f(t, \omega)]^2 \, dt$ exists. The convergence in mean-square statement was based on $\mathbb{E}\{(I[f^{(n)}] - I[f^{(2n)}])^2\} \to 0$. This is the same as $\mathbb{E}\{(I[f^{(n)} - f^{(2n)}]^2)\} \to 0$ which is the same as the squared norm of $I[f^{(n)} - f^{(2n)}] \to 0$. Stochastic integration is said to be a norm (or distance) preserving operation.

Subinterval	$f^{(2)}$	$f^{(4)}$	$f^{(2)} - f^{(4)}$	$I[f^{(2)} - f^{(4)}]$
[0, 1)	$B(0)$	$B(0)$	0	0
[1, 2)	$B(0)$	$B(1)$	$[B(0) - B(1)]$	$[B(0) - B(1)].[B(2) - B(1)]$
[2, 3)	$B(2)$	$B(2)$	0	0
[3, 4)	$B(2)$	$B(3)$	$[B(2) - B(3)]$	$[B(2) - B(3)].[B(4) - B(3)]$

Figure 3.6   Discrete integral of difference between approximations

Full mathematical details of the approximations and convergence in this section can be found in *Brzeźniak/Zastawniak* Section 7.1, *Kuo* Section 4.4, *Capasso/Bakstein* Section 3.1, and *Korn/Korn*.

3.5 PROPERTIES OF AN ITŌ STOCHASTIC INTEGRAL

It can be shown that the properties of the discrete stochastic integral for random step-functions carry over to the continuous time limit, the stochastic integral $I(f) = \int_{t=0}^{T} f(t, \omega)\, dB(t)$.

(i) *Linearity property*

$$\int_{t=0}^{T} \alpha f(t, \omega)\, dB(t) + \int_{t=0}^{T} \beta g(t, \omega)\, dB(t)$$
$$= \int_{t=0}^{T} [\alpha f(t, \omega) + \beta g(t, \omega)]\, dB(t)$$

(ii) *Distribution Properties*   The probability distribution of $I(f)$ cannot be identified in general. In the special case where f is non-random the discrete integral property carries over so $I(f)$ is then normally distributed. This is also shown directly in Section 3.7. The expected value of an Itō stochastic integral is zero.

$$\mathbb{E}[I(f)] = \mathbb{E}\left[\int_{t=0}^{T} f(t, \omega)\, dB(t)\right] = 0$$

The variance of an Itō stochastic integral is an ordinary integral

$$\mathbb{V}\mathrm{ar}[I(f)] = \mathbb{E}[I(f)^2] = \underbrace{\mathbb{E}\left[\left\{\int_{t=0}^{T} f(t, \omega)\, dB(t)\right\}^2\right] = \int_{t=0}^{T} \mathbb{E}[f(t, \omega)^2]\, dt}_{\text{Isometry}}$$

(iii) *Martingale Property*   Random variable $I(T)$ is a function of upper integration limit T. For different T it is a random process which is a *martingale*

$$\mathbb{E}\left[\int_{t=0}^{T} f(t, \omega)\, dB(t)|\Im(S)\right] = \int_{t=0}^{S} f(t, \omega)\, dB(t) \quad \text{for times } T > S$$

Proofs of the martingale property can be found in *Brzeźniak/Zastawniak* Section 7.3, *Kuo* Section 4.6, *Capasso/Bakstein* Section 3.2, and *Klebaner* p. 100.

(iv) **Product Property**

$$\mathbb{E}[I(f)I(g)] = \int_{t=0}^{T} \mathbb{E}[f(t, \omega)g(t, \omega)] \, dt$$

This property can also be shown directly by making use of the following. For any numbers a and b

$$(a+b)^2 = a^2 + 2ab + b^2 \quad \text{so} \quad ab = \tfrac{1}{2}(a+b)^2 - \tfrac{1}{2}a^2 - \tfrac{1}{2}b^2$$

Applying the latter to $a = I(f)$ and $b = I(g)$ gives

$$I(f)I(g) = \tfrac{1}{2}\{I(f) + I(g)\}^2 - \tfrac{1}{2}I(f)^2 - \tfrac{1}{2}I(g)^2$$

Using the *linearity* property, $I(f) + I(g) = \int_{t=0}^{T}\{f(t, \omega) + g(t, \omega)\} \, dB(t)$ gives

$$I(f)I(g) = \tfrac{1}{2}\left\{\int_{t=0}^{T}\{f(t, \omega) + g(t, \omega)\} \, dB(t)\right\}^2 - \tfrac{1}{2}I(f)^2 - \tfrac{1}{2}I(g)^2$$

Taken the expected value, and applying the *isometry* property to each term, gives

$$\begin{aligned}
\mathbb{E}[I(f)I(g)] &= \tfrac{1}{2}\int_{t=0}^{T} \mathbb{E}[\{f(t, \omega) + g(t, \omega)\}^2] \, dt \\
&\quad - \tfrac{1}{2}\mathbb{E}[I(f)^2] - \tfrac{1}{2}\mathbb{E}[I(g)^2] \\
&= \tfrac{1}{2}\int_{t=0}^{T} \mathbb{E}[f(t, \omega)^2] \, dt + \tfrac{1}{2}\int_{t=0}^{T} \mathbb{E}[g(t, \omega)^2] \, dt \\
&\quad + \int_{t=0}^{T} \mathbb{E}[f(t, \omega)g(t, \omega)] \, dt \\
&\quad - \tfrac{1}{2}\int_{t=0}^{T} \mathbb{E}[f(t, \omega)^2] \, dt - \tfrac{1}{2}\int_{t=0}^{T} \mathbb{E}[g(t, \omega)^2] \, dt \\
&= \int_{t=0}^{T} \mathbb{E}[f(t, \omega)g(t, \omega)] \, dt
\end{aligned}$$

The Itō integral viewed as a function of its upper integration limit is a random process which has infinite first variation and non-zero quadratic variation. Its paths are nowhere differentiable. This is discussed in *Klebaner* pp. 101, 102.

Evaluating a stochastic integral by using its limit definition is cumbersome, as can be experienced by working exercise [3.9.5]. Instead, it is normally done by a stochastic calculus rule known as Itō's formula, which is the subject of Chapter 4. The situation is the same as in ordinary calculus where a Riemann integral is not evaluated by using its definition, but by using a set of ordinary calculus rules. Although conceptually the Itō stochastic integral is a random variable, an explicit expression for it can only be found in a few special cases. An often quoted example is $\int_{t=0}^{T} B(t) \, dB(t)$. This is first evaluated as the limit of a discrete integral in exercise [3.9.5] and then by stochastic calculus in Section 4.4.4. When no closed form expression can be found, the probability distribution of the integral can be approximated by simulation.

The integrand used in the discrete Itō stochastic integral is fixed at the left endpoint of each time interval. Section 3.1 showed that this is natural in a finance context as the composition of the portfolio has to be fixed before the change in the values of the assets become known. This makes the stochastic integral a martingale. What the situation would be if a different endpoint were used is shown in the next section.

3.6   SIGNIFICANCE OF INTEGRAND POSITION

The significance of the position of the integrand f is now illustrated for the case where f is a random step-function which approximates Brownian motion. The expected value of a discrete stochastic integral is computed using different endpoints for f. These integrals are under different definitions.

Using the *left* endpoint t_k, the discrete stochastic integral is $L_n \stackrel{\text{def}}{=} \sum_{k=0}^{n-1} B(t_k) [B(t_{k+1}) - B(t_k)]$ where $B(t_k)$ and $[B(t_{k+1}) - B(t_k)]$ are *independent*. The value of $B(t_k)$ will be known before the value of $[B(t_{k+1}) - B(t_k)]$ is generated. The general term in the approximation is $B(t_k)[B(t_{k+1}) - B(t_k)]$. Its expected value, taken at time t_k, is $\mathbb{E}\{B(t_k)[B(t_{k+1}) - B(t_k)]\}$. Conditioning on the history $\Im_k$ of the process through time t_k, this is written as the expected value of a conditional expectation $\mathbb{E}\{\mathbb{E}(B(t_k)[B(t_{k+1}) - B(t_k)]|\Im_k)\}$. Position $B(t_k)$ is independent of $[B(t_{k+1}) - B(t_k)]$, and when $\Im_k$ is given, it is a known number (a realization of a random variable). So the above becomes $\mathbb{E}\{B(t_k)\mathbb{E}[B(t_{k+1}) - B(t_k)|\Im_k]\}$. With $\mathbb{E}[B(t_{k+1}) - B(t_k)|\Im_k] = 0$ the result is $\mathbb{E}\{B(t_k)0\} = 0$. Therefore using the *left* endpoint, the expected value of the stochastic integral equals zero. (Using the result from Section 3.3 with $f = B$ would, of course, have given the same.)

Using the *right* endpoint t_k, define a discrete stochastic integral as $R_n \stackrel{\text{def}}{=} \sum_{k=0}^{n-1} B(t_{k+1})[B(t_{k+1}) - B(t_k)]$. The integrand $B(t_{k+1})$ is now not non-anticipating; it is not the Itō type integral and could not be used in the setting of Section 3.1. Here $B(t_{k+1})$ and integrator $[B(t_{k+1}) - B(t_k)]$ are *no longer independent*. The general term in the approximation is now $B(t_{k+1})[B(t_{k+1}) - B(t_k)]$. To evaluate $\mathbb{E}B(t_{k+1})[B(t_{k+1}) - B(t_k)]$, *decompose* $B(t_{k+1})$ into the known value $B(t_k)$ and the random variable $B(t_{k+1}) - B(t_k)$, as in the analysis of martingales in Chapter 2. That gives

$$\mathbb{E}\{(B(t_k) + [B(t_{k+1}) - B(t_k)])[B(t_{k+1}) - B(t_k)]\}$$
$$= \mathbb{E}\{B(t_k)[B(t_{k+1}) - B(t_k)]\} + \mathbb{E}\{[B(t_{k+1}) - B(t_k)]^2\}$$

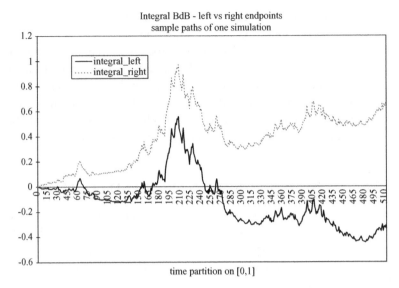

Figure 3.7 Simulated path of different integral definitions

The first term was shown above to equal zero. So the resulting expression is $\mathbb{E}\{[B(t_{k+1}) - B(t_k)]^2\} = \Delta t$. Summing over all n subintervals then gives $n\Delta t = n(T/n) = T$. Thus using the *right* endpoint, the expected value of that stochastic integral equals T. Indeed, the difference $R_n - L_n = \sum_{k=0}^{n-1}[B(t_{k+1}) - B(t_k)][B(t_{k+1}) - B(t_k)]$ is shown in Annex C to converge in mean-square to T.

The difference is illustrated in Figure 3.7 by simulation over the time period [0, 1] using 512 subintervals.

The terminal value at time 1 was recorded for 200 simulations. The results are presented in Figure 3.8.

Kuo introduces the stochastic integral in Section 4.1 by asking: If one wants the stochastic integral to be a martingale, should the integrand be taken at the left endpoint or at the right endpoint? This is discussed for the example $\int B(t)\, dB(t)$.

	integral_left	integral_right
sample_mean	-0.037770	0.964537
exact_mean	0	1
sample_variance	0.492436	0.498138
exact_variance	0.5	0.5

Figure 3.8 Simulation results left endpoint versus right endpoint

3.7   ITŌ INTEGRAL OF NON-RANDOM INTEGRAND

The special case where the integrand is a non-random function of time, $\sigma(t)$ say, arises frequently, for example in interest rate modelling. The stochastic integral $I(t) = \int_{s=0}^{t} \sigma(s)\,dB(s)$ has

$$\mathbb{E}[I(t)] = 0 \quad \text{and} \quad \mathbb{V}\text{ar}[I(t)] = \mathbb{E}[I(t)^2] = \int_{s=0}^{t} \mathbb{E}[\sigma(s)^2]\,ds$$

$$= \int_{s=0}^{t} \sigma(s)^2\,ds$$

What is its probability distribution? The discrete stochastic integral of which this $I(t)$ is the limit, is a sum of many independent Brownian motion increments each weighted by a coefficient which is a value of the non-random function σ. This discrete stochastic integral is a linear combination of independent normal random variables, and is thus normal. The intuition is therefore that $I(t)$ is normally distributed. To verify this, recall that the type of distribution of a random variable is uniquely specified by its moment generating function (mgf). Thus the mgf of $I(t)$ needs to be derived. The mgf of a normally distributed random variable Z which has mean zero and variance σ^2 is $\mathbb{E}[\exp(\theta Z)] = \exp(\frac{1}{2}\theta^2\sigma^2)$ where θ denotes the dummy parameter. The mgf of $I(t)$ is defined as $\mathbb{E}[\exp\{\theta I(t)\}]$. If this equals $\exp\{\frac{1}{2}\theta^2 \mathbb{V}\text{ar}[I(t)]\}$, then $I(t)$ is normal with mean 0 and variance $\mathbb{V}\text{ar}[I(t)]$, and $\mathbb{E}[\exp\{\theta I(t) - \frac{1}{2}\theta^2\mathbb{V}\text{ar}[I(t)]\}] = 1$. So it needs to be verified whether this expression holds. For convenience let the exponent be named X

$$X(t) \stackrel{\text{def}}{=} \theta I(t) - \tfrac{1}{2}\theta^2\mathbb{V}\text{ar}[I(t)] \qquad X(0) = 0$$

$$X(t) = \theta \int_{s=0}^{t} \sigma(s)\,dB(s) - \frac{1}{2}\theta^2 \int_{s=0}^{t} \sigma(s)^2\,ds$$

Thus it needs to be verified whether $\mathbb{E}[\exp\{X(t)\}] = 1 = \exp\{0\} = \exp\{X(0)\}$. If this holds then the random process $\exp(X)$ is a martingale. Introduce the notation $Y(t) \stackrel{\text{def}}{=} \exp\{X(t)\}$. To verify whether Y is a martingale, the method of Chapter 2 can be used. Alternatively, use is made here of Chapter 4 and Section 5.1.1, and it is checked whether the stochastic differential of Y is without drift.

$$dY = \frac{dY}{dX}dX + \frac{1}{2}\frac{d^2Y}{dX^2}(dX)^2$$

Using

$$\frac{dY}{dX} = Y \qquad \frac{d^2Y}{dX^2} = Y$$

$$dX(t) = -\tfrac{1}{2}\theta^2\sigma(t)^2\,dt + \theta\sigma(t)\,dB(t) \qquad [dX(t)]^2 = \theta^2\sigma(t)^2\,dt$$

gives

$$dY = Y\left[-\tfrac{1}{2}\theta^2\sigma(t)^2\,dt + \theta\sigma(t)\,dB(t)\right] + \tfrac{1}{2}Y\theta^2\sigma(t)^2\,dt$$
$$= Y\theta\sigma(t)\,dB(t)$$

This is indeed without drift, so Y is a martingale. It has thus been shown that an Itō integral with a non-random integrand is a normally distributed random variable with mean zero and variance given by an ordinary integral.

3.8   AREA UNDER A BROWNIAN MOTION PATH

Integral $\int_{t=0}^{T} B(t)\,dt$ often occurs in applications, for example in interest rate modelling. The previous sections explained the construction of the integral of a random process *with respect to Brownian*, the Itō stochastic integral. The integral here is an integral of a specific Brownian motion path *with respect to time*. This is *not a stochastic integral*. If a Brownian motion path is generated from time 0 to time T then the above integral is the area under that particular path. As the path is continuous, it is Riemann integrable. As the path is random, so is the integral. For a specific T the integral is a random variable. It turns out that this random variable is normally distributed with mean 0 and variance $\tfrac{1}{3}T^3$. The derivation is along the lines of the construction of the ordinary Riemann integral. As usual, discretize the time period $[0, T]$ by partitioning into n intervals of equal length $\Delta t \overset{\text{def}}{=} T/n$ with discrete time-points $t_k = k\Delta t$, $k = 0 \ldots n$. For an interval from time-point t_{k-1} to t_k, the area under the Brownian motion path (Figure 3.9) is approximated by the rectangle $\Delta t\, B(t_k)$. So

$$\int_{t=0}^{T} B(t)\,dt = \lim_{n\to\infty}\sum_{k=1}^{n} \Delta t\, B(t_k)$$

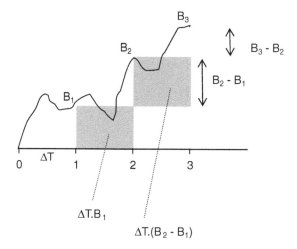

ΔT.B$_1$

ΔT.(B$_2$ - B$_1$)

all have level B$_1$
all but one have level B$_2$
etc

Figure 3.9   Area under Brownian motion path

The rectangles are:

    1st $\Delta t\, B(t_1)$
    2nd $\Delta t\, B(t_2) = \Delta t\, B(t_1) + \Delta t[B(t_2) - B(t_1)]$
    ⋮   ⋮
    nth $\Delta t\, B(t_n) = \Delta t\, B(t_1) + \cdots + \Delta t[B(t_k) - B(t_{k-1})] + \cdots$
                    $+ \Delta t[B(t_n) - B(t_{n-1})]$

Position $B(t_k)$ is the sum of the increments during the preceding intervals. The sum of the n rectangles, denoted S_n, equals

$$S_n \stackrel{\text{def}}{=} \Delta t\{n B(t_1) + (n-1)[B(t_2) - B(t_1)] + \cdots$$
$$+ k[B(t_k) - B(t_{k-1})] + \cdots + 1[B(t_n) - B(t_{n-1})]$$

This is the sum of independent normally distributed random variables, so S_n is normal with $\mathbb{E}[S_n] = 0$ and variance

$$\mathbb{V}\text{ar}[S_n] = (\Delta t)^2\{\mathbb{V}\text{ar}[n B(t_1)] + \cdots + \mathbb{V}\text{ar}[k[B(t_k) - B(t_{k-1})]]$$
$$+ \cdots + \mathbb{V}\text{ar}[1[B(t_n) - B(t_{n-1})]]\}$$
$$= (\Delta t)^2\{n^2 \Delta t + \cdots + k^2 \Delta t + \cdots + 1^2 \Delta t\}$$

$$= (\Delta t)^3 \sum_{k=1}^{n} k^2$$

$$= (\Delta t)^3 \frac{1}{6} n(n+1)(2n+1)$$

$$= \frac{T^3}{n^3} \frac{1}{6} n(n+1)(2n+1)$$

$$= T^3 \frac{1}{6} \left(1 + \frac{1}{n}\right) \left(2 + \frac{1}{n}\right)$$

Using $\int_{t=0}^{T} B(t)\, dt = \lim_{n \to \infty} S_n$

$$\mathbb{E}\left[\int_{t=0}^{T} B(t)\, dt \right] = \mathbb{E}[\lim_{n \to \infty} S_n] = \lim_{n \to \infty} \mathbb{E}[S_n] = 0$$

$$\mathbb{V}\mathrm{ar}\left[\int_{t=0}^{T} B(t)\, dt \right] = \mathbb{V}\mathrm{ar}[\lim_{n \to \infty} S_n] = \lim_{n \to \infty} \mathbb{V}\mathrm{ar}[S_n]$$

$$= \lim_{n \to \infty} T^3 \frac{1}{6} \left(1 + \frac{1}{n}\right) \left(2 + \frac{1}{n}\right) = \frac{1}{3} T^3$$

Therefore

$$\int_{t=0}^{T} B(t)\, dt \sim N\left(0, \tfrac{1}{3} T^3\right)$$

As the standard deviation of the integral is $\sqrt{\frac{1}{3} T^3}$, it has the same distribution as $\sqrt{\frac{1}{3} T^3 Z}$, where $Z \sim N(0, 1)$. Writing it as $\int_{t=0}^{T} B(t)\, dt \sim \sqrt{\frac{1}{3} T^3} Z$ can be convenient in any further analysis which uses this integral. This method is based on *Epps* and *Klebaner*. The integral is rederived in Section 4.8 using stochastic calculus.

3.9   EXERCISES

The subject of Exercises [3.9.1], [3.9.2], and [3.9.10], is the discrete Itō stochastic integral of a random step-function. As usual, the integral is over the time period $[0, T]$ partitioned into n subintervals of equal length T/n. The time-points in the partition are denoted $t_k \stackrel{\text{def}}{=} k(T/n)$, $t_0 = 0$ and $t_n = T$. The random step-function integrand corresponding to this partition is denoted $f^{(n)}$ and has level $B(t_k)$ on the left-closed right-open subinterval $[t_k, t_{k+1})$.

$$f^{(n)} = B(t_0) 1_{[t_0,t_1)} + B(t_1) 1_{[t_1,t_2)} + \cdots + B(t_k) 1_{[t_k,t_{k+1})} + \cdots$$
$$+ B(t_{n-1}) 1_{[t_{n-1},t_n)}$$
$$= \sum_{k=0}^{n-1} B(t_k) 1_{[t_k,t_{k+1})}$$

[3.9.1] (a) Give the expression for the discrete Itō stochastic integral of $f^{(n)}$, call it $I^{(n)}$, and show that it is a martingale.

(b) Compute the variance of $I^{(n)}$ (use the results derived earlier; no need to rederive from scratch).

(c) As n tends to infinity, determine the limiting value of this variance.

[3.9.2] Refine the partition by halving only the first subinterval $[0, t_1]$. This adds just one time-point to the partition, $t_{\frac{1}{2}} \overset{\text{def}}{=} \frac{1}{2}(T/n)$. The refined partition has $(n - 1)$ subintervals of length T/n, and two subintervals of length $\frac{1}{2}(T/n)$, in total $(n + 1)$ subintervals.

(a) Write the expression for $f^{(n+1)}$, the random step-function corresponding to the refined partition.

(b) Write the expression for the norm of the difference between $f^{(n+1)}$ and $f^{(n)}$, that is $|| f^{(n+1)} - f^{(n)} ||$.

(c) As n tends to infinity, show that the expression in (b) tends to zero.

(d) Write the expression for the discrete stochastic integral corresponding to $f^{(n+1)}$; call it $I^{(n+1)}$.

(e) Write the expression for the norm of the difference of the two integrals $I^{(n+1)}$ and $I^{(n)}$, that is $|| I^{(n+1)} - I^{(n)} ||$.

(f) As n tends to infinity, show that the expression in (e) tends to zero.

(g) Having done all of the above, what has been demonstrated?

[3.9.3] Compute the expected value of the square of $I(T) \overset{\text{def}}{=} \int_{t=0}^{T} B(t) \, dB(t)$ in two ways

(a) Using the property of Itō stochastic integrals.

(b) Using the closed form expression $\int_{t=0}^{T} B(t) \, dB(t) = \frac{1}{2} B(T)^2 - \frac{1}{2} T$.

(c) Compute the variance of $I(T)$.

[3.9.4] Find a closed form expression for $\int_{s=t_{i-1}}^{t_i} [\int_{y=t_{i-1}}^{s} dB(y)] \, dB(s)$. This expression is used in the Milstein simulation scheme for stochastic differential equations.

[3.9.5] Specify the discrete stochastic integral which converges to $\int_{t=0}^{T} B(t) \, dB(t)$. Show that the sequence of these discrete stochastic integrals converges in mean-square to $\frac{1}{2} B(T)^2 - \frac{1}{2} T$. To rearrange the expressions, make use of the identity $ab = \frac{1}{2}[(a+b)^2 - (a^2 + b^2)]$ which comes from rearranging $(a+b)^2 = a^2 + 2ab + b^2$.

[3.9.6] Derive the variance of $TB(T) - \int_{t=0}^{T} B(t) \, dt$.

[3.9.7] Derive the variance of $\int_{t=0}^{T} \sqrt{|B(t)|} \, dB(t)$.

[3.9.8] Derive the variance of $\int_{t=0}^{T} [B(t) + t]^2 \, dB(t)$.

[3.9.9] (a) Construct a simulation of a discrete stochastic integral for non-random step-functions.
(b) Construct a simulation of a discrete stochastic integral for random step-functions.

[3.9.10] Construct a simulation and analyse the convergence of $I^{(n)}$ as defined in Exercise [3.9.2] for the time period $[0, 1]$. Initially use the partition into $n = 2^8 = 256$ subintervals. Run at least 1000 simulations of the discrete stochastic integral for this n. Show a chart of two typical sample paths. Show a histogram of the terminal value of the integral, and compute the mean and variance of the simulated values and compare these to the theoretical values. Repeat the above by doubling n to $2^{10} = 1024$ then to 2048. Compare the results against the mean and variance of

$$\int_{t=0}^{1} B(t) \, dB(t) = \frac{1}{2} B(1)^2 - \frac{1}{2}.$$

[3.9.11] Construct a simulation for the area under the path of a Brownian motion. Show a cumulative histogram of the results and compare this with the probability distribution of the area. Compare the sample mean and variance with the theoretical values.

3.10   SUMMARY

This chapter defined the *discrete* stochastic integral of the non-anticipating *random* process f with respect to Brownian motion as $\sum_{k=0}^{n-1} f(t_k, \omega)[B(t_{k+1}) - B(t_k)]$. The key feature is that the value of f is taken at t_k before the Brownian motion increment $[B(t_{k+1}) - B(t_k)]$ over the subsequent time interval becomes known. If $\int_{t=0}^{T} \mathbb{E}[f(t, \omega)^2] \, dt$ exists, then any random process $f(t, \omega)$ can be approximated arbitrarily closely by random step-functions, here denoted $f^{(n)}$, in the sense that

$$\int_{t=0}^{T} \mathbb{E}\big[f(t, \omega) - f^{(n)}(t, \omega)\big]^2 \, dt \to 0 \quad \text{as } n \to \infty$$

The value of integral $\int_{t=0}^{T} \mathbb{E}[f(t, \omega) - f^{(n)}(t, \omega)]^2 \, dt$ depends on n. Repeating this calculation for a larger value of n (a finer partition), gives another value of the corresponding integral. Making n larger and larger gives a sequence of values which converges to zero. This approximation can also be written in terms of the difference between two random step-functions, one based on n intervals, the other based on $2n$ intervals,

$$\int_{t=0}^{T} \mathbb{E}\big[f^{(n)}(t, \omega) - f^{(2n)}(t, \omega)\big]^2 \, dt$$

The expected value of the square of the difference between the *discrete stochastic integrals* corresponding to $f^{(n)}$ and $f^{(2n)}$ is $\mathbb{E}[\{I[f^{(n)}] - I[f^{(2n)}]\}^2]$, which equals $\int_{t=0}^{T} \mathbb{E}[f^{(n)}(t, \omega) - f^{(2n)}(t, \omega)]^2 \, dt$. Since by linearity $I[f^{(n)}] - I[f^{(2n)}] = I[f^{(n)}] - f^{(2n)}]$. This goes to zero as $n \to \infty$. That means the sequence $I(f^{(n)})$ has a limit, and this limit is called the Itō stochastic integral. The stochastic integration concept is based on *closeness of random quantities in the mean-square sense*. A numerical illustration is given in Figure 3.10 for $\int_{t=0}^{1} B(t) \, dB(t)$. The simulation compares the values of the discrete stochastic integral for doubled values of n, and also compares them to the simulated values of the exact expression.

   This text only covers Itō integrals with respect to Brownian motion in one dimension, where the random integrand process f is such that $\int_{t=0}^{T} \mathbb{E}[f(t, \omega)^2] \, dt$ is finite. The martingale property of the stochastic integral stems from the use of integrands which have no knowledge of the

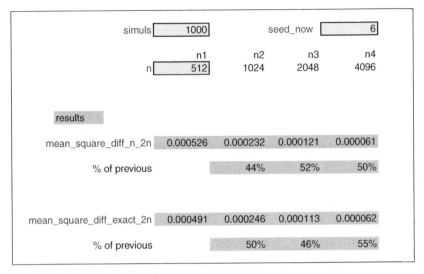

Figure 3.10 Simulation of $\int_{t=0}^{1} B(t)\, dB(t)$

future (non-anticipating); these arise naturally in a finance context. When the integrand is non-random, the stochastic integral is a normally distributed random variable[8]; this has many applications. The area under the path of a Brownian motion is not a stochastic integral. Extensions to integrals based on weaker conditions for the integrand are discussed in *Kuo*. That book also covers stochastic integrals with respect to a continuous time martingale which can have jumps, and multi-dimensional Brownian motion integrals. *Klebaner*'s book uses the concept of convergence in probability and is mathematically more advanced; *Capasso/Bakstein* first use convergence in mean square, thereafter convergence in probability.

3.11   A TRIBUTE TO KIYOSI ITŌ

More on Itō and his work can be found on the Internet in *Wikipedia*. The mathematical aspects are summarized here by *Protter* as *The Work of Kiyosi Itô* in NOTICES OF THE AMS JUNE/JULY 2007.

---

[8] *Kuo* starts with the construction of these integrals, in Section 2.3, so-called Wiener integrals.

The Work of Kiyosi Itô

Philip Protter

The *Notices* solicited the following article describing the work of Kiyosi Itô, recipient of the 2006 Gauss Prize. The International Mathematical Union also issued a news release, which appeared in the November 2006 issue of the *Notices*.

On August 22, 2006, the International Mathematical Union awarded the Carl Friedrich Gauss Prize at the opening ceremonies of the International Congress of Mathematicians in Madrid, Spain. The prizewinner is Kiyosi Itô. The Gauss prize was created to honor mathematicians whose research has had a profound impact not just on mathematics itself but also on other disciplines.

To understand the achievements of Itô, it is helpful to understand the context in which they were developed. Bachelier in 1900, and Einstein in 1905, proposed mathematical models for the phenomenon known as Brownian motion. These models represent the random motion of a very small particle in a liquid suspension. Norbert Wiener and collaborators showed in the 1920s that Einstein's model exists as a stochastic process, using the then-new ideas of Lebesgue measure theory. Many properties of the process were established in the 1930s, the most germane for this article being that its sample paths are of infinite variation on any compact time interval, no matter how small. This made the Riemann–Stieltjes integration theory inapplicable. Wiener wanted to use such integrals to study filtering theory and signal detection, important during the second world war. Despite these problems he developed a theory of integrals, known today as Wiener integrals, where the integrands are non-random functions. This served his purpose but was unsatisfying because it ruled out the study of stochastic differential equations, among other things.

The problem in essence is the following: how can one define a stochastic integral of the form $\int_0^t H_s dW_s$, where H has continuous sample paths and W is a Wiener process (another name for Brownian motion), as the limit of Riemann-style sums? That is, to define an integral as the limit of

Philip Protter is professor of operations research at Cornell University. His email address is pep4@cornell.edu.

sums such as $\sum_{1 \leq i \leq n} H_{\xi_i}(W_{t_{i+1}} - W_{t_i})$, with convergence for all such H. Unfortunately as a consequence of the Banach–Steinhaus theorem, W must then have sample paths of finite variation on compact time intervals. What Itô saw, and Wiener missed, was that if one restricts the class of potential integrands H to those that are adapted to the underlying filtration of sigma algebras generated by the Wiener process, and if one restricts the choice of $\xi_t \in [t_i, t_{i+1})$ to t_i, then one can use the independence of the increments of the Wiener process in a clever way to obtain the convergence of the sums to a limit. This became the stochastic integral of Itô. One should note that Itô did this in the mathematical isolation of Japan during the second world war and was one of the pioneers (along with G. Maruyama) of modern probability in Japan, which has since spawned some of the world's leading probabilists. Moreover since Jean Ville had named martingales as such only in 1939, and J.L. Doob had started developing his theory of martingales only in the 1940s, Itô was unaware of the spectacular developments in this area that were happening in the U.S., France, and the Soviet Union. Thus modern tools such as Doob's martingale inequalities were unavailable to Itô, and his creativity in the proofs, looked at today, is impressive. But the key result related to the stochastic integral was Itô's change of variables formula.

Indeed, one can argue that most of applied mathematics traditionally comes down to changes of variable and Taylor-type expansions. The classical Riemann–Stieltjes change of variables, for a stochastic process A with continuous paths of finite variation on compacts, and $f \in C^1$ is of course

$$f(A_t) = f(A_0) + \int_0^t f'(A_s)\, dA_s.$$

With the Itô integral it is different and contains a "correction term". Indeed, for $f \in C^2$ Itô proved

$$f(W_t) = f(W_0) + \int_0^t f'(W_s)\, dW_s$$
$$+ \tfrac{1}{2} \int_0^t f''(W_s)\, ds.$$

This theorem has become ubiquitous in modern probability theory and is astonishingly useful. Moreover Itô used this formula to show the existence and uniqueness of solutions of stochastic ordinary differential equations:

$$dX_t = \sigma(X_t)\, dW_t + b(X_t)\, dt;$$
$$X_0 = x_0,$$

when σ and b are Lipschitz continuous. This approach provided methods with an alternative intuition to the semigroup/partial differential equations approaches of Kolmogorov and Feller, for the study of continuous strong Markov processes, known as diffusions. These equations found applications without much delay: for example as approximations of complicated Markov chains arising in population and ecology models in biology (W. Feller), in electrical engineering where dW models white noise (N. Wiener, I. Gelfand, T. Kailath), in chemical reactions (e.g., L. Arnold), in quantum physics (P.A. Meyer, L. Accardi, etc.), in differential geometry (K. Elworthy, M. Emery), in mathematics (harmonic analysis (Doob), potential theory (G. Hunt, R. Getoor, P.A. Meyer), PDEs, complex analysis, etc.), and, more recently and famously, in mathematical finance (P. Samuelson, F. Black, R. Merton, and M. Scholes).

When Wiener was developing his Wiener integral, his idea was to study random noise, through sums of iterated integrals, creating what is now known as "Wiener chaos". However his papers on this were a mess, and the true architect of Wiener chaos was (of course) K. Itô, who also gave it the name "Wiener chaos". This has led to a key example of Fock spaces in physics, as well as in filtering theory, and more recently to a fruitful interpretation of the Malliavin derivative and its adjoint, the Skorohod integral.

Itô also turned his talents to understanding what are now known as Lévy processes, after the renowned French probabilist Paul Lévy. He was able to establish a decomposition of a Lévy process into a drift, a Wiener process, and an integral mixture of compensated compound Poisson processes, thus revealing the structure of such processes in a more profound way than does the Lévy–Khintchine formula.

In the late 1950s Itô collaborated with Feller's student H.P. McKean Jr. Together Itô and McKean published a complete description of one-dimensional diffusion processes in their classic tome, *Diffusion Processes and Their Sample Paths* (Springer-Verlag, 1965). This book was full of original research and permanently changed our understanding of Markov processes. It developed in detail such notions as local times and described essentially all of the different kinds of behavior the sample paths of diffusions could manifest. The importance of Markov processes for applications, and especially that of continuous Markov processes (diffusions), is hard to overestimate. Indeed, if one is studying random phenomena evolving through time, relating it to a Markov process is key to understanding it, proving properties of it, and making predictions about its future behavior.

Later in life, when conventional wisdom holds that mathematicians are no longer so spectacular, Itô embraced the semimartingale-based theory of stochastic integration, developed by H. Kunita, S. Watanabe, and principally P.A. Meyer and his school in France. This permitted him to integrate certain processes that were no longer adapted to the underlying filtration. Of course, this is a delicate business, due to the sword of Damocles Banach–Steinhaus theorem. In doing this, Itô began the theory of expansion of filtrations with a seminal paper and then left it to the work of Meyer's French school of the 1980s (Jeulin, Yor, etc.). The area became known as *grossissements de filtrations,* or in English as "the expansions of filtrations". This theory has recently undergone a revival, due to applications in finance to insider trading models, for example.

A much maligned version of the Itô integral is due to Stratonovich. While others were ridiculing this integral, Itô saw its potential for explaining parallel transport and for constructing Brownian motion on a sphere (which he did with D. Stroock), and his work helped to inspire the successful use of the integral in differential geometry, where it behaves nicely when one changes coordinate maps. These ideas have also found their way into other domains, for example in physics, in the analysis of diamagnetic inequalities involving Schrödinger operators (D. Hundertmark, B. Simon).

It is hard to imagine a mathematician whose work has touched so many different areas of applications, other than Isaac Newton and Gottfried Leibniz. The legacy of Kyosi Itô will live on a long, long time.

ACKNOWLEDGMENT

Permission by the American Mathematical Society and by Professor Philip Protter to include this article is gratefully acknowledged.

4
Itō Calculus

Thus far the discussion was about the dynamics of Brownian motion. Usually the quantity of interest is not Brownian motion itself, but a random process which depends on ('is driven by') Brownian motion. For example, the value of an option depends on the value of the underlying stock, which in turn is assumed to be driven by Brownian motion. To determine how a change in the value of the stock price affects the value of the option is a key question. Questions like these can be answered with Itō's formula which is the subject of this chapter. This formula is also known as Itō's lemma.

4.1 STOCHASTIC DIFFERENTIAL NOTATION

The analysis is normally done in terms of *stochastic differentials*, a notion which will now be introduced. Consider the integral of an ordinary function f of t, from $t = 0$ to x, $\int_{t=0}^{x} f(t) \, dt$. The value of this integral depends on the upper integration limit x, and this can be captured by writing it as $G(x) \overset{\text{def}}{=} \int_{t=0}^{x} f(t) \, dt$. According to the fundamental theorem of calculus,[1] differentiating $G(x)$ with respect to x gives $dG(x)/dx = f(x)$, the value of the function $f(t)$ at the upper integration limit $t = x$.

Inspired by this[2] it has become standard practice in stochastic calculus to write the Itō stochastic integral $I(T) = \int_{t=0}^{T} f(t, \omega) \, dB(t)$ in the form $dI(T) = f(T, \omega) \, dB(T)$. It can be easily remembered by *formally* performing the same differentiation operation on $I(T)$ as for the ordinary integral: the integral sign is omitted and t is replaced by the upper integration limit T – formally, because differentiation of a stochastic integral is not an existing concept. The notation $dI(T)$ is used as *shorthand* for $I(T)$ and it is not implied that some form of differentiation took place. The stochastic differential notation makes the rules of stochastic calculus easy to use. Similarly, a result stated in

---

[1] Recapped in Annex B, *Ordinary Integration*.
[2] See Annex B, *Ordinary Integration*, Section *Differential*.

differential form $dI(T)$ can be written in equivalent integral form $I(T)$ by formally writing out an integration, and where possible perform the integration. So

$$I(T) = \int_{t=0}^{T} f(t, \omega)\, dB(t) \quad \Leftrightarrow \quad dI(T) = f(T, \omega)\, dB(T)$$

4.2 TAYLOR EXPANSION IN ORDINARY CALCULUS

To set the stage, recall from ordinary calculus the concept of a Taylor expansion. Consider a function f of a smooth[3] ordinary non-random variable[4] x. Suppose that the value of the function at a particular point x_0 is the known value $f(x_0)$, and that the value of the function is to be determined at a point $x_0 + h$, a small distance h away from x_0. This value, $f(x_0 + h)$, can be approximated by using the steepness of the function at the present location x_0, as given by the slope $df(x)/dx$ at x_0, and the rate of change in the steepness, as given by $d^2f(x)/dx^2$ at x_0. For example, if $f(x_0)$ represents the position of a car at time x_0, then its position at time $x_0 + h$ depends on its present speed, $df(x)/dx$ at x_0, and its present acceleration, $d^2f(x)/dx^2$ at x_0. It also depends on the rate of change in the acceleration, $d^3f(x)/dx^3$ at x_0, but that is a refinement that will be neglected. If f is smooth, the change in value $f(x_0 + h) - f(x_0)$ can be approximated by the first two terms in a Taylor expansion as

$$f(x_0 + h) = f(x_0) + h\frac{df(x_0)}{dx} + \tfrac{1}{2}h^2\frac{d^2f(x_0)}{dx^2}$$

Here $df(x_0)/dx$ is *shorthand* for $df(x)/dx$ evaluated at x_0, and similarly for $d^2 f(x_0)/dx^2$.

It is convenient to define $\Delta f(x_0) \overset{\text{def}}{=} f(x_0 + h) - f(x_0)$ and rewrite the above as

$$\Delta f(x_0) = h\frac{df(x_0)}{dx} + \tfrac{1}{2}h^2\frac{d^2f(x_0)}{dx^2}$$

---

[3] Here this is shorthand for a function that is continuous in all variables, and has first and second derivatives in each of the variables, which are continuous.

[4] Also known as a *deterministic* variable.

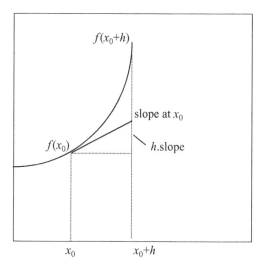

Figure 4.1 Taylor expansion in one variable

A similar approximation exists for a smooth function g of two variables x and y. Let

$$\Delta g(x_0, y_0) \overset{\text{def}}{=} g(x_0 + h, y_0 + k) - g(x_0, y_0),$$

then

$$\Delta g(x_0, y_0) =$$
$$\left.\begin{array}{l} + [\partial g(x_0, y_0)/\partial x]h \\ + [\partial g(x_0, y_0)/\partial y]k \end{array}\right\} \text{(first-order terms)}$$
$$\left.\begin{array}{l} + \tfrac{1}{2}[\partial^2 g(x_0, y_0)/\partial x^2]h^2 \\ + \tfrac{1}{2}[\partial^2 g(x_0, y_0)/\partial y^2]k^2 \end{array}\right\} \text{(second-order terms)}$$
$$+ [\partial^2 g(x_0, y_0)/\partial x \partial y]hk \quad \text{(mixed second-order terms)}$$

The standard Taylor expansion holds when the variables x and y are deterministic. There is a need for a Taylor type expansion that can also handle random variables.

4.3 ITŌ'S FORMULA AS A SET OF RULES

Typically a random process f depends explicitly on time t and on some random process X which is driven by Brownian motion, $f[t, X]$. For example f could be the value of an option which is written on stock price X. Henceforth assume that f is smooth. An expression for the change in f over a discrete time step Δt is then obtained by writing out a formal Taylor expansion which includes all second-order terms

$$\Delta f[t, X] = \frac{\partial f}{\partial t} \Delta t + \frac{\partial f}{\partial X} \Delta X + \frac{1}{2} \frac{\partial^2 f}{\partial t^2} (\Delta t)^2$$

$$+ \frac{1}{2} \frac{\partial^2 f}{\partial X^2} (\Delta X)^2 + \frac{\partial^2 f}{\partial t \partial X} \Delta t \, \Delta X$$

where $\partial f / \partial X$ is *shorthand* for the derivative of f with respect to its second argument, evaluated at the point $[t, X(t)]$, and similarly for the other derivatives. Suppose X evolves according to

$$\Delta X(t) = \mu[t, X(t)] \Delta t + \sigma[t, X(t)] \Delta B(t) \qquad (*)$$

where μ and σ are given functions of t and X. Δf has terms $(\Delta t)^2$, $\Delta t \Delta B(t)$, $[\Delta B(t)]^2$. Now take a closer look at their magnitude. For very small Δt, say 10^{-6}, $(\Delta t)^2$ is much smaller than Δt, and is therefore considered negligeable compared to Δt. $\mathbb{E}[\Delta t \, \Delta B(t)] = 0$ and $\mathbb{V}\mathrm{ar}[\Delta t \, \Delta B(t)] = (\Delta t)^2$, so for very small Δt, $\Delta t \, \Delta B(t)$ approaches 0. The variance of $[\Delta B(t)]^2$ is $2(\Delta t)^2$, which is much smaller than its expected value Δt. As $\Delta t \to 0$, $[\Delta B(t)]^2$ approaches its non-random expected value Δt. In the continuous-time limit, $(\Delta t)^2$ is written as $(dt)^2 = 0$, $\Delta t \, \Delta B(t)$ as $dt \, dB(t) = 0$, $[\Delta B(t)]^2$ as $[dB(t)]^2 = dt$. That leaves

$$df[t, X] = \left\{ \frac{\partial f}{\partial t} + \mu[t, X(t)] \frac{\partial f}{\partial X} + \frac{1}{2} \sigma[t, X(t)]^2 \frac{\partial^2 f}{\partial X^2} \right\} dt$$

$$+ \sigma[t, X(t)] \frac{\partial f}{\partial X} dB(t)$$

This is *Itō's formula*. Note that in the equivalent integral form, the first term is a time integral with a random integrand, as defined in Annex B.2.2, and the second term is the Itō integral of Chapter 3. The easiest way to use Itō's formula is to write out a second-order Taylor expansion for the function f that is being analyzed, and apply the *multiplication table* given in Figure 4.2. This is also known as *box calculus* or box algebra.

	dt	$dB(t)$
dt	0	0
$dB(t)$	0	dt

Figure 4.2 Multiplication table

Similarly, for a function of time and two processes X and Y of type (*) driven by Brownian motions B_1 and B_2 which, in general, can be correlated, Itō's formula gives

$$df = \frac{\partial f}{\partial t} dt + \frac{\partial f}{\partial X} dX + \frac{\partial f}{\partial Y} dY$$
$$+ \frac{1}{2} \frac{\partial^2 f}{\partial X^2}(dX)^2 + \frac{1}{2} \frac{\partial^2 f}{\partial Y^2}(dY)^2 + \frac{\partial^2 f}{\partial X \partial Y} dX\, dY$$

Here the term $\frac{1}{2}(\partial^2 f/\partial t^2)(dt)^2$ has already been omitted as $(dt)^2 = 0$. Similarly $(\partial^2 f/\partial t \partial X)\, dt\, dX$ and $(\partial^2 f/\partial t \partial Y)\, dt\, dY$ are not shown because dX and dY each have a dt term and a dB term, so $dt\, dX$ and $dt\, dY$ are zero.

If f does not depend on t explicitly there is no $\partial f/\partial t$ term

$$df = \frac{\partial f}{\partial X} dX + \frac{\partial f}{\partial Y} dY + \frac{1}{2} \frac{\partial^2 f}{\partial X^2}(dX)^2 + \frac{1}{2} \frac{\partial^2 f}{\partial Y^2}(dY)^2 + \frac{\partial^2 f}{\partial X \partial Y} dX\, dY$$
$$(**)$$

In the case of correlation, $dB_1(t)\, dB_2(t) = \rho\, dt$. This can be seen by using Section 1.6 and another independent $B_3(t)$.

$$\Delta B_1(t)\, \Delta B_2(t) = \Delta B_1(t)\{\rho\, \Delta B_1(t) + \sqrt{1 - \rho^2}\, \Delta B_3(t)\}$$
$$= \rho[\Delta B_1(t)]^2 + \sqrt{1 - \rho^2}\, \Delta B_1(t)\, \Delta B_3(t)$$

Taking the expected value and using the independence of B_1 and B_3 and gives

$$\mathbb{E}[\Delta B_1(t)\, \Delta B_2(t)] = \rho \mathbb{E}\{[\Delta B_1(t)]^2\} + \sqrt{1 - \rho^2}\mathbb{E}[\Delta B_1(t)]\mathbb{E}[\Delta B_3(t)]$$
$$= \rho\, \Delta t$$

Figure 4.3 shows the resulting table.

A justification of Itō's formula is given in Section 4.9. As a full rigorous proof is rather elaborate, many books just give plausibility arguments for the simple case $f[B(t)]$, which is sufficient to appreciate the basic idea. For a function of t and B, $f[t, B(t)]$, recommended references are *Brzeźniak, Shreve II, Kuo*. Technical references are *Numerical Solution of Stochastic Differential Equations* by *Kloeden/Platen*

	dt	$dB_1(t)$	$dB_2(t)$
dt	0	0	0
$dB_1(t)$	0	dt	$\rho\, dt$
$dB_2(t)$	0	$\rho\, dt$	dt

Figure 4.3   Multiplication table for correlated Brownian motions

Chapter 3, and *Korn/Korn* Excursion 3. *Stojanovic* Section 2.3 has Mathematica simulations of Itō's formula. For integration with respect to a general martingale, the corresponding Itō formula is discussed in *Kuo* Chapter 7.

4.4   ILLUSTRATIONS OF ITŌ'S FORMULA

In what follows f is a smooth function. For greater readability t is usually not shown in the derivation steps. The key to applying Itō's formula is to *first identify the variables*.

4.4.1   Frequent Expressions for Functions of Two Processes

Here equation (**) applies.

Sum Rule $f \overset{\text{def}}{=} aX(t) + bY(t)$ a and b constants
The partial derivatives are

$$\frac{\partial f}{\partial X} = a \quad \frac{\partial^2 f}{\partial X^2} = 0 \quad \frac{\partial f}{\partial Y} = b \quad \frac{\partial^2 f}{\partial Y^2} = 0$$

$$\frac{\partial^2 f}{\partial X \partial Y} = \frac{\partial}{\partial X}\left(\frac{\partial f}{\partial Y}\right) = \frac{\partial}{\partial X}(b) = 0$$

$$\frac{\partial^2 f}{\partial Y \partial X} = \frac{\partial}{\partial Y}\left(\frac{\partial f}{\partial X}\right) = \frac{\partial}{\partial Y}(a) = 0 = \frac{\partial^2 f}{\partial X \partial Y}$$

Substituting these gives the *sum rule*

$$\boxed{d[a\,X + b\,Y] = a\,dX + b\,dY}$$

Product Rule $f \overset{\text{def}}{=} X(t)\,Y(t)$
Substituting the partial derivatives

$$\frac{\partial f}{\partial X} = Y \quad \frac{\partial^2 f}{\partial X^2} = 0 \quad \frac{\partial f}{\partial Y} = X \quad \frac{\partial^2 f}{\partial Y^2} = 0$$

$$\frac{\partial^2 f}{\partial X \partial Y} = \frac{\partial}{\partial X}\left(\frac{\partial f}{\partial Y}\right) = \frac{\partial}{\partial X}(X) = 1 = \frac{\partial^2 f}{\partial Y \partial X}$$

gives the *product rule*

$$\boxed{d[XY] = Y\,dX + X\,dY + dX\,dY}$$

In integral form over $[s \le t \le u]$

$$\int_{t=s}^{u} d[X(t) Y(t)] = \int_{t=s}^{u} Y(t) \, dX(t) + \int_{t=s}^{u} X(t) \, dY(t)$$
$$+ \int_{t=s}^{u} dX(t) \, dY(t)$$

$$X(u) Y(u) - X(s) Y(s) = \int_{t=s}^{u} Y(t) \, dX(t) + \int_{t=s}^{u} X(t) \, dY(t)$$
$$+ \int_{t=s}^{u} dX(t) \, dY(t)$$

$$\int_{t=s}^{u} X(t) \, dY(t) = X(u) Y(u) - X(s) Y(s) - \int_{t=s}^{u} Y(t) \, dX(t)$$
$$- \int_{t=s}^{u} dX(t) \, dY(t)$$

This is known as the *stochastic integration by parts formula*. The first three terms on the right are the same as in ordinary calculus. The fourth term is unique to stochastic partial integration; when expressions for dX and dY are substituted it becomes an ordinary integral. The product rule can also be expressed in terms of the proportional changes in f, X, and Y.

$$\frac{d[XY]}{XY} = \frac{Y}{XY} dX + \frac{X}{XY} dY + \frac{1}{XY} dX \, dY$$

$$\boxed{\frac{d[XY]}{XY} = \frac{dX}{X} + \frac{dY}{Y} + \frac{dX}{X} \frac{dY}{Y}}$$

Ratio $f \overset{\text{def}}{=} X(t)/Y(t)$

Substituting the partial derivatives

$$\frac{\partial f}{\partial X} = \frac{1}{Y} \quad \frac{\partial f}{\partial Y} = \frac{-X}{Y^2} \quad \frac{\partial^2 f}{\partial X^2} = 0 \quad \frac{\partial^2 f}{\partial Y^2} = \frac{2X}{Y^3} \quad \frac{\partial^2 f}{\partial X \partial Y} = \frac{-1}{Y^2} = \frac{\partial^2 Z}{\partial Y \partial X}$$

gives

$$dF = \frac{1}{Y} dX + \frac{-X}{Y^2} dY + \frac{1}{2} 0 (dX)^2 + \frac{1}{2} \frac{2X}{Y^3} (dY)^2 + \frac{-1}{Y^2} dX \, dY$$

$$= \frac{X}{Y} \frac{dX}{X} - \frac{X}{Y} \frac{dY}{Y} + \frac{X}{Y} \left(\frac{dY}{Y}\right)^2 - \frac{X}{Y} \frac{dX}{X} \frac{dY}{Y}$$

$$= f \frac{dX}{X} - f \frac{dY}{Y} + f \left(\frac{dY}{Y}\right)^2 - f \frac{dX}{X} \frac{dY}{Y}$$

The *ratio rule* is

$$df = f\left[\frac{dX}{X} - \frac{dY}{Y} + \left(\frac{dY}{Y}\right)^2 - \frac{dX}{X}\frac{dY}{Y}\right]$$

This can also be expressed in terms of the proportional changes in f, X, and Y, as

$$\boxed{\frac{d[X/Y]}{X/Y} = \frac{dX}{X} - \frac{dY}{Y} + \left(\frac{dY}{Y}\right)^2 - \frac{dX}{X}\frac{dY}{Y}}$$

4.4.2   Function of Brownian Motion $f[B(t)]$

This is a function $f(x)$ of the single variable x where $x = B(t)$.

Step 1   Apply the Taylor expansion to a function of one variable, treating the random variable *as if* it were a deterministic variable, neglecting terms of order greater than 2.

$$df = \frac{df}{dx}\,dB + \frac{1}{2}\frac{d^2f}{dx^2}(dB)^2$$

Evaluate df/dx and d^2f/dx^2 at $x = B(t)$. Figure 4.4 illustrates the meaning of df/dx evaluated at $x = B(t)$. The position x where the slope is computed is given by the value of random variable $B(t)$.

Convenient shorthand for Itō's formula applied to $f[B(t)]$ is

$$df = \frac{df}{dB}\,dB + \frac{1}{2}\frac{d^2f}{dB^2}(dB)^2$$

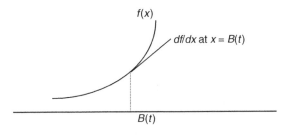

Figure 4.4   Slope at $B(t)$

Step 2   Apply the *multiplication table*
Substituting $(dB)^2 = dt$ gives

$$df = \tfrac{1}{2}\frac{d^2 f}{dB^2} dt + \frac{df}{dB} dB$$

Be aware that in this expression $d^2 f/dx^2$ and df/dx are evaluated at $x = B(t)$. In the right-hand side, the dt term is always put first in this text; some authors put the dB term first.

The simplest example is $f[B(t)] = B(t)$. Then $df/dx = 1$ and $d^2 f/dx^2 = 0$, giving

$$df = \tfrac{1}{2}0\, dt + 1\, dB(t) = dB(t)$$

as expected.

Another example is $f[B(t)] = [B(t)]^4$. Here

$$\frac{df}{dB} = 4B^3 \qquad \frac{d^2 f}{dB^2} = 12B^2 \qquad (dB)^2 = dt$$

$$df = \tfrac{1}{2}12B^2\, dt + 4B^3\, dB$$

Substituting f gives $d[B]^4 = 6B^2\, dt + 4B^3\, dB$. The equivalent integral form on $[0, T]$ is

$$[B(T)]^4 - [B(0)]^4 = 6\int_{t=0}^{T} B(t)^2\, dt + 4\int_{t=0}^{T} B(t)^3\, dB(t)$$

$$[B(T)]^4 = 6\int_{t=0}^{T} B(t)^2\, dt + 4\int_{t=0}^{T} B(t)^3\, dB(t)$$

$[B(T)]^4$ is a random variable whose expected value is

$$\mathbb{E}\{[B(T)]^4\} = 6\mathbb{E}\left\{\int_{t=0}^{T} B(t)^2\, dt\right\} + 4\mathbb{E}\left\{\int_{t=0}^{T} B(t)^3\, dB(t)\right\}$$

In the last term, $\int_{t=0}^{T} B(t)^3\, dB(t)$ is an Itō stochastic integral if $\int_{t=0}^{T} \mathbb{E}[B(t)^6]\, dt$ is finite. As B has a mgf (derived in Annex A), all its moments are finite. Here $\mathbb{E}[B(t)^6]$ is a function of t which is integrated over $[0, T]$ and is thus finite. Being an Itō stochastic integral it has an expected value of zero. In the first term, the expectation operator can be taken inside the integral according Fubini's theorem to give $6 \int_{t=0}^{T} \mathbb{E}[B(t)^2] \, dt$. As $\mathbb{E}\{B(t)^2\} = t$ this equals $6 \int_{t=0}^{T} t \, dt = 6\frac{1}{2}T^2$. So $\mathbb{E}\{B(T)^4\} = 3T^2$, which confirms the result for the fourth moment of Brownian motion derived in Annex A. It is known from probability theory that the sixth moment of a normal random variable with variance σ^2 equals $15(\sigma^2)^3$, so $\mathbb{E}[B(t)^6] = 15t^3$, but this fact is not used above.

4.4.3   Function of Time and Brownian Motion $f[t, \, B(t)]$

This is a function of two variables t and x where $x = B(t)$.

Step 1   Apply Taylor's formula for a function of two variables, treating the random variable *as if* it were a deterministic variable, neglecting terms of order greater than 2.

$$df = \frac{\partial f}{\partial t} \, dt + \frac{1}{2} \frac{\partial^2 f}{\partial t^2} (dt)^2$$
$$+ \frac{\partial f}{\partial B} \, dB + \frac{1}{2} \frac{\partial^2 f}{\partial B^2} (dB)^2$$
$$+ \frac{1}{2} \frac{\partial^2 f}{\partial t \partial B} \, dt \, dB + \frac{1}{2} \frac{\partial^2 f}{\partial B \partial t} \, dB \, dt$$

Recall from ordinary calculus that for a smooth function of two variables, the mixed second order partial derivatives are equal, $\partial^2 f / (\partial t \, \partial B) = \partial^2 f / (\partial B \, \partial t)$.

$$df = \frac{\partial f}{\partial t} \, dt + \frac{1}{2} \frac{\partial^2 f}{\partial t^2} (dt)^2 + \frac{\partial f}{\partial B} \, dB + \frac{1}{2} \frac{\partial^2 f}{\partial B^2} (dB)^2 + \frac{\partial^2 f}{\partial t \partial B} \, dt \, dB$$

Step 2   Apply the *multiplication table*
Substituting $(dt)^2 = 0$, $(dB)^2 = dt$, $dt \, dB = 0$, gives

$$df = \frac{\partial f}{\partial t} \, dt + \frac{\partial f}{\partial B} \, dB + \frac{1}{2} \frac{\partial^2 f}{\partial B^2} \, dt$$
$$= \left(\frac{\partial f}{\partial t} + \frac{1}{2} \frac{\partial^2 f}{\partial B^2} \right) dt + \frac{\partial f}{\partial B} \, dB$$

A much used expression is $f[t, B(t)] = \exp[\mu t + \sigma B(t)]$. This f is a function of two variables, t and x, $f[t, x] = \exp[\mu t + \sigma x]$, where $x = B$.

$$df = \frac{\partial f}{\partial t} dt + \frac{\partial f}{\partial B} dB + \frac{1}{2} \frac{\partial^2 f}{\partial t^2} (dt)^2 + \frac{\partial^2 f}{\partial t \partial B} dt \, dB + \frac{1}{2} \frac{\partial^2 f}{\partial B^2} (dB)^2$$

From the multiplication table, $(dt)^2 = 0$, $dt \, dB = 0$, $(dB)^2 = dt$.

$$\frac{\partial f}{\partial t} = \mu \exp[\mu t + \sigma B] \qquad \frac{\partial f}{\partial B} = \sigma \exp[\mu t + \sigma B]$$

Note that $\partial^2 f / \partial t^2$ is not needed since $(dt)^2 = 0$, $\partial^2 f / (\partial t \partial B)$ is not needed since $dt \, dB = 0$, and $\partial^2 f / \partial B^2 = \sigma^2 \exp[\mu t + \sigma B]$. Substituting these gives

$$df = \{\mu \exp[\mu t + \sigma B] + \tfrac{1}{2} \sigma^2 \exp[\mu t + \sigma B]\} \, dt + \sigma \exp[\mu t + \sigma B] \, dB$$
$$= f\{(\mu + \tfrac{1}{2}\sigma^2) \, dt + \sigma \, dB\}$$

Dividing by $f \neq 0$ gives this in the rate of return form

$$\frac{df}{f} = \left(\mu + \tfrac{1}{2}\sigma^2\right) dt + \sigma \, dB$$

4.4.4 Finding an Expression for $\int_{t=0}^{T} B(t) \, dB(t)$

Consider the stochastic integral of the random integrand $B(t)$ with respect to Brownian motion over the time period $[0, T]$, $I(T) \overset{\text{def}}{=} \int_{t=0}^{T} B(t) \, dB(t)$. This is an Itō stochastic integral because the integrand $B(t)$ is non-anticipating and

$$\int_{t=0}^{T} \mathbb{E}[B(t)^2] \, dt = \int_{t=0}^{T} t \, dt = \frac{T^2}{2}$$

is finite. The aim is to find a closed form expression for $I(T)$. This proceeds as follows. Knowing that the ordinary integral $\int x \, dx$ equals $\tfrac{1}{2} x^2$, suggests the use of $Y(T) \overset{\text{def}}{=} \tfrac{1}{2} B(T)^2$ as a trial solution. As $Y(T)$ is a function of the single variable

$$B(T), dY = \frac{dY}{dB} dB + \frac{1}{2} \frac{d^2 Y}{dB^2} (dB)^2$$

where $dY/dB = B(T)$ and $d^2Y/dB^2 = 1$. So $dY = B\,dB + \frac{1}{2}1\,dT$. In equivalent integral form,

$$\int_{t=0}^{T} dY(t) = \int_{t=0}^{T} \frac{1}{2}\,dt + \int_{t=0}^{T} B(t)\,dB(t)$$

The required integral has appeared as the second term on the right. Rearranging gives

$$\int_{t=0}^{T} B(t)\,dB(t) = Y(T) - Y(0) - \frac{1}{2}T = Y(T) - \frac{1}{2}T.$$

Substituting $Y(T) = \frac{1}{2}B(T)^2$ then gives the result

$$\int_{t=0}^{T} B(t)\,dB(t) = \frac{1}{2}B(T)^2 - \frac{1}{2}T$$

The choice of the trial solution was a lucky one because it produced $\int_{t=0}^{T} B(t)\,dB(t)$ when Itô's formula was applied. There is another way to find an expression for this integral. Suppose $\frac{1}{2}B(T)^2$ is a trial solution. If $\int_{t=0}^{T} B(t)\,dB(t)$ was equal to $\frac{1}{2}B(T)^2$ then their expected values should also be equal. But $\mathbb{E}[\int_{t=0}^{T} B(t)\,dB(t)] = 0$, whereas $\frac{1}{2}\mathbb{E}[B(T)^2] = \frac{1}{2}T$. Thus this trial solution is not the solution. Its expected value is too high by $\frac{1}{2}T$. Subtracting $\frac{1}{2}T$ gives the new trial solution $\frac{1}{2}B(T)^2 - \frac{1}{2}T$ whose expected value does equal zero, so $\frac{1}{2}B(T)^2 - \frac{1}{2}T$ is the solution.

4.4.5   Change of Numeraire

Consider $dS(t)/S(t) = r\,dt + \sigma\,dB(t)$. Let $S^*(t) \overset{\text{def}}{=} S(t)/\exp(rt)$. The denominator $\exp(rt)$ is the value at time t of a unit savings account compounded continuously at a constant rate r. $S^*(t)$ is the stock price measured in units of this savings account. The savings account is the new numeraire, and this transformation is known as a change of numeraire. The dynamics of $S^*(t)$ is now derived from the dynamics of $S(t)$ by Itô's formula. S^* is a function of the two variables t and S. Itô's formula gives

$$dS^* = \frac{\partial S^*}{\partial t} dt + \frac{\partial S^*}{\partial S} dS + \frac{1}{2} \frac{\partial^2 S^*}{\partial t^2} (dt)^2 + \frac{1}{2} \frac{\partial^2 S^*}{\partial S^2} (dS)^2$$
$$+ \frac{\partial^2 S^*}{\partial t \partial S} dt \, dS$$

$$\frac{\partial S^*}{\partial t} = -rS \exp(-rt) \qquad \frac{\partial S^*}{\partial S} = \exp(-rt)$$

$$\frac{\partial^2 S^*}{\partial S^2} = 0 \qquad (dt)^2 = 0, \text{ so} \frac{\partial^2 S^*}{\partial t^2} \text{ is not needed}$$

$$dt \, dS = dt \, (rS \, dt + \sigma S \, dB) = rS(dt)^2 + \sigma S \, dt \, dB = 0$$

Substituting the above, together with $dS = rS \, dt + \sigma S \, dB$, gives

$$dS^* = -rS \exp(-rt) \, dt + \exp(-rt)[rS \, dt + \sigma S \, dB]$$
$$\frac{dS^*(t)}{S^*(t)} = \sigma \, dB(t)$$

The drift term has dropped out as a result of the discounting. The drift of S^* is the growth relative to r. As S grows at rate r, S^* grows at rate zero. S^* is a martingale, as can be seen by integrating and taking the expected value

$$\int_{t=s}^{T} dS^*(t) = \int_{t=s}^{T} S^*(t)\sigma \, dB(t)$$

$$S^*(T) = S^*(s) + \int_{t=s}^{T} S^*(t)\sigma \, dB(t)$$

$$\mathbb{E}[S^*(T)|\mathcal{F}(s)] = S^*(s)$$

as

$$\mathbb{E}\left[\int_{t=s}^{T} S^*(t)\sigma \, dB(t)|\mathcal{F}(s)\right] = 0$$

4.4.6 Deriving an Expectation via an ODE

Let $X(t) \stackrel{\text{def}}{=} \exp[\sigma B(t)]$ so $X(0) = \exp[\sigma B(0)] = 1$. Here $\mathbb{E}[X(t)]$ is computed by solving an ordinary differential equation (ODE). The stochastic differential of X is

$$dX = \frac{dX}{dB} dB + \frac{1}{2}\frac{d^2X}{dB^2}(dB)^2$$
$$= \sigma X \, dB + \frac{1}{2}\sigma^2 X \, dt$$

In integral form over $[0, t]$

$$X(t) = X(0) + \int_{s=0}^{t} \frac{1}{2}\sigma^2 X(s) \, ds + \int_{s=0}^{t} \sigma X(s) \, dB(s)$$

Taking the expected value cancels the Itō stochastic integral term and leaves

$$\mathbb{E}[X(t)] = X(0) + \mathbb{E}\left[\int_{s=0}^{t} \frac{1}{2}\sigma^2 X(s) \, ds\right]$$

The last term is a double integral, and exchanging the order of integration is permitted (by Fubini). That moves $\mathbb{E}$ into the integrand, so

$$\mathbb{E}[X(t)] = X(0) + \int_{s=0}^{t} \frac{1}{2}\sigma^2 \mathbb{E}[X(s)] \, ds$$

For convenience introduce the notation $m(t) \stackrel{\text{def}}{=} \mathbb{E}[X(t)]$. Then

$$m(t) = X(0) + \int_{s=0}^{t} \frac{1}{2}\sigma^2 m(s) \, ds$$

This is an integral equation for the unknown function $m(t)$ which can be transformed into an ODE by differentiating with respect to the upper integration limit t, giving $dm(t)/dt = \frac{1}{2}\sigma^2 m(t)$. This type of ODE is well known and can be solved by writing $dm(t)/m(t) = \frac{1}{2}\sigma^2 \, dt$. Integration gives $\ln[m(t)] = \frac{1}{2}\sigma^2 t +$ constant; $t = 0$ gives the constant as $\ln[m(0)]$ which equals $\ln\{\mathbb{E}[X(0)]\} = \ln[X(0)] = \ln[1] = 0$ so $\ln[m(t)] = \frac{1}{2}\sigma^2 t$. Thus

$$\mathbb{E}[X(t)] = \exp[\frac{1}{2}\sigma^2 t]$$
$$\mathbb{E}\{\exp[\sigma B(t)]\} = \exp[\frac{1}{2}\sigma^2 t]$$

This agrees with applying the well-known formula for $\mathbb{E}[\exp(Y)]$ where Y is a normal random variable, $\mathbb{E}[\exp(Y)] = \exp\{\mathbb{E}[Y] + \frac{1}{2}\mathbb{V}\text{ar}[Y]$, with $\mathbb{E}[Y] = 0$ and $\mathbb{V}\text{ar}[Y] = \sigma^2 t$.

4.5 LÉVY CHARACTERIZATION OF BROWNIAN MOTION

How can one verify whether a given random process is a Brownian motion? Some special cases were the subject of exercises [1.9.1] to [1.9.3]. What follows here is more general.

Recall that Brownian motion B has the following properties:

  (i) the path of B is continuous and starts at 0
 (ii) B is a martingale, and $[dB(t)]^2 = dt$
(iii) the increment of B over time period $[s, t]$ is normal, with mean 0 and variance $(t - s)$
 (iv) the increments of B over non-overlapping time periods are independent.

Now consider a given random process M that is known to have the same first two properties as B:

(a) the path of M is continuous and starts at 0
(b) M is a martingale, and $[dM(t)]^2 = dt$

Then it turns out that M is a Brownian motion. That means that M also has the properties:

(c) the increment of M over time period $[s, t]$ is normal, with mean 0 and variance $(t - s)$
(d) the increments of M over non-overlapping time periods are independent.

Proof of Property (c) Normality with mean 0 and variance $(t - s)$ can be established by showing that the moment generating function of $M(t) - M(s)$ equals the mgf of $B(t) - B(s)$. So it has to be shown that

$$\mathbb{E}\{e^{\theta[M(t)-M(s)]}\} = e^{\frac{1}{2}\theta^2(t-s)}$$

The proof uses $f[t, M(t)] \overset{\text{def}}{=} e^{\theta M(t)-\frac{1}{2}\theta^2 t}$. The line of reasoning is that *if* f is a martingale, then

$$\mathbb{E}\{e^{\theta M(t)-\frac{1}{2}\theta^2 t}|\Im(s)\} = e^{\theta M(s)-\frac{1}{2}\theta^2 s}$$

Writing $e^{\theta M(s)}$ as $\mathbb{E}\{e^{\theta M(s)}|\Im(s)\}$ and $\mathbb{E}\{e^{-\frac{1}{2}\theta^2 t}|\Im(s)\}$ as $e^{-\frac{1}{2}\theta^2 t}$, that martingale property can be expressed as

$$\mathbb{E}\{e^{\theta[M(t)-M(s)]}|\Im(s)\} = e^{\frac{1}{2}\theta^2(t-s)} \tag{4.1}$$

Taking the expected value of this conditional expectation then gives the unconditional expectation that is required.

$$\mathbb{E}\{e^{\theta[M(t)-M(s)]}\} = \mathbb{E}[\mathbb{E}\{e^{\theta[M(t)-M(s)]}|\Im(s)\}] = e^{\frac{1}{2}\theta^2(t-s)} \tag{4.2}$$

Thus *if* f is a martingale then the increments of M have the required normality property. Showing that f is a martingale means showing that the change $f[t, M(t)] - f[s, M(s)]$ has an expected value of zero, given $\Im(s)$. An expression for this change can be found by deriving the stochastic differential of f and writing it in integral form. The Itō formula for deriving a stochastic differential presented thus far was only for a function of Brownian motion, whereas here f is not a function of B but of a general continuous path martingale. It turns out that there is also an Itō type formula for deriving the stochastic differential of $f[t, M(t)]$.[5] It is the Itō formula for B, together with its cross-multiplication table, with B replaced by M. As f is a function of the two variables t and $M(t)$, that Itō formula gives

$$df = \frac{\partial f}{\partial t}\,dt + \frac{\partial f}{\partial M}\,dM + \frac{1}{2}\frac{\partial^2 f}{\partial M^2}(dM)^2 + \frac{\partial^2 f}{\partial t\partial M}\,dt\,dM$$

where

$$\frac{\partial f}{\partial t} = -\tfrac{1}{2}\theta^2 f \qquad \frac{\partial f}{\partial M} = \theta f \qquad \frac{\partial^2 f}{\partial M^2} = \theta^2 f$$
$$(dM)^2 = dt \text{ is given} \qquad dt\,dM = 0$$

Substituting these gives $df(u) = \theta f(u)\,dM(u)$. In equivalent integral form over $[s, t]$

$$f[t, M(t)] - f[s, M(s)] = \theta \int_{u=s}^{t} f[t, M(u)]\,dM(u)$$

Taking the expected value, given $\Im(s)$, gives

$$\mathbb{E}\{f[t, M(t)|\Im(s)\} = f[s, M(s)] + \theta\mathbb{E}\left\{\int_{u=s}^{t} f[t, M(u)]\,dM(u)|\Im(s)\right\}$$

It can be shown[6] that the last term is zero, just like the expected value of a stochastic integral with respect to Brownian motion, thus f is a martingale.

---

[5] This is discussed in *Kuo* Chapter 7.
[6] See also exercise [4.10.11].

Proof of Property (d)   To show the independence of the increments of process M, use can be made of the *property* (shown in Section 2.7) that if an increment $M(t) - M(s)$ is independent of the information up to s, $\Im(s)$, then the increments of M over non-overlapping time intervals are independent. The left-hand side of expression (1) is a conditional expectation, thus a random variable, but the right-hand side is non-random, so the left-hand side is not in fact random. So (1) and (2) show that $e^{\theta[M(t)-M(s)]}$ is independent of $\Im(s)$. As $e^{\theta[M(t)-M(s)]}$ is a function of $M(t) - M(s)$ it follows that $M(t) - M(s)$ is also independent of $\Im(s)$.

This characterization of when a process is a Brownian motion was established by the French mathematician Paul Lévy who did much work on the path properties of Brownian motion. In this text it is used for simplifying combinations of Brownian motion in Section 4.6, and in the proof of the Girsanov theorem in Section 7.4.

References are *Kuo* section 8.4, *Capasso/Bakstein* section 4.3, *Lin* section 5.7. In the latter, the analysis is based directly on the exponential martingale $e^{\theta M(t) - \frac{1}{2}\theta^2 t}$.

4.6   COMBINATIONS OF BROWNIAN MOTIONS

This continues the discussion of linear combinations of Brownian motions. In Section 1.6 it was shown that the linear combination of two *independent* standard Brownian motions, $B_1(t)$ and $B_2(t)$, with constant $|\gamma| \leq 1$,

$$B_3(t) \overset{\text{def}}{=} \gamma B_1(t) + \sqrt{1 - \gamma^2}\, B_2(t)$$

is a Brownian motion. The proof used an elementary method. It will now be shown again using Lévy's characterization of a Brownian motion. This requires showing that

(a) $B_3(0) = 0$
(b) B_3 is continuous and is a martingale
(c) $[dB_3(t)]^2 = dt$

Condition (a) is clear. Regarding condition (b), as both B_1 and B_2 are continuous, B_3 is continuous. Using times $s < t$, $\mathbb{E}[B_1(t)|\Im(s)] = B_1(s)$ and $\mathbb{E}[B_2(t)|\Im(s)] = B_1(s)$. The conditional expected value of the linear combination is

$$\mathbb{E}[B_3(t)|\Im(s)] = \mathbb{E}[\gamma B_1(t) + \sqrt{1-\gamma^2} B_2(t)|\Im(s)]$$
$$= \mathbb{E}[\gamma B_1(t)|\Im(s)] + \mathbb{E}[\sqrt{1-\gamma^2} B_2(t)|\Im(s)]$$
$$= \gamma \mathbb{E}[B_1(t)|\Im(s)] + \sqrt{1-\gamma^2} \mathbb{E}[B_2(t)|\Im(s)]$$
$$= \gamma B_1(s) + \sqrt{1-\gamma^2} B_2(s)$$
$$= B_3(s)$$

Thus (b) is satisfied. For condition (c) use $dB_3(t) = \gamma\, dB_1(t) + \sqrt{1-\gamma^2}\, dB_2(t)$. Then

$$[dB_3(t)]^2 = \gamma^2[dB_1(t)]^2 + (\sqrt{1-\gamma^2})^2[dB_2(t)]^2$$
$$+ 2\gamma\sqrt{1-\gamma^2}\, dB_1(t)\, dB_2(t)$$

The last term $dB_1(t)\, dB_2(t) = 0$ due to independence, and

$$[dB_3(t)]^2 = \gamma^2\, dt + (1-\gamma^2)\, dt = dt$$

Thus B_3 is a Brownian motion.

The above case is special because the sum of the squares of the coefficients of the respective Brownian motions add to 1, which generally is not the case. Consider therefore an *arbitrary* linear combination, with constants σ_1 and σ_2, of two standard Brownian motions, $B_1(t)$ and $B_2(t)$, which are again assumed to be *independent*. The aim is to replace this by another independent Brownian motion that has a coefficient σ_3:

$$\sigma_3 B_3(t) \stackrel{\text{def}}{=} \sigma_1 B_1(t) + \sigma_2 B_2(t)$$

Coefficients σ_1 and σ_2 are given, and σ_3 has to be determined in such a way that $B_3(t)$ is a Brownian motion. Thus $B_3(t)$ must satisfy Lévy's conditions. Conditions (a) and (b) are satisfied as above. Condition (c) uses

$$\sigma_3\, dB_3(t) = \sigma_1\, dB_1(t) + \sigma_2\, dB_2(t)$$

Then, as $dB_1(t)\, dB_2(t) = 0$,

$$[dB_3(t)]^2 = \left(\frac{\sigma_1}{\sigma_3}\right)^2 dt + \left(\frac{\sigma_2}{\sigma_3}\right)^2 dt$$

This equals dt if

$$\left(\frac{\sigma_1}{\sigma_3}\right)^2 + \left(\frac{\sigma_2}{\sigma_3}\right)^2 = 1$$

So the condition that must hold is

$$\sigma_3^2 = \sigma_1^2 + \sigma_2^2$$

Then

$$B_3(t) = \frac{\sigma_1}{\sqrt{\sigma_1^2 + \sigma_2^2}} \, B_1(t) + \frac{\sigma_2}{\sqrt{\sigma_1^2 + \sigma_2^2}} \, B_2(t)$$

The role of σ_3 is now clear. It is the scaling factor that is needed to ensure that the sum of the squares of the coefficients of the respective Brownian motions add to 1, as in the first case. Linear combinations of Brownian motions appear in the stochastic differential of a ratio of random processes, as illustrated in Chapter 8. The replacement by a single Brownian motion facilitates option valuation.

In the general case the component Brownian motions are *not independent*. Consider the linear combination

$$\sigma_4 B_4(t) \stackrel{\text{def}}{=} \sigma_1 B_1(t) + \sigma_2 B_2(t)$$

where B_1 and B_2 have correlation coefficient ρ, σ_1 and σ_2 are constants, and σ_4 is to be determined. The first step is to transform this into a linear combination of two *independent* Brownian motions. To this end write B_2 as a linear combination of B_1 and another independent Brownian motion B_3, using ρ in the coefficient of B_1,

$$B_2(t) \stackrel{\text{def}}{=} \rho B_1(t) + \sqrt{1 - \rho^2} \, B_3(t)$$

Using this expression for B_2 it can be readily seen that B_1 and B_2 have correlation ρ.

$$\sigma_4 B_4(t) = \sigma_1 B_1(t) + \sigma_2 [\rho B_1(t) + \sqrt{1 - \rho^2} \, B_3(t)]$$
$$= (\sigma_1 + \sigma_2 \rho) B_1(t) + \sigma_2 \sqrt{1 - \rho^2} \, B_3(t)$$

Now applying the result for the case of independent Brownian motions

$$\sigma_4^2 = (\sigma_1 + \sigma_2 \rho)^2 + (\sigma_2 \sqrt{1 - \rho^2})^2$$
$$= \sigma_1^2 + 2\rho\sigma_1\sigma_2 + \sigma_2^2$$

In the case of a difference of Brownian motions

$$\sigma_5 B_5(t) \stackrel{\text{def}}{=} \sigma_1 B_1(t) - \sigma_2 B_2(t)$$

changing the sign of σ_2 in $B_4(t)$ gives

$$\sigma_5^2 = \sigma_1^2 - 2\rho\sigma_1\sigma_2 + \sigma_2^2$$

The above can also be applied when the Brownian motion coefficients are time dependent but non-random. Consider the linear combination of correlated Brownian motions B_1 and B_2

$$\sigma_3(t)\,dB_3(t) \stackrel{\text{def}}{=} \sigma_1(t)\,dB_1(t) \pm \sigma_2(t)\,dB_2(t)$$

Then (without reworking to a linear combination of independent Brownian motions)

$$\begin{aligned}
[\sigma_3(t)\,dB_3(t)]^2 &= [\sigma_1(t)\,dB_1(t)]^2 \pm 2\sigma_1(t)\sigma_2(t)\,dB_1(t)\,dB_2(t) \\
&\quad + [\sigma_2(t)\,dB_2(t)]^2 \\
&= \sigma_1(t)^2 dt \pm 2\sigma_1(t)\sigma_2(t)\rho dt + \sigma_2(t)^2 dt
\end{aligned}$$

So B_3 is a Brownian motion of $\sigma_3(t)^2 = \sigma_1(t)^2 \pm 2\sigma_1(t)\sigma_2(t)\rho + \sigma_2(t)^2$.

4.7   MULTIPLE CORRELATED BROWNIAN MOTIONS

Stock prices are driven by Brownian motion. Creating correlated stock prices requires correlated Brownian motions. The basic ingredients for this are correlated normal random variables. The starting point is n *un*correlated standard normal random variables $Z_1, \cdots, Z_i, \cdots, Z_n$. From these, normal random variables can be constructed which are correlated with a pre-specified correlation coefficient by linear combinations of the Z_i. If the weights are denoted by λ_{ij}, then

$$X_1 \stackrel{\text{def}}{=} \lambda_{11}Z_1 + \cdots + \lambda_{1j}Z_j + \cdots \lambda_{1n}Z_n$$
$$\vdots$$
$$X_i \stackrel{\text{def}}{=} \lambda_{i1}Z_1 + \cdots + \lambda_{ij}Z_j + \cdots \lambda_{in}Z_n$$
$$\vdots$$
$$X_n \stackrel{\text{def}}{=} \lambda_{n1}Z_1 + \cdots + \lambda_{nj}Z_j + \cdots \lambda_{nn}Z_n \qquad (*)$$

This can be conveniently put into matrix form by letting matrix L be the collection of weightings λ_{ij}, arranging the Z_i into column vector Z, and the X_i into column vector X.

$$Z \stackrel{\text{def}}{=} \begin{bmatrix} Z_1 \\ Z_i \\ Z_n \end{bmatrix} \qquad X \stackrel{\text{def}}{=} \begin{bmatrix} X_1 \\ X_i \\ X_n \end{bmatrix}$$

Then X is the matrix product $X = L \cdot Z$, where the matrix product is denoted by the big dot. The correlation of random variables X_i and X_j must be ρ_{ij}. It is the expected value of the product of the value of X_i and X_j, $\mathbb{E}[X_i X_j] = \rho_{ij}$. Collecting all these into a matrix gives the correlation matrix R

$$\mathbb{E} \begin{bmatrix} X_1 X_1 & X_1 X_j & X_1 X_n \\ X_i X_1 & X_i X_j & X_i X_n \\ X_n X_1 & X_n X_j & X_n X_n \end{bmatrix} = \begin{bmatrix} \mathbb{E}[X_1 X_1] & \mathbb{E}[X_1 X_j] & \mathbb{E}[X_1 X_n] \\ \mathbb{E}[X_i X_1] & \mathbb{E}[X_i X_j] & \mathbb{E}[X_i X_n] \\ \mathbb{E}[X_n X_1] & \mathbb{E}[X_n X_j] & \mathbb{E}[X_n X_n] \end{bmatrix}$$

$$= \begin{bmatrix} 1 & \rho_{1j} & \rho_{1n} \\ \rho_{i1} & 1 & \rho_{in} \\ \rho_{n1} & \rho_{nj} & 1 \end{bmatrix} \quad \text{which is denoted } R.$$

The matrix

$$\begin{bmatrix} X_1 X_1 & X_1 X_j & X_1 X_n \\ X_i X_1 & X_i X_j & X_i X_n \\ X_n X_1 & X_n X_j & X_n X_n \end{bmatrix}$$

can also be written as the matrix product of X and its transpose X^{T}

$$X \cdot X^{\text{T}} = (L \cdot Z) \cdot (L \cdot Z)^{\text{T}} = (L \cdot Z) \cdot (Z^{\text{T}} \cdot L^{\text{T}}) = L \cdot (Z \cdot Z^{\text{T}}) \cdot L^{\text{T}}$$

$X \cdot X^{\text{T}}$ is a collection of random variables in matrix form. Taking the expected value

$$\mathbb{E}[X \cdot X^{\text{T}}] = \mathbb{E}[L \cdot (Z \cdot Z^{\text{T}}) \cdot L^{\text{T}}] = L \cdot \mathbb{E}[Z \cdot Z^{\text{T}}] \cdot L^{\text{T}}$$

$\mathbb{E}$ is taken inside as the elements of matrix L are non-random and $\mathbb{E}$ is a linear operation. $\mathbb{E}[Z \cdot Z^{\mathrm{T}}]$ is the correlation matrix of standard normal variables Z, and, as these are uncorrelated, this equals the identity matrix I, leaving

$$\mathbb{E}[X \cdot X^{\mathrm{T}}] = L \cdot L^{\mathrm{T}}$$

Thus the matrix L of weightings must be such that

$$L \cdot L^{\mathrm{T}} = R$$

Correlation matrix R is symmetric, $\rho_{ij} = \rho_{ji}$, and positive definite. This makes it possible to decompose R into the product of a lower-triangular matrix L and its transpose L^{T}. This is known as the Cholesky decomposition.

$$R = \begin{bmatrix} 1 & \rho_{1j} & \rho_{1n} \\ \rho_{i1} & 1 & \rho_{in} \\ \rho_{n1} & \rho_{nj} & 1 \end{bmatrix}$$

$$= \begin{bmatrix} 1 & 0 & 0 & 0 & 0 \\ \lambda_{21} & \lambda_{22} & 0 & 0 & 0 \\ \lambda_{31} & \lambda_{32} & \lambda_{33} & 0 & 0 \\ \vdots & & & & \\ \lambda_{n1} & \lambda_{n2} & \lambda_{n3} & & \lambda_{nn} \end{bmatrix} \times \begin{bmatrix} 1 & \lambda_{21} & \lambda_{31} & \lambda_{n1} \\ 0 & \lambda_{22} & \lambda_{32} & \lambda_{n2} \\ 0 & 0 & \lambda_{33} & \lambda_{n3} \\ \vdots & & & \\ 0 & 0 & 0 & \lambda_{nn} \end{bmatrix}$$

$$R = L \cdot L^{\mathrm{T}}$$

Using the elements of L in (*) then gives the correlated normal random variables X_i as

$$X_1 = 1Z_1 = Z_1$$
$$X_2 = \lambda_{21}Z_1 + \lambda_{22}Z_2$$
$$X_3 = \lambda_{31}Z_1 + \lambda_{32}Z_2 + \lambda_{33}Z_3$$
$$\vdots$$
$$X_n = \lambda_{n1}Z_1 + \cdots + \lambda_{ni}Z_i + \cdots + \lambda_{nn}Z_n$$

There are efficient numerical algorithms for carrying out the Cholesky decomposition. It can be done symbolically with Mathematica. By way of example, for $n = 2$,

$$L = \begin{bmatrix} 1 & 0 \\ \rho & \sqrt{1 - \rho^2} \end{bmatrix}, \quad L^{\mathrm{T}} = \begin{bmatrix} 1 & \rho \\ 0 & \sqrt{1 - \rho^2} \end{bmatrix}, \quad L \cdot L^{\mathrm{T}} = \begin{bmatrix} 1 & \rho \\ \rho & 1 \end{bmatrix},$$

$$\lambda_{21} = \rho, \quad \lambda_{22} = \sqrt{1 - \rho^2}, \quad X_1 = Z_1, \quad X_2 = \rho Z_1 + \sqrt{1 - \rho^2} Z_2,$$

which is the construction used in Section 1.6.

Having thus computed the lambdas from the given correlation matrix R, and constructed the correlated normal random variables X_i, the correlated Brownian motions for a time step Δt are obtained by multiplying each X_i by $\sqrt{\Delta t}$. The dynamics of the correlated stock prices for option valuation are then given by the SDEs

$$\frac{dS_1}{S_1} = (r - \tfrac{1}{2}\sigma_1^2)\, dt + \sigma_1 X_1 \sqrt{\Delta t}$$

$$\frac{dS_2}{S_2} = (r - \tfrac{1}{2}\sigma_2^2)\, dt + \sigma_2 X_2 \sqrt{\Delta t}$$

$$\vdots$$

$$\frac{dS_n}{S_n} = (r - \tfrac{1}{2}\sigma_n^2)\, dt + \sigma_n X_n \sqrt{\Delta t}$$

4.8   AREA UNDER A BROWNIAN MOTION PATH – REVISITED

Section 3.6 introduced the random variable $I(T) \overset{\text{def}}{=} \int_{t=0}^{T} B(t)\, dt$ and derived its distribution. Here $I(T)$ is analysed using the product rule. For processes X and Y

$$d[X(t)Y(t)] = X(t)\, dY(t) + Y(t)\, dX(t) + dX(t)\, dY(t)$$

If $X(t) = t$ and $Y(t) = B(t)$, this becomes

$$d[t\, B(t)] = t\, dB(t) + B(t)\, dt + \underbrace{dt\, dB(t)}_{=0}$$

$$= t\, dB(t) + B(t)\, dt$$

In equivalent integral form

$$\int_{t=0}^{T} d[t\, B(t)] = \int_{t=0}^{T} t\, dB(t) + \int_{t=0}^{T} B(t)\, dt$$

$$T\, B(T) - 0\, B(0) = \int_{t=0}^{T} t\, dB(t) + \int_{t=0}^{T} B(t)\, dt$$

$$I(T) = \int_{t=0}^{T} B(t)\, dt = T\, B(T) - \int_{t=0}^{T} t\, dB(t)$$

$$= T \int_{t=0}^{T} dB(t) - \int_{t=0}^{T} t\, dB(t)$$

$$= \int_{t=0}^{T} (T - t)\, dB(t)$$

This is a stochastic integral with a non-random integrand, so random variable $I(T)$ has a normal distribution (according to Section 3.7). Being normal, it is fully specified by its mean and its variance

$$\mathbb{E}[I(T)] = 0$$

$$\mathbb{Var}[I(T)] = \mathbb{E}[I^2(T)] = \int_{t=0}^{T} \mathbb{E}[(T - t)^2]\, dt$$

$$= \int_{t=0}^{T} (T - t)^2\, dt = -\int_{t=0}^{T} (T - t)^2\, d(T - t) = \tfrac{1}{3}T^3$$

4.9   JUSTIFICATION OF ITŌ'S FORMULA[7]

The full proof of Itō's formula, when f is a function of t, and of the process $X(t)$ with dynamics $dX(t) = \mu[t, X(t)]\, dt + \sigma[t, X(t)]\, dB(t)$, is quite involved and outside the scope of this text. The coverage here is confined to a sketch of the proof for the simpler case where f is a function of Brownian motion only, $f[B(t)]$. Ignoring time as an explicit variable entails no loss of generality because the term that is special in Itō's formula comes from the Brownian motion dependence. The application of Itō's formula shown earlier gave the change in f as

$$df[B(t)] = \frac{1}{2} \left. \frac{d^2 f(x)}{dx^2} \right|_{x=B(t)} dt + \left. \frac{df(x)}{dx} \right|_{x=B(t)} dB(t)$$

---

[7] Can be skipped without loss of continuity.

Use the familiar discrete-time framework where $[0, T]$ is partitioned into n intervals of equal length $\Delta t = T/n$ and the time-points in the partition are denoted $t_k \stackrel{\text{def}}{=} k \, \Delta t$. Write $f[B(T)] - f[B(0)]$ as the sum of the changes in f over all subintervals

$$f[B(T)] - f[B(0)] = \sum_{k=0}^{n-1} (f[B(t_{k+1})] - f[B(t_k)])$$

Now write the term $f[B(t_{k+1})] - f[B(t_k)]$ as a Taylor expansion about $B[t_k]$ where the derivatives are evaluated at $x = B(t_k)$

$$f[B(t_{k+1})] - f[B(t_k)] = [B(t_{k+1}) - B(t_k)] \frac{df(x)}{dx}\bigg|_{x=B(t_k)}$$

$$+ \tfrac{1}{2}[B(t_{k+1}) - B(t_k)]^2 \frac{d^2 f(x)}{dx^2}\bigg|_{x=B(t_k)}$$

$$+ \text{remainder terms}$$

Then

$$f[B(T)] - f[B(0)] = \sum_{k=0}^{n-1} \frac{df(x)}{dx}\bigg|_{x=B(t_k)} [B(t_{k+1}) - B(t_k)]$$

$$+ \sum_{k=0}^{n-1} \frac{1}{2} \frac{d^2 f(x)}{dx^2}\bigg|_{x=B(t_k)} [B(t_{k+1}) - B(t_k)]^2$$

$$+ \sum_{k=0}^{n-1} \text{remainder terms}$$

Now it has to be shown that as $n \to \infty$ the following three properties hold:

(a) $\int_{t=0}^{T} \frac{df(x)}{dx}\big|_{x=B(t)} \, dB(t)$ is the limit of $\sum_{k=0}^{n-1} \frac{df(x)}{dx}\big|_{x=B(t_k)} [B(t_{k+1}) - B(t_k)]$ in the mean-square sense
(b) $\int_{t=0}^{T} \frac{1}{2} \frac{d^2 f(x)}{dx^2}\big|_{x=B(t)} \, dt$ is the limit of $\sum_{k=0}^{n-1} \frac{1}{2} \frac{d^2 f(x)}{dx^2}\big|_{x=B(t_k)} [B(t_{k+1}) - B(t_k)]^2$ in the mean-square sense
(c) $\sum_{k=0}^{n-1}$ remainders $\to 0$ in the mean-square. sense

Property (a) The limit as $n \to \infty$ of $\sum_{k=0}^{n-1} [df(x)/dx]\big|_{x=B(t_k)} [B(t_{k+1}) - B(t_k)]$, in the mean-square sense, is $\int_{t=0}^{T} [df(x)/dx]\big|_{x=B(t)} \, dB(t)$, by the definition of the Itō stochastic integral. The function being integrated is $df(x)/dx$ at $x = B(t)$. This is being approximated in the discrete stochastic integral by the random step-function $df(x)/dx$ at $x = B(t_k)$, over the time interval $[t_k, t_{k+1}]$. The formal mean-square convergence formulation, with $\int_{t=0}^{T}(df/dB)\,dB(t)$ denoted by I, is

$$\mathbb{E}\left[\left\{\sum_{k=0}^{n-1}\frac{df(x)}{dx}\bigg|_{x=B(t_k)}[B(t_{k+1})-B(t_k)]-I\right\}^2\right]=0 \text{ as } n\to\infty, \Delta t\to 0$$

Property (b)   Expression

$$\sum_{k=0}^{n-1}\frac{1}{2}\frac{d^2f(x)}{dx^2}\bigg|_{x=B(t_k)}[B(t_{k+1})-B(t_k)]^2$$

looks like the quadratic variation expression $\sum_{k=0}^{n-1}[B(t_{k+1})-B(t_k)]^2$ which is known to converge to T in mean-square (as shown in Annex C). Converging to T is the same as converging to $\int_{t=0}^{T}1\,dt$. This suggests that

$$\sum_{k=0}^{n-1}\frac{1}{2}\frac{d^2f(x)}{dx^2}\bigg|_{x=B(t_k)}[B(t_{k+1})-B(t_k)]^2$$

may converge to $\frac{1}{2}[d^2f(x)/dx^2]|_{x=B(t)}\,dt$. The integral $\int_{t=0}^{T}\frac{1}{2}[d^2f(x)/dx^2]|_{x=B(t)}\,dt$ is an ordinary Riemann integral and is by definition the limit, as $n \to \infty$, of $\sum_{k=0}^{n-1}\frac{1}{2}[d^2f(x)/dx^2]|_{x=B(t_k)}\Delta t$. So to verify the anticipated result it has to be shown that

$$\mathbb{E}\left[\left\{\sum_{k=0}^{n-1}\frac{1}{2}\frac{d^2f(x)}{dx^2}\bigg|_{x=B(t_k)}[B(t_{k+1})-B(t_k)]^2\right.\right.$$
$$\left.\left.-\sum_{k=0}^{n-1}\frac{1}{2}\frac{d^2f(x)}{dx^2}\bigg|_{x=B(t_k)}\Delta t\right\}^2\right]$$

converges to zero as $n \to \infty$. Expanding $\{...\}^2$ gives cross terms whose expected values are zero. That leaves

$$\mathbb{E}\left[\sum_{k=0}^{n-1}\left[\frac{1}{2}\frac{d^2f(x)}{dx^2}\bigg|_{x=B(t_k)}\right]^2\{[B(t_{k+1})-B(t_k)]^2-\Delta t\}^2\right]$$

The Brownian increment after t_k is independent of the value of $d^2f(x)/dx^2$ at $x = B(t_k)$. Taking $\mathbb{E}$ inside the sum and using this independence gives

$$\sum_{k=0}^{n-1} \mathbb{E}\left[\frac{1}{2}\frac{d^2f(x)}{dx^2}\bigg|_{x=B(t_k)}\right]^2 \mathbb{E}[\{[B(t_{k+1}) - B(t_k)]^2 - \Delta t\}^2]$$

The second part in the product term evaluates to

$$\mathbb{E}\{[B(t_{k+1}) - B(t_k)]^4 - 2[B(t_{k+1}) - B(t_k)]^2 \Delta t + [\Delta t]^2\}$$
$$= 3(\Delta t)^2 - 2\Delta t \Delta t + (\Delta t)^2 = 2(\Delta t)^2$$
$$= 2\left(\frac{T}{n}\right)^2$$

As $n \to \infty$, this $\to 0$, and the convergence has been shown.

Property (c)   The third-order term in the Taylor expansion is

$$\frac{1}{6}\frac{d^3f(x)}{dx^3}\bigg|_{x=B(t_k)} [B(t_{k+1}) - B(t_k)]^3$$

Due to the independence of $B(t_k)$ and $[B(t_{k+1}) - B(t_k)]$, its expected value is the product

$$\mathbb{E}\left\{\frac{1}{6}\frac{d^3f(x)}{dx^3}\bigg|_{x=B(t_k)}\right\} \mathbb{E}\{[B(t_{k+1}) - B(t_k)]^3\}$$

This equals zero as the third moment of Brownian motion is zero.
   The fourth-order term in the Taylor expansion is

$$\frac{1}{24}\frac{d^4f(x)}{dx^4}\bigg|_{x=B(t_k)} [B(t_{k+1}) - B(t_k)]^4$$

Its expected value is the product

$$\mathbb{E}\left\{\frac{1}{24}\frac{d^4f(x)}{dx^4}\bigg|_{x=B(t_k)}\right\} \mathbb{E}\{[B(t_{k+1}) - B(t_k)]^4\}$$

As the second expectation equals $3(\Delta t)^2$, the expected value of this fourth-order term tends to zero as $n \to \infty$. This is the pattern. All higher order terms either have expected value zero, or an expected value that tends to zero, as all odd moments of Brownian motion are zero and all even moments are powers of Δt, as can be seen in Annex A. Therefore the remainder terms have no presence in the Itō formula.

4.10 EXERCISES

Derive the stochastic differential of the functions specified in [4.10.1]–[4.10.7].

[4.10.1] $\frac{1}{3}B(t)^3$

[4.10.2] $B(t)^2 - t$

[4.10.3] $\exp[B(t)]$

[4.10.4] $\exp[B(t) - \frac{1}{2}t]$

[4.10.5] $\exp[(\mu - \frac{1}{2}\sigma^2)t + \sigma B(t)]$ where μ and σ are constants

[4.10.6] $\ln[S(t)]$ where $dS(t) = \mu S(t)\,dt + \sigma S(t)\,dB(t)$

[4.10.7] $1/S(t)$ where $S(t) = S(0)\exp[(\mu - \frac{1}{2}\sigma^2)t + \sigma B(t)]$

[4.10.8] *Exchange rate dynamics.* Let $Q(t)$ denote the exchange rate at time t. It is the price in domestic currency of one unit of foreign currency and converts foreign currency into domestic currency. A model for the dynamics of the exchange rate is $dQ(t)/Q(t) = \mu_Q\,dt + \sigma_Q\,dB(t)$. This has the same structure as the common model for the stock price. The reverse exchange rate, denoted $R(t)$, is the price in foreign currency of one unit of domestic currency $R(t) = 1/Q(t)$. Derive $dR(t)$.

[4.10.9] *Bond price dynamics.* Consider the value of a zero-coupon bond that matures at time T. A class of models treats the bond value as being fully determined by the spot interest rate. The bond value at time t is then denoted by $P(t, T)$. A general model for the dynamics of the spot rate is $dr(t) = \mu(r, t)\,dt + \sigma(r, t)\,dB(t)$. Derive dP.

[4.10.10] For a martingale M, show that $[dM(t)]^2 = dt$ means the same as $M(t)^2 - t$ being a martingale.

[4.10.11] *Stochastic integral with respect to continuous martingale M.* Following Section 3.2, the discrete stochastic integral for random step-functions f is defined as $J_n \stackrel{\text{def}}{=} \sum_{k=0}^{n-1} f_k[M(t_{k+1}) - M(t_k)]$. Show that J_n is a martingale. The continuous stochastic integral $J = \int_{t=0}^{T} f(t, \omega) \, dM(t)$ is then the limit in the mean-square sense of the sequence of J_n, and inherits the martingale property of J_n. Then informally, $\mathbb{E}[dM(t)] = 0$ and $dt \, dM(t) = 0$.

4.11   SUMMARY

This chapter explained Itō's formula for determining how quantities that are Brownian motion dependent change over time. It was illustrated by many examples. It can also be used to verify whether a given random process is a Brownian motion, the so-called Lèvy characterization. This in turn was used to combine correlated Brownian motions into a single Brownian motion.

5

Stochastic Differential Equations

5.1 STRUCTURE OF A STOCHASTIC DIFFERENTIAL EQUATION

A stochastic *differential* equation (SDE) describes the increment of a variable, say X, which is driven by one or several underlying random processes. Here these sources of randomness are Brownian motion. When there is one Brownian motion, the general specification is of the form

$$dX(t) = \mu[t, X(t)] \, dt + \sigma[t, X(t)] \, dB(t)$$

where μ and σ are known continuous functions of t and X. The initial time at which the value of the process is known is taken as $t = 0$, so $X(0)$ is known. The intuitive interpretation is that $dX(t)$ is the change in X over a 'very short' time interval dt, from t to $t + dt$, so $dX(t) = X(t + dt) - X(t)$. The SDE is the commonly used *informal shorthand notation* for the integral expression over the clearly defined period $[0, T]$

$$\int_{t=0}^{T} dX(t) = X(T) - X(0) = \int_{t=0}^{T} \mu[t, X(t)] \, dt + \int_{t=0}^{T} \sigma[t, X(t)] \, dB(t)$$

$$X(T) = X(0) + \int_{t=0}^{T} \mu[t, X(t)] \, dt + \int_{t=0}^{T} \sigma[t, X(t)] \, dB(t)$$

It is this latter integral expression that is the exact specification of the SDE. This is also known as a stochastic *integral* equation (SIE). In this text the name SDE is used when the random process is to be determined from the equation, while the name stochastic differential is used for the expression that results from applying Itō's formula to a given random process (as in Chapter 4). The term $\mu[t, X(t)] \, dt$ is called the drift, and $\sigma[t, X(t)] \, dB(t)$ the diffusion. Coefficient $\sigma[t, X(t)]$ serves as a scaling of the randomness generated by the Brownian motion. The values of X used in the coefficients may only depend on the history (non-anticipating property). The integral $\int_{t=0}^{T} \mu[t, X(t)] \, dt$ is a pathwise ordinary integral, and $\int_{t=0}^{T} \sigma[t, X(t)] \, dB(t)$ is an Itō stochastic integral.

A *solution* to a SDE is a random process that has a stochastic differential which has the same form as the SDE when the Brownian motion process is given. In the technical literature this is called a strong solution, to distinguish it from a so-called weak solution which is not discussed here. The SDE for X is often called the *dynamics* of X. When does a SDE have a solution, when is this solution unique, and how can it be found? According to Chapter 3, the stochastic integral term $\int_{t=0}^{T} \sigma[t, X(t)]\, dB(t)$ exists if $\int_{t=0}^{T} \mathbb{E}\{\sigma[t, X(t)]^2\}\, dt$ is finite. However, as the process X is not known, this check cannot be carried out and some other criteria are needed. If there was no random term, the SDE would be an ordinary differential equation (ODE) for non-random function X, $dX(t)/dt = \mu[t, X(t)]$. In the theory of ODEs the conditions for existence and uniqueness of a solution have been obtained from successive approximations to X. A similar approach has been developed for SDEs. A unique pathwise solution X to a SDE exists if there are positive constants K and L such that the following two conditions are satisfied at any $0 \le t \le T$ for arbitrary x.

$Growth$ condition: $\mu(t, x)^2 + \sigma(t, x)^2 \le K(1 + x^2)$
$Lipschitz$ condition: $|\mu(t, x_1) - \mu(t, x_2)|$
$$+ |\sigma(t, x_1) - \sigma(t, x_2)| \le L|x_1 - x_2|$$

The proof that these conditions are sufficient is not covered here as it would require an elaborate mathematical detour. A nicely motivated exposition is given in *Kuo* Chapter 10.

Some stochastic differential equations which are widely used in finance are now presented, showing how the solutions are found. The uniqueness of each solution is verified in exercise [5.12.10].

5.2   ARITHMETIC BROWNIAN MOTION SDE

Arithmetic Brownian motion (ABM) is the name for a random process, say X, specified by the SDE

$$dX(t) = \mu\, dt + \sigma\, dB(t) \quad \mu \text{ and } \sigma \text{ known constants, and } \sigma > 0$$

In this model, the drift coefficient $\mu[t, X(t)] = \mu$, and the diffusion coefficient $\sigma[t, X(t)] = \sigma$, are both constant. In equivalent integral form

$$\int_{t=0}^{T} dX(t) = \int_{t=0}^{T} \mu\, dt + \int_{t=0}^{T} \sigma\, dB(t)$$

which can be written as

$$X(T) - X(0) = \mu[T - 0] + \sigma[B(T) - B(0)]$$
$$X(T) = X(0) + \mu T + \sigma B(T)$$

As there is no unknown in the right hand side, this is the solution, as can be readily verified by deriving dX. Solution $X(T)$ equals a non-random term, $X(0) + \mu T$, plus a constant times the normally distributed random variable $B(T)$, so it is normally distributed, and can take on negative values. The distribution parameters are

$$\mathbb{E}[X(T)] = \mathbb{E}[X(0) + \mu T + \sigma B(T)] = X(0) + \mu T + \sigma \mathbb{E}[B(T)]$$
$$= X(0) + \mu T$$
$$\mathbb{V}\text{ar}[X(T)] = \mathbb{V}\text{ar}[X(0) + \mu T + \sigma B(T)] = \mathbb{V}\text{ar}[\sigma B(T)] = \sigma^2 T$$

The mean and the variance of the position at T increase linearly with T. This model can be a suitable specification for an economic variable that grows at a constant rate and is characterized by increasing uncertainty. But as the process can take on negative values it is not suitable as a model for stock prices, since limited liability prevents stock prices from going negative.

5.3   GEOMETRIC BROWNIAN MOTION SDE

Geometric Brownian motion (GBM) is a model for the change in a random process, $dX(t)$, *in relation to the current value*, $X(t)$. This *proportional change $dX(t)/X(t)$*, or rate of return, is modelled as an ABM. The SDE is

$$\frac{dX(t)}{X(t)} = \mu \, dt + \sigma \, dB(t) \quad \mu \text{ and } \sigma \text{ known constants, and } \sigma > 0$$

Multiplying through by $X(t)$ gives the SDE for the change in X itself as

$$dX(t) = \mu X(t) \, dt + \sigma X(t) \, dB(t)$$

Drift coefficient $\mu X(t)$ and diffusion coefficient $\sigma X(t)$ are both proportional to the latest *known* value $X(t)$, and thus change continuously. The higher the latest X, the greater the drift coefficient. The diffusion coefficient is then also greater and the random term is thus generated by a Brownian motion with greater variance, so a greater random increment

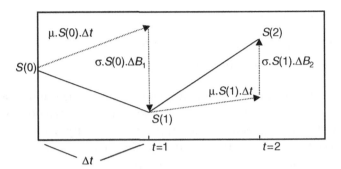

Figure 5.1 Structure of GBM

is more likely. This is the standard model for the stock price process; that random process is often denoted S.

A schematic illustration of the dynamics is now given in a discretized time setting. Starting at time 0 with the known value $S(0)$, visualize the increments as taking place in two steps, firstly the known growth, secondly the random effect. $S(0)$ moves up at a rate of $\mu S(0)$ per unit of time. Thereafter, the Brownian motion causes a random effect, with scaling coefficient $\sigma S(0)$, shown vertically in Figure 5.1. The resulting net increment is shown as a solid line. At time 1 the drift coefficient is somewhat smaller than previously because $S(1)$ is less than $S(0)$. The diffusion term can be anything.

To find the solution to the GBM, a trial solution is postulated, and Itō's formula is applied to derive the corresponding stochastic differential. If that matches the GBM, then this trial solution is the definitive solution. Notation S is now used. The intuition for the trial solution is that if S were deterministic, $dS(t)/S(t)$ would be the derivative of $\ln[S(t)]$ with respect to S. This suggests to find an expression for the stochastic differential of $\ln[S(t)]$, a function of the single random variable $S(t)$.

$$d\ln[S] = \frac{d\ln[S]}{dS} dS + \frac{1}{2}\frac{d^2\ln[S]}{dS^2}(dS)^2$$

where

$$\frac{d\ln[S]}{dS} = \frac{1}{S} \qquad \frac{d^2\ln[S]}{dS^2} = \frac{-1}{S^2} \qquad (dS)^2 = \sigma^2 S^2\, dt$$

Substituting these, together with dS, gives

$$d \ln[S] = \frac{1}{S}(\mu S\, dt + \sigma S\, dB) + \frac{1}{2}\frac{-1}{S^2}\sigma^2 S^2\, dt$$
$$= \mu\, dt + \sigma\, dB - \tfrac{1}{2}\sigma^2\, dt$$
$$= (\mu - \tfrac{1}{2}\sigma^2)\, dt + \sigma\, dB$$

In integral form

$$\int_{t=0}^{T} d \ln[S(t)] = \int_{t=0}^{T} \left(\mu - \tfrac{1}{2}\sigma^2\right) dt + \int_{t=0}^{T} \sigma\, dB(t)$$

$$\ln[S(T)] - \ln[S(0)] = \left(\mu - \tfrac{1}{2}\sigma^2\right)T + \sigma B(T)$$

$$\ln\left[\frac{S(T)}{S(0)}\right] = \left(\mu - \tfrac{1}{2}\sigma^2\right)T + \sigma B(T)$$

Thus $\ln[S(T)/S(0)]$ has a normal distribution with parameters

$$\mathbb{E}\left\{\ln\left[\frac{S(T)}{S(0)}\right]\right\} = \mathbb{E}[(\mu - \tfrac{1}{2}\sigma^2)T + \sigma B(T)] = (\mu - \tfrac{1}{2}\sigma^2)T$$

$$\mathbb{V}\mathrm{ar}\left\{\ln\left[\frac{S(T)}{S(0)}\right]\right\} = \mathbb{V}\mathrm{ar}[(\mu - \tfrac{1}{2}\sigma^2)T + \sigma B(T)] = \sigma^2 T$$

It is the ln of the stock price that is modelled as a normal random variable. Both mean and variance increase linearly with time. This is the same as in the ABM but here for the ln. Taking exponentials gives the final expression for $S(T)$:

$$\frac{S(T)}{S(0)} = \exp[(\mu - \tfrac{1}{2}\sigma^2)T + \sigma B(T)]$$

$$S(T) = S(0)\exp[(\mu - \tfrac{1}{2}\sigma^2)T + \sigma B(T)]$$

Being an exponential, $S(T)$ *cannot become negative*. At $S(T) = x$, it has the lognormal density

$$\frac{1}{xv\sqrt{2\pi}}\exp\left\{-\frac{1}{2}\left[\frac{\ln(x) - m}{v}\right]^2\right\}$$

where $m \overset{\text{def}}{=} \mathbb{E}\{\ln[S(T)]\} = \ln[S(0)] + (\mu - \tfrac{1}{2}\sigma^2)T$ and $v \overset{\text{def}}{=}$ Stdev $\{\ln[S(T)]\} = \sigma\sqrt{T}$. So when a process is modelled as a GBM it cannot become negative. It can be a suitable specification for an economic variable which cannot assume negative values and whose variability depends linearly on the level of the variable.

GBM is the traditional model for the stock price. It can be written as $S(t) = S(0)e^X$ where X is normal with mean $(\mu - \frac{1}{2}\sigma^2)t$ and variance $\sigma^2 t$. Analyses of actual stock prices have led to other models in which X is not normal. A nice empirical example in given in *Benth*, Chapter 2; other examples are in *McLeish*, Chapter 4; extensive coverage is in *Epps*, Chapters 8 and 9.

5.4   ORNSTEIN–UHLENBECK SDE

Probably the earliest recorded SDE is

$$dX(t) = -\lambda X(t)\,dt + \sigma\,dB(t)$$

λ and σ known constants, both positive

It was postulated by Ornstein–Uhlenbeck (OU) as a description of the acceleration of a pollen particle in a liquid subject to bombardments by molecules. $X(t)$ represents the velocity of the particle in one dimension, $dX(t)$ is the change in velocity per unit of time, its acceleration. This acceleration is modelled as being retarded by a frictional force proportional to the velocity, $-\lambda X(t)$, plus a random perturbation $B(t)$ with intensity σ caused by the bombardments. Extensions of this SDE are used in interest rate modelling.

The method of solution for this SDE uses a technique from ordinary differential equations to eliminate the drift. It is the transformation $Y(t) \stackrel{\text{def}}{=} X(t)\exp(\lambda t)$. Applying Itō's formula to Y, as a function of X and t, gives the dynamics of Y as

$$dY = \frac{\partial Y}{\partial t}\,dt + \frac{\partial Y}{\partial X}\,dX + \frac{1}{2}\frac{\partial^2 Y}{\partial X^2}(dX)^2 + \frac{\partial^2 Y}{\partial X \partial t}\,dt\,dX$$

where

$$\frac{\partial Y}{\partial t} = X\exp(\lambda t)\lambda = Y\lambda$$

$$\frac{\partial Y}{\partial X} = \exp(\lambda t) \quad \frac{\partial^2 Y}{\partial X^2} = 0 \quad \text{so } (dX)^2 \text{ is not needed}$$

$$dt\,dX = dt\,(-\lambda X\,dt + \sigma\,dB) = 0 \quad \text{so } \frac{\partial^2 Y}{\partial X \partial t} \text{ is not needed}$$

Substituting the above gives

$$dY = Y\lambda \, dt + \exp(\lambda t) \, dX$$

Substituting dX gives

$$dY = Y\lambda \, dt + \exp(\lambda t)(-\lambda X \, dt + \sigma \, dB)$$
$$= Y\lambda \, dt - \lambda Y \, dt + \sigma \exp(\lambda t) \, dB$$

Term $Y\lambda \, dt$ cancelled by design. The result is the driftless SDE

$$dY(t) = \sigma \exp(\lambda t) \, dB(t)$$

In integral form

$$Y(T) = Y(0) + \sigma \int_{t=0}^{T} \exp(\lambda t) \, dB(t)$$

To express this in terms of X, use $X(T) = \exp(-\lambda T) Y(T)$

$$X(T) = \exp(-\lambda T)[X(0) + \sigma \int_{t=0}^{T} \exp(\lambda t) \, dB(t)]$$

$$= \exp(-\lambda T) X(0) + \exp(-\lambda T)\sigma \int_{t=0}^{T} \exp(\lambda t) \, dB(t)$$

The stochastic integral in the last term is normally distributed because its integrand is non-random (according to Section 3.7). It has the same distribution as

$$\text{Stdev}\left[\int_{t=0}^{T} \exp(\lambda t) \, dB(t) \right] Z$$

where Z is a standard normal random variable

As the first right-hand term is non-random, $X(T)$ has a normal distribution. Given $X(0)$, its mean is

$$\mathbb{E}[X(T)] = \mathbb{E}[\exp(-\lambda T)X(0)] + \mathbb{E}[\exp(-\lambda T)\sigma \int_{t=0}^{T} \exp(\lambda t) \, dB(t)]$$

As the Itō stochastic integral has mean zero

$$\mathbb{E}[X(T)] = \exp(-\lambda T)X(0)$$

For large T the expected value $\mathbb{E}[X(T)]$ approaches zero. Its variance is

$$\mathbb{V}\text{ar}[X(T)] = \mathbb{E}\{[X(T) - \text{mean}]^2\}$$

Using

$$X(T) - \text{mean} = \exp(-\lambda T)\sigma \int_{t=0}^{T} \exp(\lambda t)\, dB(t)$$

gives

$$\mathbb{V}\text{ar}[X(T)] = \mathbb{E}\left\{\left[\exp(-\lambda T)\sigma \int_{t=0}^{T} \exp(\lambda t)\, dB(t)\right]^2\right\}$$

$$= [\exp(-\lambda T)\sigma]^2 \mathbb{E}\left\{\left[\int_{t=0}^{T} \exp(\lambda t)\, dB(t)\right]^2\right\}$$

$$= [\exp(-\lambda T)\sigma]^2 \int_{t=0}^{T} \mathbb{E}[\exp(\lambda t)^2]\, dt$$

and as $\exp(\lambda t)^2$ is non-random

$$= \exp(-2\lambda T)\sigma^2 \int_{t=0}^{T} \exp(2\lambda t)\, dt$$

$$= \exp(-2\lambda T)\sigma^2 \frac{1}{2\lambda} \exp(2\lambda t)|_{t=0}^{T}$$

$$= \exp(-2\lambda T)\sigma^2 \frac{1}{2\lambda}[\exp(2\lambda T) - 1]$$

$$= \sigma^2 \frac{1}{2\lambda}[1 - \exp(-2\lambda T)]$$

For large T the variance approaches $\sigma^2(1/2\lambda)$.

5.5   MEAN-REVERSION SDE

Mean-reversion (MR) is used for modelling random processes which fluctuate about a mean level. The prime example is the (continuously compounded) interest rate. Although random processes are often denoted by upper case letters, the lower case notation r is well established and used here. Its SDE is

$$dr(t) = -\lambda[r(t) - \bar{r}]\, dt + \sigma\, dB(t)$$
λ, σ, and $\bar{r}$ known constants, all positive

The drift coefficient $-\lambda[r(t) - \bar{r}]$ varies with the latest actual interest rate $r(t)$. If $r(t) < \bar{r}$, the drift coefficient is positive, and there is an upward drift; alternatively, if $r(t) > \bar{r}$, the drift coefficient is negative and there is a downward drift. It is shown below that $\bar{r}$ is the long-term mean, so r reverts to the mean, hence the name mean-reversion. The parameter λ controls how fast the drift coefficient changes. This SDE has the structure of the OU SDE. It can be turned into an OU SDE by the transformation $X(t) \stackrel{\text{def}}{=} r(t) - \bar{r}$, which is a function of the single variable r. By Itō's formula $dX = dr = -\lambda[r - \bar{r}]\,dt + \sigma\,dB$. Substituting X gives $dX = -\lambda X\,dt + \sigma\,dB$ which has the solution

$$X(T) = \exp(-\lambda T)X(0) + \exp(-\lambda T)\sigma \int_{t=0}^{T} \exp(\lambda t)\,dB(t)$$

Using $r(T) = X(T) + \bar{r}$ and $X(0) = r(0) - \bar{r}$ gives

$$
\begin{aligned}
r(T) &= \exp(-\lambda T)[r(0) - \bar{r}] + \exp(-\lambda T)\sigma \int_{t=0}^{T} \exp(\lambda t)\,dB(t) + \bar{r} \\
&= r(0)\exp(-\lambda T) + \bar{r}[1 - \exp(-\lambda T)] \\
&\quad + \exp(-\lambda T)\sigma \int_{t=0}^{T} \exp(\lambda t)\,dB(t)
\end{aligned}
$$

The stochastic integral is normally distributed as its integrand is non-random (according to Section 3.7). And as the other terms in the $r(T)$ expression are non-random, $r(T)$ is normally distributed. The expected value is

$$\mathbb{E}[r(T)] = r(0)\exp(-\lambda T) + \bar{r}[1 - \exp(-\lambda T)]$$

For large T the long-run expected value approaches $\bar{r}$. So $\bar{r}$ in the SDE for $r(t)$ is the long-run expected value to which $r(t)$ is attracted.

The variance is

$$\mathbb{V}\text{ar}[r(T)] = \mathbb{V}\text{ar}[X(T) + \bar{r}] = \mathbb{V}\text{ar}[X(T)] \qquad \text{since } \bar{r} \text{ is constant}$$

Copying the variance results from the OU SDE gives

$$\mathbb{V}\text{ar}[r(T)] = \sigma^2 \frac{1}{2\lambda}[1 - \exp(-2\lambda T)]$$

For large T the variance approaches $\sigma^2(1/2\lambda)$. Note that the interest rate *can become negative* under this model.

5.6 MEAN-REVERSION WITH SQUARE-ROOT DIFFUSION SDE

To exclude the possibility of the interest rate ever becoming negative, the diffusion coefficient in the previous mean-reversion model is modified to $\sigma \sqrt{r(t)}$. The SDE is then

$$dr(t) = -\lambda[(r(t) - \bar{r}]\,dt + \sigma\sqrt{r(t)}\,dB(t)$$

Now both the drift coefficient and the diffusion coefficient change continuously. It is important to realize that because a Brownian motion path is continuous, the interest rate path must be continuous. Jumps in the value of r are thus not possible under this model. Therefore, if r presently has a small positive value, then further negative Brownian motion increments cannot make it negative because r has to decrease in a continuous fashion to zero, and then the diffusion has become zero so it cannot go any lower. The drift coefficient is then $-\lambda[0 - \bar{r}] = \lambda\bar{r}$ which is positive and gives an upward drift. There is no expression for $r(t)$ but its probability distribution can be derived; it is a non-central chi-square distribution. That can then be used to value products which are based on r, such as bonds and options on bonds.

It is noted that some expositions specify the drift coefficient in the above mean-reversion SDEs in the form $[b - ar(t)]$. Taking $-a$ outside gives $-a[r(t) - b/a]$ which is the earlier specification with a as λ and b/a as $\bar{r}$. Analysis of the square-root diffusion SDE is given in *Cairns, Epps, Lamberton/Lapeyre*, and *Shreve II*. It is generally known as the CIR model, after its authors: Cox, Ingersoll, Ross.

5.7 EXPECTED VALUE OF SQUARE-ROOT DIFFUSION PROCESS

The square root process for X is

$$dX(t) = -\lambda[X(t) - \overline{X}]\,dt + \sigma\sqrt{X(t)}\,dB(t)$$

with $X(0)$ known

It is possible to determine the expected value of $X(t)$ without knowledge of its distribution, as will now be shown. In integral form over the time period $0 \leq s \leq t$

$$\int_{s=0}^{t} dX(s) = -\int_{s=0}^{t} \lambda[X(s) - \overline{X}]\,ds + \int_{s=0}^{t} \sigma\sqrt{X(s)}\,dB(s)$$

$$X(t) = X(0) - \int_{s=0}^{t} \lambda[X(s) - \overline{X}]\,ds + \int_{s=0}^{t} \sigma\sqrt{X(s)}\,dB(s)$$

Take the expected value conditional upon the known initial value $X(0)$; for greater readability this condition is not included in the notation

$$\mathbb{E}[X(t)] = X(0) - \mathbb{E}\left\{\int_{s=0}^{t} \lambda[X(s) - \overline{X}]\,ds\right\} + \mathbb{E}\left[\int_{s=0}^{t} \sigma\sqrt{X(s)}\,dB(s)\right]$$

As the expected value of the Itō stochastic integral equals zero

$$\mathbb{E}[X(t)] = X(0) - \mathbb{E}\left\{\int_{s=0}^{t} \lambda[X(s) - \overline{X}]\,ds\right\}$$

Applying $\mathbb{E}$ on the right-hand side is integrating, so that term is a double integral. The order of integration can be reversed[1] and moves $\mathbb{E}$ inside the integral, giving

$$\mathbb{E}[X(t)] = X(0) - \int_{s=0}^{t} \lambda\{\mathbb{E}[X(s)] - \overline{X}\}\,ds$$

This is an *integral equation* where the unknown function is the expected value of X. Let this unknown function be denoted $m(t)$. Then

$$m(t) = X(0) - \int_{s=0}^{t} \lambda[m(s) - \overline{X}]\,ds$$

To remove the integral, differentiate with respect to upper integration limit t. This gives the *ordinary differential equation*

$$\frac{dm(t)}{dt} = -\lambda[m(t) - \overline{X}] \quad \text{or} \quad \frac{dm(t)}{dt} + \lambda m(t) = \lambda\overline{X}$$

This is a well-known ODE. The first step in solving it is to multiply both sides of the equation by $\exp(\lambda t)$, a so-called integrating factor, giving

$$\exp(\lambda t)\frac{dm(t)}{dt} + \exp(\lambda t)\lambda m(t) = \exp(\lambda t)\lambda\overline{X}$$

Now the left-hand side can be written as the derivative with respect to t of $\exp(\lambda t)m(t)$

$$\frac{d[\exp(\lambda t)m(t)]}{dt} = \exp(\lambda t)\lambda\overline{X}$$

---

[1] By Fubini's theorem; see *Epps* Section 2.1, *Klebaner* p. 53.

This can be integrated. Using s as integration variable

$$\int_{s=0}^{t} d[\exp(\lambda s)m(s)] = \int_{s=0}^{t} \overline{X}\exp(\lambda s)\,d(\lambda s)$$
$$\exp(\lambda t)m(t) - \exp(\lambda 0)m(0) = \overline{X}[\exp(\lambda t) - \exp(\lambda 0)]$$
$$\exp(\lambda t)m(t) - m(0) = \overline{X}[\exp(\lambda t) - 1]$$
$$m(t) = \exp(-\lambda t)m(0) + \overline{X}[1 - \exp(-\lambda t)]$$

In terms of the original notation

$$\mathbb{E}[X(t)] = \exp(-\lambda t)m(0) + \overline{X}[1 - \exp(-\lambda t)]$$

As $m(0) = \mathbb{E}[X(0)] = X(0)$

$$\mathbb{E}[X(t)] = X(0)\exp(-\lambda t) + \overline{X}[1 - \exp(-\lambda t)]$$

Note that no use has been made of the specific form of the integrand of the stochastic integral. The result therefore applies not only to the square-root diffusion interest rate model in Section 5.6 but also to the model in Section 5.5.

5.8   COUPLED SDEs

The stock price model discussed earlier had a constant σ. There is empirical evidence that this is a rather strong simplification of reality, and that it would be more appropriate to model the volatility σ itself as a random process, for example of the mean-reverting type. The full model then comprises two coupled SDEs, driven by two Brownian motions $B_1(t)$ and $B_2(t)$ which can be correlated. A well-known model of this type is

$$\frac{dS(t)}{S(t)} = \mu(t)\,dt + \sqrt{v(t)}\,dB_1(t)$$
$$dv(t) = -\alpha[v(t) - \overline{v}]\,dt + \beta\sqrt{v(t)}\,dB_2(t)$$

where $\overline{v}$ is the long-run average of $v(t)$. Numerous interest rate models are coupled SDEs. There is usually no analytical pathwise solution for $S(t)$ but its distribution can be readily simulated. Recommended further reading on this is *Epps* Chapter 8, and *Martingale Methods in Financial Modelling*, 2nd edition by *Musiela/Rutkowski* Chapter 7.

5.9   CHECKING THE SOLUTION OF A SDE

Having derived that $dS(t) = \mu S(t)\,dt + \sigma S(t)\,dB(t)$ has the solution

$$S(t) = S(0)\exp[(\mu - \tfrac{1}{2}\sigma^2)t + \sigma B(t)]$$

it is now verified whether this expression for $S(t)$ indeed satisfies the SDE. To this end, derive an expression for $dS(t)$, treating $S(t)$ as a function of the two variables t and $B(t)$

$$dS(t) = \frac{\partial S}{\partial t}\,dt + \frac{\partial S}{\partial B}\,dB + \frac{1}{2}\frac{\partial^2 S}{\partial B^2}(dB)^2$$

The second-order term involving $(dt)^2$ and the cross term involving $dt\,dB$ have been omitted since these are negligible. The partial derivatives are

$$\frac{\partial S}{\partial t} = S(0)\exp[(\mu - \tfrac{1}{2}\sigma^2)t + \sigma B(t)](\mu - \tfrac{1}{2}\sigma^2) = S(t)(\mu - \tfrac{1}{2}\sigma^2)$$

$$\frac{\partial S}{\partial B} = S(0)\exp[(\mu - \tfrac{1}{2}\sigma^2)t + \sigma B(t)]\sigma = S(t)\sigma$$

$$\frac{\partial^2 S}{\partial B^2} = \frac{\partial}{\partial B}\left\{\frac{\partial S}{\partial B}\right\} = \frac{\partial}{\partial B}\{S(t)\sigma\} = \frac{\partial S}{\partial B}\sigma = S(t)\sigma\sigma = S(t)\sigma^2$$

Substituting these derivatives, together with $[dB(t)]^2 = dt$ gives

$$dS(t) = S(t)(\mu - \tfrac{1}{2}\sigma^2)\,dt + S(t)\sigma\,dB + \tfrac{1}{2}S(t)\sigma^2\,dt$$

Grouping dt terms then gives

$$dS(t) = \mu S(t)\,dt + \sigma S(t)\,dB(t)$$

It has been shown that $S(t)$ solves the SDE. Checking the solution of a mean-reversion SDE is requested in exercise [5.12.1].

5.10   GENERAL SOLUTION METHODS FOR LINEAR SDEs

The SDE $dS(t)/S(t) = \mu\,dt + \sigma\,dB(t)$, where μ and σ are constants, was solved by taking $\ln[S(t)]$ and deriving its stochastic differential $d\ln[S(t)]$. It turned out that this transformed the SDE in such a way that the right-hand side contained only t and $B(t)$, and the unknown $S(t)$ on the left-hand side could be written as an expression in t and $B(t)$. This was a lucky choice, which happened to work. Generally it is not easy to find a 'trial solution' that produces a right-hand side with only t and $B(t)$. If the unknown still appears in the right-hand side of the stochastic differential of a trial solution, then the trial solution has to be discarded. There is a need for a more general approach.

Consider a SDE which is linear in the unknown process X and of the general form

$$dX(t) = [\mu_{1X}(t) + \mu_{2X}(t)X(t)]\,dt + [\sigma_{1X}(t) + \sigma_{2X}(t)X(t)]\,dB(t)$$

where $X(0)$ is a known constant and $\mu_{1X}(t)$, $\mu_{2X}(t)$, $\sigma_{1X}(t)$, $\sigma_{2X}(t)$, are known non-random functions of time. If $\mu_{1X}(t) = 0$ and $\sigma_{1X}(t) = 0$, then X is a geometric Brownian motion. On the other hand, if $\mu_{2X}(t) = 0$ and $\sigma_{2X}(t) = 0$, then X is an arithmetic Brownian motion. This is highlighted by writing the SDE as

$$dX(t) = \underbrace{[\mu_{1X}(t)\,dt + \sigma_{1X}(t)\,dB(t)]}_{\text{arithmetic}}$$
$$+ \underbrace{[\mu_{2X}(t)X(t)\,dt + \sigma_{2X}(t)X(t)\,dB(t)]}_{\text{geometric}}$$

This suggests expressing X as the product of two processes, geometric Brownian motion Y and arithmetic Brownian motion Z. Process Y is specified as

$$\frac{dY(t)}{Y(t)} = \mu_Y(t)\,dt + \sigma_Y(t)\,dB(t) \qquad Y(0) = 1$$

This SDE can be solved because its drift coefficient and diffusion coefficient are assumed to be known. It is done as shown previously by deriving the stochastic differential of $\ln[Y(t)]$.

$$Y(t) = Y(0)\exp\left\{\int_{u=0}^{t}[\mu_Y(u) - \tfrac{1}{2}\sigma_Y(u)^2]\,du + \int_{u=0}^{t}\sigma_Y(u)\,dB(u)\right\}$$

Process Z is specified as

$$dZ = \mu_Z(t)\,dt + \sigma_Z(t)\,dB(t) \qquad Z(0) = X(0) \text{ a known constant}$$

where $\mu_Z(t)$ and $\sigma_Z(t)$ can be random processes. In integral form

$$Z(t) = Z(0) + \int_{u=0}^{t}\mu_Z(u)\,du + \int_{u=0}^{t}\sigma_Z(u)\,dB(u)$$

The drift and diffusion coefficient of Z have to be determined such that $X(t) = Y(t)Z(t)$. To this end write the stochastic differential of the process $Y(t)Z(t)$, and equate the drift and diffusion to that of the original X SDE:

$$dX(t) = Y(t)\,dZ(t) + Z(t)\,dY(t) + dY(t)\,dZ(t)$$

The last term is

$$dY(t)\,dZ(t) = [\sigma_Y(t)Y(t)\,dB(t)][\sigma_Z(t)\,dB(t)] = \sigma_Y(t)\sigma_Z(t)Y(t)\,dt$$

and

$$
\begin{aligned}
dX(t) &= Y(t)[\mu_Z(t)\,dt + \sigma_Z(t)\,dB(t)] + \\
&\quad + Z(t)[\mu_Y(t)Y(t)\,dt + \sigma_Y(t)Y(t)\,dB(t)] + \sigma_Y(t)\sigma_Z(t)Y(t)\,dt \\
&= Y(t)[\mu_Z(t)\,dt + \sigma_Z(t)\,dB(t)] + \\
&\quad + [\mu_Y(t)X(t)\,dt + \sigma_Y(t)X(t)\,dB(t)] + \sigma_Y(t)\sigma_Z(t)Y(t)\,dt \\
&= [\underbrace{Y(t)\mu_Z(t) + \sigma_Y(t)\sigma_Z(t)Y(t)}_{\mu_{1X}(t)} + \underbrace{\mu_Y(t)X(t)}_{\mu_{2X}(t)}]\,dt + \\
&\quad + [\underbrace{Y(t)\sigma_Z(t)}_{\sigma_{1X}(t)} + \underbrace{\sigma_Y(t)X(t)}_{\sigma_{2X}(t)}]\,dB(t)
\end{aligned}
$$

Equating the drift coefficients gives

$$Y(t)\mu_Z(t) + \sigma_Y(t)\sigma_Z(t)Y(t) = \mu_{1X}(t) \qquad (5.1)$$

$$\boxed{\mu_Y(t) = \mu_{2X}(t)}$$

Equating the diffusion coefficients gives

$$Y(t)\sigma_Z(t) = \sigma_{1X}(t) \qquad (5.2)$$

$$\boxed{\sigma_Y(t) = \sigma_{2X}(t)}$$

Substituting (5.2) into (5.1) gives

$$Y(t)\mu_Z(t) + \sigma_Y(t)\sigma_{1X}(t) = \mu_{1X}(t)$$

so

$$\boxed{\mu_Z(t) = [\mu_{1X}(t) - \sigma_Y(t)\sigma_{1X}(t)]/Y(t)}$$

Equation (5.2) gives

$$\boxed{\sigma_Z(t) = \frac{\sigma_{1X}(t)}{Y(t)}}$$

Coefficients $\mu_Z(t)$ and $\sigma_Z(t)$ have now been expressed in terms of known coefficients, $\mu_{1X}(t)$, $\sigma_{1X}(t)$, $\sigma_Y(t)$, and solution $Y(t)$.

Example 5.10.1: *Ornstein–Uhlenbeck SDE*

$$dX(t) = -\lambda X(t)\, dt + \sigma\, dB(t) \qquad \text{where } X(0) \text{ is known}$$

Here

$$\mu_{1X}(t) = 0 \qquad \mu_{2X}(t) = -\lambda \qquad \sigma_{1X}(t) = \sigma \qquad \sigma_{2X}(t) = 0$$

and

$$\mu_Z(t) = 0 \qquad \sigma_Z(t) = \sigma/Y(t)$$

The equation for Y is then

$$dY(t) = -\lambda Y(t)\, dt \qquad \text{where } Y(0) = 1$$

or

$$\frac{dY(t)}{Y(t)} = -\lambda\, dt \qquad Y(0) = 1$$

which is an ODE with solution $Y(t) = \exp(-\lambda t)$. The expression for $Z(t)$ becomes

$$Z(t) = X(0) + \int_{u=0}^{t} \sigma \exp(\lambda u)\, dB(u)$$

The product of $Y(t)$ and $Z(t)$ is the solution

$$X(t) = \exp(-\lambda t)\left[X(0) + \int_{u=0}^{t} \sigma \exp(\lambda u)\, dB(u) \right]$$

which is the same as in Section 5.4.

Example 5.10.2: *Brownian bridge SDE*

For t in the time interval $[0, T]$

$$dX(t) = -\frac{1}{T-t} X(t)\, dt + dB(t) \qquad \text{where } X(0) = 0$$

Here

$$\mu_{1X}(t) = 0 \qquad \mu_{2X}(t) = -1/(T-t) \qquad \sigma_{1X}(t) = 1 \qquad \sigma_{2X}(t) = 0$$

and

$$\mu_Z(t) = 0 \qquad \sigma_Z(t) = [1/Y(t)]$$

The equation for Y is then

$$dY(t) = -\frac{1}{T-t}Y(t)\,dt$$

or

$$\frac{dY(t)}{dt} = -\frac{1}{T-t}Y(t)$$

which is an ODE. The solution is found by separating Y and t, and integrating

$$\int \frac{dY(t)}{Y(t)} = \int -\frac{1}{T-t}\,dt = \int \frac{d(T-t)}{T-t}$$

$$\ln[Y(t)] = \ln(T-t) + c \qquad \text{where } c \text{ is a constant}$$

To find c, use initial condition $Y(0) = 1$, giving

$$\ln[Y(0)] = \ln(1) = 0 = \ln(T) + c \quad c = -\ln(T)$$
$$\ln[Y(t)] = \ln(T-t) - \ln(T)$$
$$Y(t) = (T-t)/T$$

The SDE for Z is

$$dZ(t) = 0\,dt + \frac{1}{Y(t)}\,dB(t)$$

so

$$Z(t) = X(0) + \int_{u=0}^{t} \frac{T}{T-u}\,dB(u)$$

The product of $Y(t)$ and $Z(t)$ is the solution for all t excluding T

$$X(t) = \frac{T-t}{T}\left[X(0) + \int_{u=0}^{t} \frac{T}{T-u}\,dB(u)\right]$$

$$= \frac{T-t}{T}X(0) + (T-t)\int_{u=0}^{t} \frac{1}{T-u}\,dB(u)$$

As X is a stochastic integral and $X(0)$ is constant

$$\mathbb{E}[X(t)] = \frac{T-t}{T}X(0)$$

$$\mathbb{V}\text{ar}[X(t)] = (T-t)^2 \int_{u=0}^{t} \left(\frac{1}{T-u}\right)^2 du$$

The integral equals

$$-\int_{u=0}^{t} \frac{1}{(T-u)^2} d(T-u) = \frac{1}{T-u} \qquad \text{between } u = t \text{ and } 0$$

$$= \frac{1}{T-t} - \frac{1}{T} = \frac{t(T-t)}{T}$$

$$\mathbb{V}\text{ar}[X(t)] = t\left(1 - \frac{t}{T}\right)$$

which goes to zero as t goes to T. The terminal position is non-random, $X(T) = 0$.

5.11   MARTINGALE REPRESENTATION

Suppose a random process X follows the driftless SDE $dX(t) = \varphi(t, \omega)\, dB(t)$, where $X(0)$ is known, and $\varphi(t, \omega)$ is a random process (highlighted by ω in the notation). The corresponding integral expression over time period $s \leq t \leq u$ is

$$\int_{t=s}^{u} dX(t) = \int_{t=s}^{u} \varphi(t, \omega)\, dB(t)$$

$$X(u) = X(s) + \int_{t=s}^{u} \varphi(t, \omega)\, dB(t)$$

Taking the expected value conditional upon history $\Im(s)$ gives

$$\mathbb{E}[X(u)|\Im(s)] = X(s) + \mathbb{E}\left[\int_{t=s}^{u} \varphi(t, \omega)\, dB(t)|\Im(s)\right]$$

$$= X(s)$$

as the Itō stochastic integral has expected value zero. Thus a *driftless SDE describes a martingale*.

It turns out that there is also a reverse relationship. If Brownian motion is the only source of randomness, then a continuous *martingale*, *say M, can be expressed as a driftless SDE* driven by Brownian motion, $dM(t) = h(t, \omega) \, dB(t)$. Often M is written as a stochastic integral

$$M(t) = \mathbb{E}[M(t)] + \int_{s=0}^{t} h(\omega, s) \, dB(s)$$

which comes from

$$\int_{s=0}^{t} dM(s) = \int_{s=0}^{t} h(s, \omega) \, dB(s)$$

$$M(t) = M(0) + \int_{s=0}^{t} h(s, \omega) \, dB(s)$$

and using $\mathbb{E}[M(t)] = M(0)$. It is known as the *martingale representation property*. This only establishes the existence of the random process h. Its proof is rather elaborate and will only be outlined later. First, some examples.

Example 5.11.1 Earlier it was shown that the random process $M(t) \overset{\text{def}}{=} B(t)^2 - t$ is a martingale. It was also shown that $\int_{s=0}^{t} B(s) \, dB(s) = \frac{1}{2}[B(t)^2 - t] = \frac{1}{2}M(t)$. Thus $M(t)$ can be represented in terms of an Itō stochastic integral as $\int_{s=0}^{t} 2B(s) \, dB(s)$, and the random process h is $2B(t)$.

Example 5.11.2 For fixed T, the area under a Brownian motion path, $I(T) = \int_{t=0}^{T} B(t) \, dt$, is a random variable. For variable T it is a random process which is a martingale (why?). Earlier it has been shown that $I(T) = \int_{t=0}^{T}(T - t) \, dB(t)$, so that is its martingale representation, and h is the deterministic function $(T - t)$.

Example 5.11.3 If the random process M is known to be a martingale which is a function of t and random process X, whose SDE is of the general form $dX(t) = \mu(t, X) \, dt + \sigma(t, X) \, dB(t)$, then an expression for h can be derived as follows. Itō's formula gives, in shorthand

$$dM = \frac{\partial M}{\partial t} dt + \frac{\partial M}{\partial X} dX + \frac{1}{2}\frac{\partial^2 M}{\partial X^2}(dX)^2 + \frac{\partial^2 M}{\partial t \partial X} dt \, dX$$

$$= \left[\frac{\partial M}{\partial t} + \mu(t, X)\frac{\partial M}{\partial X} + \frac{1}{2}\sigma(t, X)^2 \frac{\partial^2 M}{\partial X^2} \right] dt$$

$$+ \sigma(t, X)\frac{\partial M}{\partial X} dB(t)$$

As M is known to be a martingale, the drift term in its SDE must be zero, leaving

$$dM(t) = \sigma(t, X)\frac{\partial M}{\partial X}\,dB(t)$$

Thus

$$h(t, \omega) = \sigma(t, X)\frac{\partial M}{\partial X}$$

This has its application in the martingale method for option valuation, where the existence of a process h implies the existence of a replicating portfolio.

Idea of proof   The full proof of the martingale representation property is highly technical, and beyond the scope of this text. It is based on the property that any random variable which depends only on Brownian motion, say X, can be expressed in terms of linear combinations of random variables of the form

$$Z(t) = \exp\left[-\frac{1}{2}\int_{s=0}^{t}\phi(s, \omega)^2\,ds + \int_{s=0}^{t}\phi(s, \omega)\,dB(s)\right]$$

which are the solution to the SDE $dZ(t) = \phi(t, \omega)Z(t)\,dB(t)$ with $Z(0) = 1$. In integral form

$$\int_{s=0}^{t}dZ(s) = \int_{s=0}^{t}\phi(s, \omega)Z(s)\,dB(s)$$

$$Z(t) = Z(0) + \int_{s=0}^{t}\phi(s, \omega)Z(s)\,dB(s)$$

$$= 1 + \int_{s=0}^{t}\phi(s, \omega)Z(s)\,dB(s)$$

which is of martingale representation form, with $h(t, \omega) = \phi(t, \omega)Z(t)$. Taking the conditional expectation of random variable X turns it into a random process which is a martingale (as shown in Section 2.4).

Although the martingale representation seems intuitive, its proof requires a major excursion into probability territory. Introductory references are: *Bass, Lin* Section 5.8, *Capasso/Bakstein* Section 3.6, *Björk* Section 11.1, the source of Example 5.11.3.

5.12   EXERCISES

[5.12.1] *Verifying the OU solution* Show that

$$X(t) = \exp(-\lambda t)\left[X(0) + \sigma \int_{s=0}^{t} \exp(\lambda s)\, dB(s)\right] \quad \text{solves the SDE}$$

$dX(t) = -\lambda X(t)\, dt + \sigma\, dB(t)$ where $X(0)$ is known

Hint: Introduce $Z(t) \stackrel{\text{def}}{=} \int_{s=0}^{t} \exp(\lambda s)\, dB(s)$. Then X is a function of the variables ...

[5.12.2]   *Deriving the Brownian bridge SDE*

(a) Random process X is specified for $0 \le t < T$ (strictly) as

$$X(t) \stackrel{\text{def}}{=} (T - t) \int_{s=0}^{t} \frac{1}{T - s}\, dB(s) \qquad \text{so } X(0) = 0$$

Derive the SDE for X.
(b) Random process Y is specified for $0 \le t < T$ (strictly) as

$$Y(t) \stackrel{\text{def}}{=} a\left(1 - \frac{t}{T}\right) + b\frac{t}{T} + (T - t)\int_{s=0}^{t} \frac{1}{T - s}\, dB(s) \quad \text{so}$$
$$Y(0) = a$$

Derive the SDE for Y.

[5.12.3]   *Solving the HW SDE.*   Hull and White developed a model for the short-term interest rate r in which the long-run mean is specified via the non-random time-dependent function $b(t)$

$$dr(t) = [b(t) - ar(t)]\, dt + \sigma\, dB(t) \text{ where } r(0) \text{ is known.}$$

Derive the solution by first simplifying the drift coefficient via the transformation $X(t) \stackrel{\text{def}}{=} \exp(at)r(t)$.

[5.12.4]   *Transforming the CIR volatility.*   Consider a function g of interest rate r specified by the CIR model.

(a) Derive dg
(b) Choose dg/dr so the diffusion coefficient of dg equals 1.
(c) Integrate dg/dr to get an expression for g in terms of r; write r in terms of g, and use this to write dg entirely in terms of g.

[5.12.5]   *Solving the BDT SDE.*   Black, Derman, and Toy developed a discrete time model for the short-term interest rate r. Its equivalent in continuous time is

$$d \ln[r(t)] = \theta(t)\,dt + \sigma\,dB(t)$$

Derive the SDE for r in terms of r.

[5.12.6]   *Solving and verifying the solution*

$$dX(t) = X(t)\,dt + dB(t) \quad X(0) \text{ known constant}$$

[5.12.7]   *Solving and verifying the solution*
$$dX(t) = -X(t)\,dt + \exp(-t)\,dB(t) \quad X(0) \text{ known constant}$$

[5.12.8]   *Solving and verifying the solution*
$$dX(t) = m\,dt + \sigma X(t)\,dB(t) \quad X(0), m \text{ and } \sigma \text{ known constants}$$

[5.12.9]   *Finding the variance of a square-root process.*   Let X be specified by the CIR SDE. Determine the variance of X by deriving the dynamics of $X(t)^2$ and using the method and the results of Section 5.7.

[5.12.10]   Verify that the growth conditions hold for all the SDEs discussed in this chapter, and that the Lipschitz condition holds for all except the square-root diffusion SDE.

[5.12.11]   Construct a simulation program for the mean-reversion SDE of Section 5.5.

[5.12.12]   Construct a simulation program for the square-root diffusion SDE of Section 5.6.

[5.12.13]   Let $S(t)$ be the standard stock price expression $S(0)$ $\exp[(\mu - \frac{1}{2}\sigma^2)t + \sigma B(t)]$. Verify that $\int_{t=0}^{1} \sigma S(t)\,dB(t)$ is an Itō stochastic integral.

5.13   SUMMARY

This chapter introduced and solved the basic stochastic differential equations used in finance. A general solution method was given for SDEs with linear coefficients. But there are few SDEs for which a closed form solution exists. A comprehensive overview of these is *Numerical Solution of Stochastic Differential Equations* by *Kloeden/Platen* Chapter 3. The distribution of SDEs can be readily analysed by simulation. A comprehensive reference for this is *Glasserman*.

6

Option Valuation

The purpose of this chapter is to illustrate how stochastic calculus is used in the valuation of an option. The concept of an option contract was outlined in the Preface of this text, and further specification is given below.[1] Firstly, the so-called *partial differential equation method* and, secondly, the so-called *martingale method* (also known as the *risk-neutral method*) are discussed. Thereafter the link between these two methods is explained: the so-called *Feynman–Kač representation*. The key concepts used in option valuation are the self-financing replicating portfolio and the martingale probability (also known as the risk-neutral probability). The essence of the martingale method can be best understood in the simplified discrete-time one-period binomial framework that was introduced in Chapter 2. The exposition in a continuous-time framework is a logical more technical continuation. Here the notation is

$S(t)$ price of underlying stock at time t

K strike price at which underlying can be bought or sold

T maturity time of option contract

r risk-free interest rate

$V(t)$ option value at time t

$S^*(t)$ stock price discounted by savings account, $\exp(-rt)S(t)$

$\widehat{\mathbb{P}}$ probability distribution under which $S^*(t)$ is a martingale

$\mathbb{E}_{\widehat{\mathbb{P}}}$ expected value under $\widehat{\mathbb{P}}$.

In a financial market, a so-called arbitrage opportunity is a situation where an investor can make a guaranteed profit without incurring any risk. Such situations arise regularly and there are people who specialize in spotting them, using sophisticated communication technology. They then immediately initiate a trade which changes the supply-demand situation, and restores the market price to equilibrium. Arbitrage opportunities are therefore very short lived. The fundamental condition for establishing the price of an option is that it should not permit an arbitrage

---

[1] As this is not a book about options per se, it may be useful to consult an options book for further terminology, such as the elementary introduction by *Benth*.

opportunity. Absence of arbitrage is a highly realistic assumption, and more tangible than the equilibrium assumptions in economics.

6.1   PARTIAL DIFFERENTIAL EQUATION METHOD

The value of an option on a stock is modelled as a function of two variables – calender time t, and stock price $S(t)$ – and is denoted $V(t)$. The stock price is assumed to evolve according to

$$\frac{dS(t)}{S(t)} = \mu \, dt + \sigma \, dB(t) \qquad \text{where } \mu \text{ and } \sigma \text{ are constants}$$

Itō's formula gives the change in the option value resulting from the change in time, dt, and the change in S over dt, as

$$dV = \frac{\partial V}{\partial t} \, dt + \frac{\partial V}{\partial S} \, dS + \frac{1}{2} \frac{\partial^2 V}{\partial S^2} (dS)^2$$

Using $dS = \mu S \, dt + \sigma S \, dB$ and $dS^2 = \sigma^2 S^2 \, dt$ gives

$$dV = \frac{\partial V}{\partial t} \, dt + \frac{\partial V}{\partial S} [\mu S \, dt + \sigma S \, dB] + \frac{1}{2} \frac{\partial^2 V}{\partial S^2} \sigma^2 S^2 \, dt$$

$$= \left[\frac{\partial V}{\partial t} + \mu S \frac{\partial V}{\partial S} + \tfrac{1}{2} \sigma^2 S^2 \frac{\partial^2 V}{\partial S^2} \right] dt + \sigma S \frac{\partial V}{\partial S} \, dB$$

The partial differential equation method of option valuation is based on the insight that the option and the stock on which it is written have the *same source of randomness*. Thus, by taking opposite positions in the option and the stock, the randomness of the one asset can offset the randomness of the other. It is therefore possible to form a portfolio of stock and options in such proportion that the overall randomness of this portfolio is zero. Moreover, if the proportion of stock and options in this portfolio is changed as the value of the stock changes, this portfolio can be maintained riskless at all times.

At time t, form a portfolio that is *long* λ shares and *short* 1 option. The value P of this portfolio at time t is

$$P(t) = \lambda S(t) - V(t)$$

The minus sign comes from the fact that the option is not owned but owed. The value of this portfolio changes according to

$$dP = \lambda \, dS - dV$$

$$= \lambda [\mu S \, dt + \sigma S \, dB] - \left[\frac{\partial V}{\partial t} + \mu S \frac{\partial V}{\partial S} + \tfrac{1}{2} \sigma^2 S^2 \frac{\partial^2 V}{\partial S^2} \right] dt - \sigma S \frac{\partial V}{\partial S} \, dB$$

$$= \left[\lambda \mu S - \frac{\partial V}{\partial t} - \mu S \frac{\partial V}{\partial S} - \tfrac{1}{2} \sigma^2 S^2 \frac{\partial^2 V}{\partial S^2} \right] dt + \left[\lambda \sigma S - \sigma S \frac{\partial V}{\partial S} \right] dB$$

The random term can be made to disappear by choosing λ in such a way that the coefficient

$$\left[\lambda \sigma S - \sigma S \frac{\partial V}{\partial S} \right] = 0$$

$$\lambda = \frac{\partial V}{\partial S}$$

That leaves

$$dP = \left[\lambda \mu S - \frac{\partial V}{\partial t} - \mu S \frac{\partial V}{\partial S} - \frac{1}{2} \sigma^2 S^2 \frac{\partial^2 V}{\partial S^2} \right] dt$$

The *portfolio* is then *riskless* so its value must increase in accordance with the risk-free interest rate, otherwise there would be an arbitrage opportunity. The interest accrued on 1 unit of money over a time interval of length dt is $1r\,dt$. The value of the portfolio thus grows by $Pr\,dt$ over dt. Equating the two expressions for the change in the value of P gives

$$\left[\lambda \mu S - \frac{\partial V}{\partial t} - \mu S \frac{\partial V}{\partial} S - \frac{1}{2} \sigma^2 S^2 \frac{\partial^2 V}{\partial S^2} \right] dt = Pr\,dt$$

$$\lambda \mu S - \frac{\partial V}{\partial t} - \mu S \frac{\partial V}{\partial S} - \frac{1}{2} \sigma^2 S^2 \frac{\partial^2 V}{\partial S^2} = Pr$$

Substituting $P = \lambda S - V$ and $\lambda = \partial V / \partial S$ gives

$$\frac{\partial V}{\partial S} \mu S - \frac{\partial V}{\partial t} - \mu S \frac{\partial V}{\partial S} - \frac{1}{2} \sigma^2 S^2 \frac{\partial^2 V}{\partial S^2} = \left(\frac{\partial V}{\partial S} S - V \right) r$$

The term $\mu S(\partial V / \partial S)$ cancels, leaving

$$-\frac{\partial V}{\partial t} - \frac{1}{2} \sigma^2 S^2 \frac{\partial^2 V}{\partial S^2} = rS \frac{\partial V}{\partial S} - rV$$

which rearranges to

$$\frac{1}{2} \sigma^2 S^2 \frac{\partial^2 V}{\partial S^2} + rS \frac{\partial V}{\partial S} + \frac{\partial V}{\partial t} = rV$$

This is a second-order partial differential equation (PDE) in the unknown function V. The fact that this PDE does not contain the growth rate μ of the stock price may be surprising at first sight. The PDE must be accompanied by the specification of the option value at the time of exercise, the so-called option payoff. For the actual PDE solution method, the reader is referred to *Wilmott* and *Jiang*. The PDE method was developed by *Black and Scholes using key insights by Merton*. Merton and Scholes were awarded the 1997 'Nobel prize' in Economics for this seminal work; Black had died in 1995.

The PDE derivation given here is attractive for its clarity. There are, however, other derivations which are considered to be more satisfactory in a technical mathematical sense.

6.2   MARTINGALE METHOD IN ONE-PERIOD BINOMIAL FRAMEWORK

The approach is to form a so-called replicating portfolio which comprises shares of the stock on which the option contract is based and an amount of risk-free borrowing. If this portfolio can replicate the value of the option at all times then the initial value of the portfolio must be the value of the option. If that was not the case then there would be an arbitrage opportunity in the market and the price would not be stable. The value of the replicating portfolio is a random process whose value is denoted V. The initial portfolio consists of α *shares* of initial price $S(0) = S$ and a *loan* of β, where α and β are to be determined.

$$V(0) = \alpha S + \beta$$

During the period $[0, T]$, using discrete compounding at rate r for the period, the amount β grows to $\beta(1 + r)$, and S becomes uS or dS. The terminal value of the portfolio, $V(T)$, is then

in the up-state: $\alpha uS + (1 + r)\beta = V(T)_{up}$
in the down-state: $\alpha dS + (1 + r)\beta = V(T)_{down}$

As $V(T)$ must replicate the value of the option at time T, its values will be known in terms of the values of the underlying stock. For example, in the case of a standard European call option with strike price K

$$V(T)_{up} = \max[uS - K, 0]$$
$$V(T)_{down} = \max[dS - K, 0]$$

The two equations are linear in the unknowns α and β, and have the unique solution

$$\alpha = \frac{V(T)_{up} - V(T)_{down}}{(u - d)S} \qquad \beta = \frac{V(T)_{down}u - V(T)_{up}d}{(1 + r)(u - d)}$$

Substituting α and β gives initial portfolio value $V(0)$, after regrouping terms, as

$$V(0) = \frac{(1+r)-d}{u-d}\frac{V(T)_{\text{up}}}{1+r} + \frac{u-(1+r)}{u-d}\frac{V(T)_{\text{down}}}{1+r}$$

Note that $\frac{u-(1+r)}{u-d}$ can also be written as $1 - \frac{(1+r)-d}{u-d}$. If $d < 1+r < u$ then both $\frac{(1+r)-d}{u-d}$ and $1 - \frac{(1+r)-d}{u-d}$ lie between 0 and 1. They can therefore be interpreted as probabilities q and $1-q$

$$q \overset{\text{def}}{=} \frac{(1+r)-d}{u-d} \qquad 1-q = \frac{u-(1+r)}{u-d}$$

So $V(0)$ can be expressed as the expected value of the terminal option values discounted by the risk-free rate

$$V(0) = q\frac{V(T)_{\text{up}}}{1+r} + (1-q)\frac{V(T)_{\text{down}}}{1+r}$$

Thus the *discounted* option value process is a discrete martingale when probability q is used. For this reason q is also known as the *martingale probability*. Probability q is an *artificial probability* and its sole use is in the valuation of the option. It is not the real probability of an up-movement of the stock price. Multiplying both sides of the expression for q by S and rearranging gives $(1+r)S = q(uS) + (1-q)(dS)$. The left-hand side is the terminal value when an amount S is invested in a risk-free savings account. The right-hand side is the expected value of the stock price if the amount S is used to purchase a stock. The equation says that the investor is, in expected value terms, indifferent to whether the amount S is invested in a risk-free savings account or whether it is invested in a stock. Because of this interpretation, q is also called the *risk-neutral probability*. It is *as if* the investor is indifferent to the risk of the stock price increment, when probability q is used in the valuation of the option. That, of course, is not the true attitude of an investor towards risk. The option value computation turns out to be the correct one if the investor is treated as being risk-neutral, hence the name risk-neutral probability. Probability q was determined without using investors' views on the probability of an up-movement. If the expectation is taken under a value that is different from q then the result is an option value that permits arbitrage. So the risk-neutral probability is linked to the absence of arbitrage. It is now shown that the aforementioned condition $d < 1+r < u$ follows from the absence of arbitrage.

No-Arbitrage Condition   Assume, without loss of generality, that $S = 1$. If α and β can be chosen such that initial investment $V(0)$ is zero, and terminal values $V(T)_{up}$ and $V(T)_{down}$ are both non-negative, but either $V(T)_{up}$ or $V(T)_{down}$ is strictly positive, then this is a scheme which produces a non-negative return for certain without any down-side risk. That is an *arbitrage opportunity*. Restated in terms of inequalities in the (β, α) plane

$$V(0) = \alpha + \beta = 0 \qquad\qquad \alpha = -\beta$$
$$V(T)_{up} = \alpha u + \beta(1 + r) \geqq 0 \qquad \alpha \geq -\frac{(1+r)}{u}\beta$$
$$V(T)_{down} = \alpha d + \beta(1 + r) \geqq 0 \qquad \alpha \geq -\frac{(1+r)}{d}\beta$$

If $(1 + r) \geq u$ then the slopes of the three lines relate as

$$-\frac{(1+r)}{d} \leq -\frac{(1+r)}{u} \leq -1,$$

as illustrated in Figure 6.1. The line $\alpha = -\beta$ lies on or above the other lines, and strictly above one of them (the area where the inequality holds

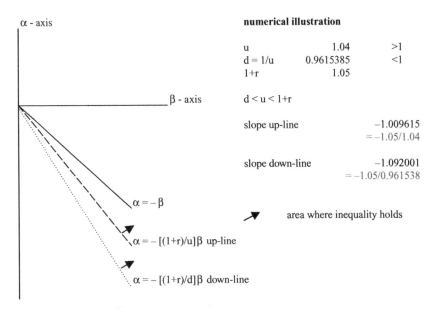

α values on line α = −β are above or on the other lines
negative α is short-selling

Figure 6.1   Arbitrage opportunity

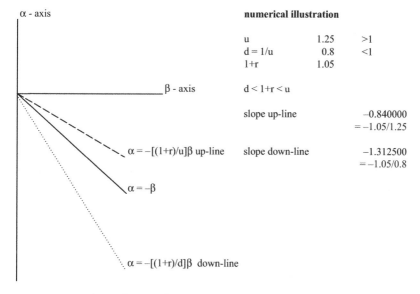

α - axis

numerical illustration

u	1.25	>1
d = 1/u	0.8	<1
1+r	1.05	

β - axis   d < 1+r < u

slope up-line   −0.840000
                                 = −1.05/1.25

α = −[(1+r)/u]β up-line   slope down-line   −1.312500
                                 = −1.05/0.8

α = −β

α = −[(1+r)/d]β down-line

α values on line α = −β are between other lines

Figure 6.2   No-arbitrage

is indicated by an arrow). The *no-arbitrage* condition is that $\alpha = -\beta$ must lie *between* the other two lines, as shown in Figure 6.2.

That is the *convex cone* spanned by the other two lines. The *no-arbitrage condition* in terms of slopes is thus

$$-\frac{(1+r)}{d} \leq -1 \leq -\frac{(1+r)}{u} \quad \text{or} \quad d \leq (1+r) \leq u$$

The last inequality says that $(1 + r)$ is then a convex combination of d and u, which can be written uniquely with weightings $0 < q < 1$ and $(1 - q)$ as $1 + r = qu + (1 - q)d$. This gives the familiar expression for the unique binomial martingale probability

$$q = \frac{(1+r) - d}{u - d}$$

So it has been shown that the no-arbitrage condition implies $d < 1 + r < u$, which in turn implies the existence of q.

Knowing that the discounted value process is a martingale makes it possible to determine the initial value of the option from the terminal values of the option which are known. So the *martingale concept is used in reverse*. That means that the valuation process can be carried out the other way around, namely by first finding a risk-neutral probability, and then taking the expected value under that probability of the discounted terminal option value. That is the way it is done in continuous time. One other feature has to be mentioned: the replicating portfolio must be self-financing, which means that the change in portfolio value over time should only come from the change in the value of the stock and the change in borrowing. No money is withdrawn or added freely.

The above showed that a *martingale probability arises in a completely natural way*. It can also be derived from the condition that the discounted stock price process must be a martingale under discrete probability q. The expectation of the terminal discounted stock price under q is

$$q \frac{uS}{1+r} + (1-q) \frac{dS}{1+r}$$

Equating this to S and solving for q gives

$$q = \frac{(1+r) - d}{u - d}$$

Using the notation $\mathbb{E}_q$ for an expectation using probability q, the martingale expression is

$$\mathbb{E}_q \left[\frac{S(T)}{1+r} \middle| S \right] = S$$

The discounted value of the self-financing replicating portfolio is also a martingale under q, as can be seen as follows. The *discounted* terminal portfolio value is

$$\text{in the up-state: } \frac{V(T)_{\text{up}}}{1+r} = \alpha u \frac{S}{1+r} + \beta$$

$$\text{in the down-state: } \frac{V(T)_{\text{down}}}{1+r} = \alpha d \frac{S}{1+r} + \beta$$

The expectation, under probability q, of the *discounted* terminal portfolio value is

$$\mathbb{E}_q \left[\frac{V(T)}{1+r} \right] = q \left[\frac{\alpha uS}{1+r} + \beta \right] + (1-q) \left[\frac{\alpha dS}{1+r} + \beta \right]$$

$$= q \frac{\alpha uS}{1+r} + (1-q) \frac{\alpha dS}{1+r} + \beta$$

$$= q \left[\frac{\alpha uS}{1+r} - \frac{\alpha dS}{1+r} \right] + \frac{\alpha dS}{1+r} + \beta$$

$$= q\alpha \frac{u-d}{1+r} S + \alpha \frac{d}{1+r} S + \beta$$

Substituting $q = [(1 + r) - d]/(u - d)$ gives

$$\mathbb{E}_q\left[\frac{V(T)}{(1 + r)}\right] = \frac{(1 + r) - d}{u - d}\alpha\frac{u - d}{1 + r}S + \alpha\frac{d}{1 + r}S + \beta$$

$$= \alpha\frac{(1 + r) - d}{1 + r}S + \alpha\frac{d}{1 + r}S + \beta$$

$$= \alpha S + \beta$$

$$= V(0)$$

Since the terminal option values $V(T)$ are *known* in every terminal state of the market, the unknown initial value of the replicating portfolio, $V(0)$, and thus of the option, can be determined by turning the martingale expression around

> initial option value $= \mathbb{E}_q[\text{terminal option value}/(1 + r)]$

In this simple binomial model there are two risky assets, the stock and the option. The expected rate of return is defined as *(expected terminal value − initial value)/initial value*. Under q both risky assets in the model have an expected rate of return equal to the risk-free rate.

The binomial tree for option valuation is of such conceptual and computational importance that entire books have been devoted to it: *Roman* and *Shreve I* are particularly good introductions. The graphical explanation of the relationship between the absence of arbitrage and the existence of a martingale probability is based on *McLeish*; an alternative exposition is in *Epps* Chapter 5.

6.3 MARTINGALE METHOD IN CONTINUOUS-TIME FRAMEWORK

In the continuous time setting, start from the standard SDE for the stock price

$$\frac{dS(t)}{S(t)} = \mu\,dt + \sigma\,dB(t)$$

Introduce the discounted stock price

$$S^*(t) \stackrel{\text{def}}{=} \frac{S(t)}{\exp(rt)}$$

This expresses the stock price value in terms of the savings account as the numeraire. Itō's formula, for S^* as a function of t and S, gives

$$\frac{dS^*(t)}{S^*(t)} = (\mu - r)\,dt + \sigma\,dB(t)$$

The expected growth rate of S^* is $(\mu - r)$, which is r less than the growth rate of S, because it is measured against the riskless growth rate r of the savings account. As this SDE has a drift, the process S^* is not a martingale under the probability distribution of Brownian motion B. In the replicating portfolio of the above discrete-time model, the discounted stock price was a martingale. That led to the discounted replicating portfolio also being a martingale. The same recipe will be followed here. A probability distribution can be found under which S^* is a martingale. This goes as follows. Rewrite the SDE as

$$\frac{dS^*(t)}{S^*(t)} = \sigma\left[\frac{\mu - r}{\sigma}\,dt + dB(t)\right] = \sigma[\varphi\,dt + dB(t)]$$

where $\varphi \overset{\text{def}}{=} (\mu - r)/\sigma$. The probability density of $B(t)$ at $B(t) = x$ is

$$\frac{1}{\sqrt{t}\sqrt{2\pi}}\exp\left[-\frac{1}{2}\left(\frac{x}{\sqrt{t}}\right)^2\right]$$

This can be decomposed into the product of two terms

$$\frac{1}{\sqrt{t}\sqrt{2\pi}}\exp\left[-\frac{1}{2}\left(\frac{\varphi t + x}{\sqrt{t}}\right)^2\right]\exp\left[\tfrac{1}{2}\varphi^2 t + \varphi x\right]$$

With $y \overset{\text{def}}{=} \varphi t + x$ the first term can be written as

$$\frac{1}{\sqrt{t}\sqrt{2\pi}}\exp\left[-\frac{1}{2}\left(\frac{y}{\sqrt{t}}\right)^2\right]$$

which is the probability density of another Brownian motion, say $\widehat{B}(t)$, at $\widehat{B}(t) = y$. It defines $\widehat{B}(t) \overset{\text{def}}{=} \varphi t + B(t)$, so $d\widehat{B}(t) = \varphi\,dt + dB(t)$. This is further discussed in Chapter 7. Substituting the latter into the SDE for S^* gives

$$\frac{dS^*(t)}{S^*(t)} = \sigma\,d\widehat{B}(t) \quad \text{and} \quad \frac{dS(t)}{S(t)} = r\,dt + \sigma\,d\widehat{B}(t)$$

This says that under the probability distribution of Brownian motion $\widehat{B}$, S^* is a martingale. Let this probability be denoted $\mathbb{P}$.

The value of the replicating portfolio at time t is denoted $V(t)$. The portfolio consists of a quantity $\alpha(t)$ of stock $S(t)$ and an amount $\beta(t)$ of risk-free borrowing. The evolution of $\beta(t)$ is specified by the ordinary differential equation $d\beta(t) = \beta(t)r\,dt$; in rate of return form, $d\beta(t)/\beta(t) = r\,dt$.

$$V(t) = \alpha(t)S(t) + \beta(t) \quad \text{where} \quad \beta(t) = \exp(rt) \qquad \beta(0) = 1$$

The replicating portfolio must be *self-financing*, so the change in the value of the portfolio must only come from the change in the value of the stock and the change in the value of the borrowing. This condition is represented by

$$dV(t) = \alpha(t)\,dS(t) + d\beta(t)$$

The discounted value of the portfolio

$$V^*(t) \stackrel{\text{def}}{=} \frac{V(t)}{\exp(rt)}$$

is a function of t and V. Itō's formula gives

$$dV^* = \frac{\partial V^*}{\partial t}\,dt + \frac{\partial V^*}{\partial V}\,dV + \frac{1}{2}\frac{\partial^2 V^*}{\partial V^2}(dV)^2 + \frac{\partial^2 V^*}{\partial t\partial V}\,dt\,dV$$

Substituting

$$\frac{\partial V^*}{\partial t} = -rV^* \qquad \frac{\partial V^*}{\partial V} = \exp(-rt) \qquad \frac{\partial^2 V^*}{\partial V^2} = 0 \qquad dt\,dV = 0$$

gives

$$dV^* = -rV^*\,dt + \exp(-rt)\,dV$$

Substituting V^* and dV then gives

$$dV^* = -r[\exp(-rt)V]\,dt + \exp(-rt)[\alpha\,dS + d\beta]$$

Finally substituting the expression for V gives

$$\begin{aligned} dV^* &= -r\exp(-rt)[\alpha S + \beta]\,dt + \exp(-rt)[\alpha\,dS + r\beta\,dt] \\ &= -r\exp(-rt)\alpha S\,dt - r\exp(-rt)\beta\,dt + \exp(-rt)\alpha\,dS \\ &\quad + r\exp(-rt)\beta\,dt \end{aligned}$$

The term with β cancels, leaving

$$
\begin{aligned}
dV^* &= -r \exp(-rt)\alpha S \, dt + \exp(-rt)\alpha \, dS \\
&= \alpha[-r \exp(-rt)S \, dt + \exp(-rt) \, dS] \\
&= \alpha[-r \exp(-rt)S \, dt + \exp(-rt)rS \, dt + \exp(-rt)\sigma S \, d\widehat{B}] \\
&= \alpha \exp(-rt)\sigma S \, d\widehat{B}
\end{aligned}
$$

So the final SDE is

$$ dV^*(t) = \alpha(t) \, dS^*(t) $$

As there is no drift term, random process $V^*(t)$ is a martingale under $\widehat{\mathbb{P}}$. Thus its expected value at future time T equals its value at time 0, $\mathbb{E}_{\widehat{\mathbb{P}}}[V^*(T)|V^*(0)] = V^*(0) = V(0)$. Writing this the other way around, the initial value of the option is determined from

$$ \boxed{V(0) = \mathbb{E}_{\widehat{\mathbb{P}}}[V^*(T)|S(0)] = \mathbb{E}_{\widehat{\mathbb{P}}}\left[\frac{V(T)}{\exp(rT)}\Big|S(0)\right]} $$

In Section 6.5 this is illustrated for the valuation of some European options.

On $0 \le t \le T$ the SDE $dS^*(t)/S^*(t) = \sigma \, d\widehat{B}(t)$ has the solution

$$ S^*(T) = S^*(0) \exp\left[-\tfrac{1}{2}\sigma^2 T + \sigma \widehat{B}(T)\right] $$

and using $S(t) = \exp(rt)S^*(t)$ gives

$$ S(T) = S(0) \exp\left[(r - \tfrac{1}{2}\sigma^2)T + \sigma \widehat{B}(T)\right] $$

The existence of a replicating portfolio is guaranteed by the martingale representation (of Section 5.11), and is shown in *Epps* pp 265–266. The actual expressions for α and β are derived in *Lamberton/Lapeyre* Section 4.3.3.

6.4   OVERVIEW OF RISK-NEUTRAL METHOD

In both the discrete- and the continuous-time framework the *discounted stock price is forced to become a martingale*. That produces a new probability distribution, which is called the *martingale probability* or *risk-neutral probability*. Using this new probability distribution, the expected value of the discounted stock price at some future time equals the known discounted stock price at an earlier time. Then a *self-financing* portfolio is formed which replicates the value of an option at all times in all possible states of the market. The discounted value of this self-financing replicating portfolio is also found to be a martingale under this new probability. Thus the expectation of the discounted value of the replicating portfolio, at the time of exercise of the option, equals the present value of the portfolio. As all possible values of the option at the time of exercise are known, and the martingale probability is also known, their expected value can be computed. At the present time, the discounted value of the replicating portfolio equals the value of the option that is to be determined. Hence the fact that the discounted portfolio value is a martingale makes it possible to compute this present value of the option. The steps in the methodology are summarized below.

Table 6.1

	Stock price	Portfolio
Value at time t	$S(t)$	$V(t) = \alpha(t)S(t) + \beta(t)$
Dynamics under $B(t)$	$\frac{dS(t)}{S(t)} = \mu \, dt + \sigma \, dB(t)$	self-financing $dV(t) = \alpha(t) \, dS(t) + d\beta(t)$
Discounted value at time t	$S^*(t) = \exp(-rt)S(t)$	$V^*(t) = \exp(-rt)V(t)$
Dynamics of discounted value under $B(t)$	$\frac{dS^*(t)}{S^*(t)} = (\mu - r) \, dt + \sigma \, dB(t)$	$dV^*(t) = \alpha(t) \, dS^*(t)$
Property	$S^*(t)$ *not* martingale	$V^*(t)$ *not* martingale
Dynamics of discounted value under $\widehat{B}(t)$	$\frac{dS^*(t)}{S^*(t)} = \sigma \, d\widehat{B}(t)$	$dV^*(t) = \alpha(t) \, dS^*(t)$ $= \alpha(t)\sigma \, S^*(t) \, d\widehat{B}(t)$
Property	$S^*(t)$ *martingale* $\mathbb{E}_{\widehat{p}}[S^*(T)\|S(0)] = S(0)$	$V^*(t)$ *martingale* $\mathbb{E}_{\widehat{p}}[V^*(T)\|V(0)] = V(0)$

6.5   MARTINGALE METHOD VALUATION OF SOME EUROPEAN OPTIONS

The illustration of the martingale method in continuous time starts with two examples which only require simple integration.

6.5.1   Digital Call

The payoff of this option at exercise time T is a fixed amount (here set at 1), or nothing,

$$c(T) = \begin{cases} 1 & \text{if } S(T) \geq K \\ 0 & \text{otherwise} \end{cases}$$

The discounted call value process $c(t)/\exp(rt)$ is a martingale under probability $\widehat{\mathbb{P}}$, so at time 0

$$\frac{c(0)}{\exp(0T)} = \mathbb{E}_{\widehat{\mathbb{P}}}\left[\frac{c(T)}{\exp(rT)}\right] = \exp(-rT)\mathbb{E}_{\widehat{\mathbb{P}}}[1_{\{S(T)\geq K\}}]$$

where $1_{\{.\}}$ is the indicator function which has value 1 when the condition in the curly brackets is satisfied, and 0 otherwise. As the expected value is taken of a random process that is a martingale, the payoff condition $S(T) \geq K$ is now expressed in terms of that martingale S^* as $S^*(T) \geq \exp(-rT)K$, and

$$c(0) = \exp(-rT)\mathbb{E}_{\widehat{\mathbb{P}}}[1_{\{S^*(T)\geq \exp(-rT)K\}}]$$
$$= \exp(-rT)\widehat{\mathbb{P}}[S^*(T) \geq \exp(-rT)K]$$

as the expected value of an indicator function equals the probability of the indicator event. Under $\widehat{\mathbb{P}}$

$$S^*(T) = S(0)\exp\left[-\tfrac{1}{2}\sigma^2 T + \sigma\widehat{B}(T)\right]$$

Now determine the values of $\widehat{B}(T)$ for which $S^*(T) \geq \exp(-rT)K$, or equivalently

$$\ln[S^*(T)] \geq -rT + \ln[K]$$
$$\ln[S(0)] - \tfrac{1}{2}\sigma^2 T + \sigma\widehat{B}(T) \geq -rT + \ln[K]$$
$$\widehat{B}(T) \geq \underbrace{-\left\{\ln\left[\frac{S(0)}{K}\right] + \left(r - \tfrac{1}{2}\sigma^2\right)T\right\}\bigg/\sigma}_{a}$$

$$\widehat{B}(T) \geq a$$

Then

$$\widehat{\mathbb{P}}[S^*(T) \geq \exp(-rT)K] = \widehat{\mathbb{P}}[\widehat{B}(T) \geq a]$$
$$= \int_{x=a}^{\infty} \underbrace{\frac{1}{\sqrt{T}\sqrt{2\pi}}\exp\left[-\frac{1}{2}\left(\frac{x}{\sqrt{T}}\right)^2\right]dx}_{\text{density of }\widehat{B}(T)\text{ at }x}$$

The integrand is now transformed to the standard normal density by the change of variable $y = x/\sqrt{T}$, lower integration limit $x = a$ becomes

$y = a/\sqrt{T}, dy = dx/\sqrt{T}$, and

$$\widehat{\mathbb{P}}[S^*(T) \geq \exp(-rT)K] = \int_{y=a/\sqrt{T}}^{\infty} \frac{1}{\sqrt{2\pi}} \exp\left[-\tfrac{1}{2}y^2\right] dy$$

$$= 1 - N\left(\frac{a}{\sqrt{T}}\right) = N\left(-\frac{a}{\sqrt{T}}\right)$$

where $N(d)$ denotes the cumulative standard normal $\int_{y=-\infty}^{d} \frac{1}{\sqrt{2\pi}}$ $\exp\left[-\tfrac{1}{2}y^2\right] dy$.

$$c(0) = \exp(-rT)N\left(-\frac{a}{\sqrt{T}}\right)$$

The expression

$$\frac{-a}{\sqrt{T}} = \frac{\ln\left[\frac{S(0)}{K}\right] + \left(r - \tfrac{1}{2}\sigma^2\right)T}{\sigma\sqrt{T}}$$

is usually denoted d_2. The final result is

$$\boxed{c(0) = \exp(-rT)N(d_2)}$$

6.5.2   Asset-or-Nothing Call

This call has a payoff equal to the value of the underlying at maturity, if that value is not below the strike price, and zero otherwise,

$$c(T) = S(T) 1_{\{S(T) \geq K\}}$$

As before

$$c(0) = \exp(-rT) \mathbb{E}_{\widehat{\mathbb{P}}}[c(T)]$$
$$= \exp(-rT) \mathbb{E}_{\widehat{\mathbb{P}}}\left[S(T) 1_{\{S(T) \geq K\}}\right]$$
$$= \mathbb{E}_{\widehat{\mathbb{P}}}\left[S^*(T) 1_{\{S^*(T) \geq \exp(-rT)K\}}\right]$$

This expected value can be computed by using the lognormal probability density of random variable $S^*(T)$, but it is easier to view $S^*(T)$ as a function of random variable $\widehat{B}(T)$ and use its density.

$$c(0) = \mathbb{E}_{\widehat{\mathbb{P}}}[S(0) \exp[-\tfrac{1}{2}\sigma^2 T + \sigma \widehat{B}(T)] \, 1_{\{\widehat{B}(T) \geq a\}}]$$

$$= \int_{x=a}^{\infty} S(0) \exp[-\tfrac{1}{2}\sigma^2 T + \sigma x] \underbrace{\frac{1}{\sqrt{T}\sqrt{2\pi}} \exp\left[-\frac{1}{2}\left(\frac{x}{\sqrt{T}}\right)^2\right]}_{\text{density of } \widehat{B}(T) \text{ at } x} dx$$

Change of variable $y \stackrel{\text{def}}{=} x/\sqrt{T}$ gives

$$c(0) = \int_{y=a/\sqrt{T}}^{\infty} S(0) \exp\left[-\tfrac{1}{2}\sigma^2 T + \sigma\sqrt{T}y\right] \frac{1}{\sqrt{2\pi}} \exp\left[-\tfrac{1}{2}y^2\right] dy$$

The exponent of the integrand can be rearranged to $-\tfrac{1}{2}(y - \sigma\sqrt{T})^2$

$$c(0) = S(0) \int_{y=a/\sqrt{T}}^{\infty} \frac{1}{\sqrt{2\pi}} \exp\left[-\tfrac{1}{2}(y - \sigma\sqrt{T})^2\right] dy$$

This integrand is transformed to the standard normal density by $z \stackrel{\text{def}}{=} y - \sigma\sqrt{T}$

$$c(0) = S(0) \int_{z=z_{\text{low}}}^{\infty} \frac{1}{\sqrt{2\pi}} \exp\left[-\tfrac{1}{2}z^2\right] dz$$

where $z_{\text{low}} \stackrel{\text{def}}{=} a/\sqrt{T} - \sigma\sqrt{T}$.

$$c(0) = S(0)[1 - N(z_{\text{low}})]$$
$$= S(0)N(-z_{\text{low}})$$

The expression for $-z_{\text{low}}$ is commonly denoted

$$d_1 \stackrel{\text{def}}{=} \frac{\ln[S(0)/K] + (r + \tfrac{1}{2}\sigma^2)T}{\sigma\sqrt{T}}$$

and

$$\boxed{c(0) = S(0)N(d_1)}$$

6.5.3 Standard European Call

The derivation steps are the same as for the above calls. Only the integral for the expected value is somewhat more involved. As above, the derivation is done directly in terms of the martingale S^*. That makes the methodology compatible with option valuation when a different numeraire than the savings account is used, as discussed in Chapter 8. For the standard European call

$$c(T) = \max[S(T) - K, 0]$$
$$= \exp(rT)\max[S^*(T) - \exp(-rT)K, 0]$$

and

$$c(0) = \exp(-rT)\mathbb{E}_{\widehat{\mathbb{P}}}[c(T)]$$
$$= \mathbb{E}_{\widehat{\mathbb{P}}}\{\max[S^*(T) - \exp(-rT)K, 0]\}$$
$$= \mathbb{E}_{\widehat{\mathbb{P}}}\{\max[S^*(T) - K^*, 0]\}$$

where $K^* \stackrel{\text{def}}{=} \exp(-rT)K$. The expression for $c(0)$ is of the same form as for the asset-or-nothing call, but now there is an additional term K^* in the integrand. The lower integration limit a for positive payoff is the same as before.

$$c(0) = \int_{x=a}^{\infty} \left\{ S(0)\exp\left[-\tfrac{1}{2}\sigma^2 T + \sigma x\right] - K^* \right\}$$
$$\times \underbrace{\frac{1}{\sqrt{T}\sqrt{2\pi}}\exp\left[-\tfrac{1}{2}\left(\tfrac{x}{\sqrt{T}}\right)^2\right]dx}_{\text{density of } \widehat{B}(T) \text{ at } x}$$

It is convenient to write this as the difference of the two integrals

$$I_1 \stackrel{\text{def}}{=} \int_{x=a}^{\infty} S(0)\exp\left[-\tfrac{1}{2}\sigma^2 T + \sigma x\right]\frac{1}{\sqrt{T}\sqrt{2\pi}}\exp\left[-\frac{1}{2}\left(\frac{x}{\sqrt{T}}\right)^2\right]dx$$

$$I_2 \stackrel{\text{def}}{=} \int_{x=a}^{\infty} K^* \frac{1}{\sqrt{T}\sqrt{2\pi}}\exp\left[-\frac{1}{2}\left(\frac{x}{\sqrt{T}}\right)^2\right]dx$$

Integral I_1 has already been evaluated above as $S(0)N(d_1)$. Integral I_2 transforms to

$$
\begin{aligned}
I_2 &= K^* \int_{x=a/\sqrt{T}}^{\infty} \frac{1}{\sqrt{2\pi}} \exp\left[-\tfrac{1}{2}y^2\right] dy \\
&= K^* \left[1 - N\left(\frac{a}{\sqrt{T}}\right)\right] \\
&= K^* N\left(-\frac{a}{\sqrt{T}}\right) \\
&= \exp(-rT)K N(d_2)
\end{aligned}
$$

The final result is

$$
\boxed{c(0) = S(0)N(d_1) - \exp(-rT)KN(d_2)}
$$

This is the famous *Black–Scholes pricing expression for a standard European call.*

Setting $K = 0$ recovers the pricing expression for the *asset-or-nothing call.* Note that the second term in the Black–Scholes expression equals K times the digital call price, and that the first term equals the all-or-nothing call price. Thus a standard European call can be seen as a portfolio that is long K European digital calls, and short one all-or-nothing call.

6.6 LINKS BETWEEN METHODS

6.6.1 Feynman-Kač Link between PDE Method and Martingale Method

The Black–Scholes PDE is an example of a so-called parabolic PDE. Such PDEs can be solved by a variety of classical analytical or numerical methods, but there is also an altogether different method which produces a solution in the form of an expected value. This is known as the Feynman–Kač representation. It links the PDE method for option valuation to the martingale method in continuous time.

Let S be the standard stock price process with SDE $dS(t)/S(t) = \mu \, dt + \sigma \, dB(t)$. Let V be the value of an option on S which matures at time T. It is a function of time t and $S(t)$, but to keep the notation simple, its value is denoted as $V(t)$. As derived previously,

$$
dV = \left[\frac{\partial V}{\partial t} + \mu S \frac{\partial V}{\partial S} + \tfrac{1}{2}\sigma^2 S^2 \frac{\partial^2 V}{\partial S^2}\right] dt + \sigma S \frac{\partial V}{\partial S} dB
$$

Consider the discounted value of V, $Z(t) \overset{\text{def}}{=} V(t)\exp(-rt)$. It is a function of the two variables t and V, but to keep the notation simple, its value is denoted as $Z(t)$. By Itô's formula

$$dZ = \frac{\partial Z}{\partial t} dt + \frac{\partial Z}{\partial V} dV + \frac{1}{2}\frac{\partial^2 Z}{\partial V^2}(dV)^2 + \frac{\partial^2 Z}{\partial t\,\partial V} dt\,dV$$

Substituting

$$\frac{\partial Z}{\partial t} = V\exp(-rt)(-r) \quad \frac{\partial Z}{\partial V} = \exp(-rt) \quad \frac{\partial^2 Z}{\partial V^2} = 0 \quad dt\,dV = 0$$

together with the expression for dV, gives

$$dZ = \exp(-rt)(-r)V\,dt +$$
$$+ \exp(-rt)\left[\frac{\partial V}{\partial t} + \mu S\frac{\partial V}{\partial S} + \frac{1}{2}\sigma^2 S^2\frac{\partial^2 V}{\partial S^2}\right]dt$$
$$+ \exp(-rt)\sigma S\frac{\partial V}{\partial S}\,dB$$
$$= \exp(-rt)\left[-rV + \frac{\partial V}{\partial t} + \mu S\frac{\partial V}{\partial S} + \frac{1}{2}\sigma^2 S^2\frac{\partial^2 V}{\partial S^2}\right]dt +$$
$$+ \exp(-rt)\sigma S\frac{\partial V}{\partial S}\,dB$$

If V *is such that its drift coefficient equals zero,*

$$-rV + \frac{\partial V}{\partial t} + \mu S\frac{\partial V}{\partial S} + \frac{1}{2}\sigma^2 S^2\frac{\partial^2 V}{\partial S^2} = 0 \qquad (*)$$

then $dZ = \exp(-rt)\sigma S\dfrac{\partial V}{\partial S}\,dB$. In integral form over the time period $[0, T]$

$$Z(T) = Z(0) + \int_{t=0}^{T} \exp(-rt)\sigma S(t)\frac{\partial V}{\partial S}\,dB(t)$$

At time 0, $Z(0)$ is a non-random value. Taking the expected value of both sides of the equation, gives

$$\mathbb{E}[Z(T)] = Z(0) + \mathbb{E}\left\{\int_{t=0}^{T} \exp(-rt)\sigma S(t)\frac{\partial V}{\partial S}\,dB(t)\right\}$$

As $\mathbb{E}$ of the stochastic integral equals zero, $Z(0) = \mathbb{E}[Z(T)]$. With $Z(0) = \exp(-r0)V(0) = V(0)$ and $Z(T) = \exp(-rT)V(T)$, the result is

$$V(0) = \mathbb{E}[\exp(-rT)V(T)]$$

This says that the initial value of the option can be determined as the expected value of the discounted random terminal value of the option. It was shown that the value V of an option on stock price S satisfies the Black–Scholes PDE

$$-rF + \frac{\partial V}{\partial t} + rS\frac{\partial V}{\partial S} + \tfrac{1}{2}\sigma^2 S^2 \frac{\partial^2 V}{\partial S^2} = 0$$

This is of the form (*) with $\mu = r$. Thus the S dynamics to be used in the expected value expression has r as the drift coefficient instead of μ

$$\frac{dS(t)}{S(t)} = r\,dt + \sigma\,dB(t)$$

This is the same SDE as used in the martingale method.

   The Feynman–Kač link is that

- the coefficient of the first-order partial derivative $\partial V/\partial S$ in the PDE is the drift coefficient in the martingale dynamics $dS(t) = rS(t)\,dt + \sigma S(t)\,dB(t)$
- the coefficient of the second-order partial derivative $\partial^2 V/\partial S^2$ in the PDE contains the square of the diffusion coefficient in the martingale dynamics dS.

So when the martingale dynamics for the underlying asset of the option are given, the corresponding PDE for the option value can be written down without any further derivation.

6.6.2   Multi-Period Binomial Link to Continuous

By dividing the time to maturity T into n time-steps $\Delta t = T/n$, and repeating the binomial stock price increment, a tree like that shown in Figure 1.2 is produced with a stock price at each node; $n+1$ stock prices at maturity. As for the random walk, it is assumed that the stock price increments over successive time-steps are independent. There are several choices for the tree parameters u and d. The original, and most widely known, is

$$u = \exp(\sigma\sqrt{\Delta t}) \qquad d = \exp(-\sigma\sqrt{\Delta t})$$

It can be shown that for $n \to \infty$:

- the binomial scheme converges to the Black–Scholes PDE
- the terminal stock price converges to the standard continuous time terminal stock price
- the binomial value of an option converges to the Black–Scholes value.

The binomial option value is highly non-linear in n. The convergence pattern depends greatly on the ratio K/S, and for the above choice of u and d it is highly irregular when $K \neq S$. This makes it difficult to fix a suitable n. A better choice is

$$u = \exp\left[\sigma\sqrt{\Delta t} + \tfrac{1}{n}\ln\left(\tfrac{K}{S}\right)\right] \qquad d = \exp\left[-\sigma\sqrt{\Delta t} + \tfrac{1}{n}\ln\left(\tfrac{K}{S}\right)\right]$$

Now the convergence pattern is gradual for all K/S. The convergence from binomial to continuous is well described in *Roman*.

6.7 EXERCISE

[6.7.1] *Futures option.*   The object of this exercise is to use the replicating portfolio method to determine the initial value of an option on a futures contract.

A futures contract is an agreement between two parties in which the holder of the contract agrees to trade a specified quantity of an asset with the writer of the contract, at a delivery date and place specified in the contract, for a delivery price that is set when the contract is entered into. The contract holder has the *obligation* to trade. Payment takes place at delivery. This delivery price is set such that the *value of the futures contract* is zero to both parties when the contract is initiated. If $F(t)$ denotes the futures price at time t of a futures contract on stock price $S(t)$ for delivery at T^*, then $F(t) = \exp[r(T^* - t)]S(t)$, the compounded value of S. There is no initial cost when buying a futures contract. The predetermined *delivery price* is also known as the *initial futures price*.Under this price there exists no arbitrage opportunity. A futures contract is traded and administered on exchanges. The trading produces futures price at all subsequent times. As time progresses, the asset which the holder of the future has contracted, changes in value over time due to market forces. The price for which the same asset can be bought at the original delivery date thus changes. Hence the futures price changes, but the delivery price stays as fixed originally. The exchange on which the contract is traded keeps track of these changes. An *account* is kept for the value of the contract. Initially the value of the account is zero. When a new market price for the futures contract becomes available on the exchange, the account value is increased (decreased) by the increase (decrease) of the current futures price from the previous futures price. It is essential to be clear about the distinction between the *initial futures price* (at which the asset will be delivered), the subsequent *futures prices,* and the *value of the account.* An *option on a futures contract* gives its holder the option to exercise the futures contract. It removes the obligation at the cost of the option premium. The underlying of this option is a futures contract of equal or later maturity than the option. At option maturity date T the payoff to the option holder is the excess of the futures price $F(T)$ over the strike price K of the option, plus a long position in a newly opened futures contract. As this new futures contract has zero value, the option payoff can be specified as $\max[F(T) - K, 0]$.

6.8   SUMMARY

This chapter described how stochastic calculus is used in option valuation. Two methodologies were presented, and the link between them. Fully worked out examples were given for the martingale approach. The stage was set for the detailed exposition of the methodology for a change in probability in Chapter 7. The connection between the absence of arbitrage assumption and the existence of a martingale probability was shown for the discrete-time model of Section 6.2. That model was for a market where every terminal payoff can be replicated by a self-financing portfolio. Such a market is said to be complete, and the martingale probability is unique. A similar connection exists in the continuous-time framework but is much more technical and outside the scope of this text. An introduction is given in *Dana/Jeanblanc* Section 3.2.

The reason for discounting the stock price in option valuation deserves some further comment. In the option valuation set up there is a risk-free savings account, a stock price, and a replicating portfolio. The savings account has a rate of return equal to the risk-free rate r. If there is to be no arbitrage opportunity, then the stock price process and the value of the replicating portfolio must also have the same rate of return. That is, when the value of the stock price is discounted by the savings account its rate of return must be zero, and is thus a martingale. The same applies to the value of the replicating portfolio. This discounting has the effect of creating an effective interest rate of zero. This is very clear in the one-period binomial model for option valuation. The only assumption that is being made there is the absence of arbitrage. The discounting in that model is not something that is imposed externally, but follows logically from solving the two linear equations. The calculations produce completely naturally a probability, there named q, under which the stock price when discounted is a martingale. When setting out the methodology for option valuation in a continuous-time framework, *the basic idea of using a discounted stock price is borrowed from the one-period binomial derivation.* That is why that method in continuous-time starts by discounting the stock price. But the steps can also be carried out in a different way, as follows.

Start from $dS(t)/S(t) = \mu\,dt + \sigma\,dB(t)$ and decompose the drift into its risk-free component r and the excess over the risk-free rate $(\mu - r)$

$$\frac{dS(t)}{S(t)} = r\,dt + \underbrace{(\mu - r)\,dt + \sigma\,dB(t)}_{\text{Girsanov}}$$

Now apply a Girsanov transformation (as described in full in Chapter 7) to create a new probability distribution $\widehat{\mathbb{P}}$:

$$\sigma\,\widehat{B}(t) = (\mu - r)t + \sigma\,B(t)$$
$$d\widehat{B}(t) = \frac{\mu - r}{\sigma}\,dt + dB(t)$$

$\widehat{B}$ has picked up the excess return $(\mu - r)$ per unit of risk σ. The result is

$$\frac{dS(t)}{S(t)} = r\,dt + \sigma\,d\widehat{B}(t)$$

This has a drift, so under $\widehat{\mathbb{P}}$ the process S is *not* a martingale. But discounting to $S^*(t) \stackrel{\text{def}}{=} S(t)/\exp(rt)$ gives

$$\frac{dS^*(t)}{S^*(t)} = \sigma\,d\widehat{B}(t)$$

which is driftless, so the discounted stock price is a martingale under $\widehat{\mathbb{P}}$. That the drift coefficient of S for option valuation must be r can also be seen from the Feynman–Kač representation. Discounting by the savings account means using the savings account as the numeraire. One can also choose another numeraire, for example the stock price. That is not a logical choice but the methodology and the resulting option value are the same. For each numeraire there is a corresponding probability distribution under which all assets in the model are martingales when expressed in terms of that numeraire. This is discussed with several examples in Chapter 8.

Further information on the 1997 Nobel prize in Economics is given on the Internet at *Nobelprize.org*. This includes the lectures given by Merton and Scholes at the prize ceremony.

7
Change of Probability

It was shown in Chapter 6 that if the discounted underlying asset of an option is a martingale, then the value of the option can be computed as the expected value of the discounted option payoff. The real world randomness of the underlying asset is not a martingale, but it can be transformed into a martingale by the change of probability technique applied to the driving Brownian motion. The change of probability technique is ubiquitous in financial mathematics. In order to prepare for the general exposition of the concept, the basic idea is first illustrated for a discrete distribution. This is followed by a continuous distribution using the normal distribution as an example.

7.1 CHANGE OF DISCRETE PROBABILITY MASS

Consider an experiment with outcomes recorded numerically by random variable X. Let the possible values of X be $x_1 = 3, x_2 = 7, x_3 = 11$, with equal probability $p_i = \frac{1}{3}$.

The mean of X is

$$\mathbb{E}[X] = 3\tfrac{1}{3} + 7\tfrac{1}{3} + 11\tfrac{1}{3} = 7$$

The variance of X is

$$\mathbb{V}\text{ar}[X] = \mathbb{E}[X^2] - \{\mathbb{E}[X]\}^2 = 3^2\tfrac{1}{3} + 7^2\tfrac{1}{3} + 11^2\tfrac{1}{3} - 7^2 = 10\tfrac{2}{3}$$

Now change the mean, while keeping the variance the same, by changing the discrete probabilities (p_1, p_2, p_3) to (q_1, q_2, q_3). The new probabilities have to satisfy the following three equations:

(i) $q_1 + q_2 + q_3 = 1$
(ii) $\mathbb{E}[X] = 3q_1 + 7q_2 + 11q_3 = \mu$ (new mean)
(iii) $\mathbb{V}\text{ar}[X] = 3^2 q_1 + 7^2 q_2 + 11^2 q_3 - \mu^2 = 10\tfrac{2}{3}$ (original variance)

In matrix notation

$$
\begin{bmatrix} 1 & 1 & 1 \\ 3 & 7 & 11 \\ 9 & 49 & 121 \end{bmatrix} \cdot \begin{bmatrix} q_1 \\ q_2 \\ q_3 \end{bmatrix} = \begin{bmatrix} 1 \\ \mu \\ \mu^2 + 10\frac{2}{3} \end{bmatrix}
$$

There are three linear equations in three unknowns, so there exists a unique solution. But for the q's to be probabilities they have to be no less than 0 and no greater than 1. For $\mu = 9$ the solution is

$$
\begin{bmatrix} q_1 \\ q_2 \\ q_3 \end{bmatrix} = \begin{bmatrix} 0.2083 \\ 0.0833 \\ 0.7083 \end{bmatrix}
$$

The ratios of probabilities q to p are shown below and denoted z_i.

$$
\begin{bmatrix} q_1/p_1 \\ q_2/p_2 \\ q_3/p_3 \end{bmatrix} = \begin{bmatrix} 0.2083/0.3333 \\ 0.0833/0.3333 \\ 0.7083/0.3333 \end{bmatrix} = \begin{bmatrix} 0.625 \\ 0.25 \\ 2.125 \end{bmatrix} = \begin{bmatrix} z_1 \\ z_2 \\ z_3 \end{bmatrix}
$$

The first feature is that the values of z_i are positive. This must be so because they are ratios of probabilities and probabilities are positive. The second feature is that if Z is a random variable which can take the values z_1, z_2, z_3, then the expected value of Z under the original probabilities p_i always equals 1.

$$
\mathbb{E}_p[Z] = \frac{q_1}{p_1}p_1 + \frac{q_2}{p_2}p_2 + \frac{q_3}{p_3}p_3 = q_1 + q_2 + q_3 = 1
$$

For the above example

$$
\mathbb{E}_p[Z] = z_1 p_1 + z_2 p_2 + z_3 p_3 = (0.625)\tfrac{1}{3} + (0.25)\tfrac{1}{3} + (2.125)\tfrac{1}{3} = 1
$$

The role of Z is to *redistribute* the original *probability mass*. For example, the amount of probability originally concentrated at $X = 3$ was 0.3333 and has been changed to $(0.3333)z_1 = (0.3333)(0.625) = 0.2083$. The values of X where the probability mass is located are not changed by this change of probability. Whether there exists a solution depends on the target mean μ that is chosen. For example $\mu = 10$ gives

$$
\begin{bmatrix} q_1 \\ q_2 \\ q_3 \end{bmatrix} = \begin{bmatrix} 0.2346 \\ -0.2292 \\ 0.9896 \end{bmatrix}
$$

but q_2 is not a probability. The reason that there is no solution in this case is that there is no q weighting of the values 3, 7, 11, that can produce the value 10.

Now consider a function h of X. This is a random variable. To get from an expected value of h under the original probability p, to an expected value of h under the new probability q, multiply the original values by $z_i = q_i/p_i$ and use the original probability. A subscript is now used to refer to the probability under which the expectation is taken.

$$\mathbb{E}_q[h(X)] = \mathbb{E}_p[Z\,h(X)]$$

left-hand side: $h(x_1)q_1 + h(x_2)q_2 + h(x_3)q_3$

right-hand side: $z_1 h(x_1)p_1 + z_2 h(x_2)p_2 + z_3 h(x_3)p_3$

Substituting $z_i = q_i/p_i$ then shows the equality of the two expectations $\mathbb{E}_q$ and $\mathbb{E}_p$.

7.2   CHANGE OF NORMAL DENSITY

In the standard normal probability density, $(1/\sqrt{2\pi})\exp[-\frac{1}{2}x^2]$, the exponent $-\frac{1}{2}x^2$ can be decomposed as $-\frac{1}{2}(x-\mu)^2 - \mu x + \frac{1}{2}\mu^2$. Thus the standard normal density can be written as

$$\frac{1}{\sqrt{2\pi}} \exp\left[-\tfrac{1}{2}(x-\mu)^2\right] \exp\left[-\mu x + \tfrac{1}{2}\mu^2\right]$$

which is the product of a normal density with mean μ and standard deviation 1, and the factor $\exp[-\mu x + \frac{1}{2}\mu^2]$. This decomposition is the key to a change of probability for normal distributions, and hence for Brownian motions. Consider a function h of random variable $X \sim N(0, 1)$.

$$\mathbb{E}[h(X)] = \int_{x=-\infty}^{\infty} h(x)\frac{1}{\sqrt{2\pi}} \exp\left[-\tfrac{1}{2}x^2\right] dx \text{ where } x \text{ is a value of } X$$

Using the above decomposition, this can also be written as

$$\int_{x=-\infty}^{\infty} h(x)\frac{1}{\sqrt{2\pi}} \exp\left[-\tfrac{1}{2}(x-\mu)^2\right] \exp\left[-\mu x + \tfrac{1}{2}\mu^2\right] dx$$

or as

$$\int_{x=-\infty}^{\infty} \left\{h(x)\exp[-\mu x + \tfrac{1}{2}\mu^2]\right\} \frac{1}{\sqrt{2\pi}} \exp\left[-\tfrac{1}{2}(x-\mu)^2\right] dx$$

Now view $\{...\}$ as a new function $g(y) \stackrel{\text{def}}{=} h(y)\exp[-\mu y + \frac{1}{2}\mu^2]$. Then the above can also be written as

$$\int_{y=-\infty}^{\infty} g(y)\frac{1}{\sqrt{2\pi}}\exp\left[-\tfrac{1}{2}(y-\mu)^2\right]dy$$

which is the expected value of the function g of Y where $Y \sim N(\mu, 1)$. The probability distribution $N(0, 1)$ will be referred to as $\mathbb{P}$, the probability distribution $N(\mu, 1)$ as $\widehat{\mathbb{P}}$. So there are now two expressions for the expected value, one based on $\widehat{\mathbb{P}}$, the other based on $\mathbb{P}$. To distinguish them, the subscripts $\mathbb{P}$ and $\widehat{\mathbb{P}}$ are used.

$$\mathbb{E}_{\mathbb{P}}[h(X)] = \int_{x=-\infty}^{\infty} h(x)\frac{1}{\sqrt{2\pi}}\exp\left[-\tfrac{1}{2}x^2\right]dx$$

$$\mathbb{E}_{\widehat{\mathbb{P}}}[g(Y)] = \int_{y=-\infty}^{\infty} g(y)\frac{1}{\sqrt{2\pi}}\exp\left[-\tfrac{1}{2}(y-\mu)^2\right]dy$$

$$\mathbb{E}_{\widehat{\mathbb{P}}}[g(Y)] = \mathbb{E}_{\mathbb{P}}[h(X)]$$

The left uses the Y density

$$\frac{1}{\sqrt{2\pi}}\exp\left[-\tfrac{1}{2}(y-\mu)^2\right] \text{ at } Y = y$$

and the integrand is

$$g(y) = h(y)\exp\left[-\mu y + \tfrac{1}{2}\mu^2\right]$$

It can also be written as $\mathbb{E}_{\widehat{\mathbb{P}}}\{h(Y)\exp[-\mu Y + \tfrac{1}{2}\mu^2]\}$. So an alternative way to compute the expected value of $h(X)$ under $\widehat{\mathbb{P}}$ is to multiply the value of $h(Y)$ by $\exp[-\mu Y + \tfrac{1}{2}\mu^2]$, then use $Y \sim N(\mu, 1)$. Note that $\mathbb{E}_{\widehat{\mathbb{P}}}$ is applied to the product of two functions of the same random variable Y, so there is no need for a joint probability distribution. The right uses the X density $(1/\sqrt{2\pi})\exp[-\tfrac{1}{2}x^2]$ at $X = x$ and the integrand is $h(x)$. Using $\widehat{\mathbb{P}}$, the computations are based on a random variable, Y, that has its mean shifted relative to X.

7.3   CHANGE OF BROWNIAN MOTION

The probability density of Brownian motion at $B(t) = x$ is

$$\frac{1}{\sqrt{t}\sqrt{2\pi}}\exp\left[-\frac{1}{2}\left(\frac{x}{\sqrt{t}}\right)^2\right]$$

Using the decomposition of the previous section and replacing x by $x/\sqrt{t}$ and μ by $-\varphi\sqrt{t}$, where φ is a constant, this can be written as

$$\frac{1}{\sqrt{t}\sqrt{2\pi}}\exp\left[-\tfrac{1}{2}\left(\tfrac{x+\varphi t}{\sqrt{t}}\right)^2\right]\exp\left[\varphi x + \tfrac{1}{2}\varphi^2 t\right]$$

which is the product of a normal density and the factor $\exp[\varphi x + \tfrac{1}{2}\varphi^2 t]$. The ratio of the second density to the original density at x is

$$\frac{1}{\sqrt{t}\sqrt{2\pi}}\exp\left[-\tfrac{1}{2}\left(\tfrac{x+\varphi t}{\sqrt{t}}\right)^2\right] \Big/ \frac{1}{\sqrt{t}\sqrt{2\pi}}\exp\left[-\tfrac{1}{2}\left(\tfrac{x}{\sqrt{t}}\right)^2\right]$$
$$= \exp[-\varphi x - \tfrac{1}{2}\varphi^2 t]$$

For different values x of $B(t)$ the right-hand side is a random variable, say Z.

$$Z(t) \stackrel{\text{def}}{=} \exp\left[-\varphi B(t) - \tfrac{1}{2}\varphi^2 t\right]$$

This ratio is called the Radon–Nikodym derivative with respect to $B(t)$, and is further explained in Section 7.10. Now introduce the new random variable $\widehat{B}(t) \stackrel{\text{def}}{=} B(t) + \varphi t$. Then a value $B(t) = x$ corresponds to a value $x + \varphi t$ of $\widehat{B}(t)$. If values taken by $\widehat{B}(t)$ are denoted y, then $y = x + \varphi t$. The expression

$$\frac{1}{\sqrt{t}\sqrt{2\pi}}\exp\left[-\tfrac{1}{2}\left(\tfrac{x+\varphi t}{\sqrt{t}}\right)^2\right] \quad \text{can then be written as}$$

$$\frac{1}{\sqrt{t}\sqrt{2\pi}}\exp[-\tfrac{1}{2}(\tfrac{y}{\sqrt{t}})^2]$$

which is the density of a Brownian motion. The random variable $\widehat{B}(t) = B(t) + \varphi t$ has *induced a new probability* under which it is a Brownian motion. Under the original probability, $\mathbb{E}_{\text{orig}}[\widehat{B}(t)] = \varphi t$ and $\mathbb{V}\text{ar}_{\text{orig}} = t$. Under the new probability, $\mathbb{E}_{\text{new}}[\widehat{B}(t)] = 0$ and $\mathbb{V}\text{ar}_{\text{new}} = t$. By the change of probability, the expected value has changed but the variance has not.

7.4   GIRSANOV TRANSFORMATION

Let φ be a constant, and define the random process

$$Z(t) \stackrel{\text{def}}{=} \exp\left[-\varphi B(t) - \frac{1}{2}\varphi^2 t\right] \quad \text{for } 0 < t \le T.$$

It starts at $Z(0) = \exp(0) = 1$, and its values are non-negative. The application of Itō's formula readily gives $dZ(t) = -\varphi Z(t) \, dB(t)$, which is driftless so the process Z is a martingale; values greater than 1, and values between 0 and 1, average to $Z(0) = 1$. For any t, $\mathbb{E}[Z(t)] = 1$. The previous section showed that by using $Z(t)$ a new probability $\widehat{\mathbb{P}}$ could be created. Now define a new random process $\widehat{B}$ as a drift change to Brownian motion B for all t, under probability $\mathbb{P}$.

$$\widehat{B}(t) \stackrel{\text{def}}{=} B(t) + \varphi t$$

It turns out that $\widehat{B}(t)$ is a Brownian motion under the new probability distribution $\widehat{\mathbb{P}}$ which is created by Z from the original probability $\mathbb{P}$. This construction is known as the *Girsanov transformation*, after the Russian mathematician.

The proof uses the Lévy characterization of Brownian motion. It must thus be shown that:

(i) $\widehat{B}(0) = 0$ and $\widehat{B}$ has a continuous path
(ii) $\widehat{B}(t)$ is a martingale under $\widehat{\mathbb{P}}$
(iii) $[d\widehat{B}(t)]^2 = dt$ under $\widehat{\mathbb{P}}$. This is equivalent to $\widehat{B}(t)^2 - t$ being a martingale under $\widehat{\mathbb{P}}$, as shown below in the subsection "Equivalent Expression".

Proof

(i) $\widehat{B}(0) = B(0) + \varphi 0 = 0$, and the transformation from B to $\widehat{B}$ is continuous in t.
(ii) Use $\mathbb{E}_{\widehat{\mathbb{P}}}[\widehat{B}] = \mathbb{E}_{\mathbb{P}}[Z\widehat{B}]$. If it is shown that $Z(t)\widehat{B}(t)$ is a martingale under $\mathbb{P}$, then $\widehat{B}(t)$ is a martingale under $\widehat{\mathbb{P}}$[1]. To verify the former, use the stochastic differential of $Z\widehat{B}$ to derive an expression for the change in $Z\widehat{B}$ over an arbitrary time period, and check if the expected value of this change is zero. For convenience introduce the notation $Y(t) \stackrel{\text{def}}{=} Z(t)\widehat{B}(t)$. The product rule gives $dY(t) = Z(t) \, d\widehat{B}(t) + \widehat{B}(t) \, dZ(t) + dZ(t) \, d\widehat{B}(t)$, under $\mathbb{P}$. Substituting $dZ(t) = -\varphi Z(t) \, dB(t)$ and $d\widehat{B}(t) = dB(t) + \varphi(t) \, dt$ gives

$$\begin{aligned}
dY &= Z[dB + \varphi \, dt] + \widehat{B}[-\varphi Z \, dB] + [-\varphi Z \, dB][dB + \varphi \, dt] \\
&= Z[1 - \widehat{B}\varphi] \, dB \\
&= Z[1 - \{B + \varphi t\}\varphi] \, dB
\end{aligned}$$

---

[1] This relationship is derived in *Kuo* p 142 and in *Shreve II* p 212.

Introduce the simplified notation

$$f(t) \stackrel{\text{def}}{=} Z(t)[1 - \{B(t) + \varphi t\}\varphi]$$
$$= \exp\left[-\varphi B(t) - \tfrac{1}{2}\varphi^2 t\right][1 - \{B(t) + \varphi t\}\varphi]$$

This is non-anticipating as its value can be determined at any time t. The corresponding integral expression over $[t, t + u]$ is then

$$Y(t + u) - Y(t) = \int_{s=t}^{t+u} f(t)) \, dB(s)$$

It can be shown that $\int_{s=t}^{t+u} \mathbb{E}[f(t)^2] \, dt$ is finite [2], so $\int f \, dB$ is an Itō stochastic integral. Thus $\mathbb{E}_{\mathbb{P}}[Y(t + u) - Y(t)|\Im(t)] = 0$ and Y is a martingale under $\mathbb{P}$, as was to be shown.

(iii) Using the same approach as in (ii), introduce the notation $X(t) \stackrel{\text{def}}{=} \widehat{B}(t)^2 - t$, and verify whether XZ is a martingale under $\mathbb{P}$. The product rule gives $d[XZ] = X \, dZ + Z \, dX + dX \, dZ$. By Itō's formula

$$dX = \frac{\partial X}{\partial t} dt + \frac{\partial X}{\partial \widehat{B}} d\widehat{B} + \frac{1}{2} \frac{\partial^2 X}{\partial \widehat{B}^2} (d\widehat{B})^2 + \frac{\partial^2 X}{\partial t \partial \widehat{B}} dt \, d\widehat{B}$$
$$= -1 \, dt + 2\widehat{B} \, d\widehat{B} + \tfrac{1}{2} 2(d\widehat{B})^2 + 0 \, dt \, d\widehat{B}$$

Under probability $\mathbb{P}$, $d\widehat{B}(t) = dB(t) + \varphi \, dt$, so $[d\widehat{B}(t)]^2 = [dB(t)]^2 = dt$. *Beware* that one cannot write $[d\widehat{B}(t)]^2 = dt$ without going via B because this is under $\mathbb{P}$. Similarly, $dt \, d\widehat{B} = dt \, (dB + \varphi \, dt) = dt \, dB + \varphi \, dt \, dt = 0$, although that term is not needed. That gives $dX = 2\widehat{B} \, d\widehat{B}$. Then

$$d[XZ] = X(-\varphi Z \, dB) + Z(2\widehat{B} \, d\widehat{B}) + (2\widehat{B} \, d\widehat{B})(-\varphi Z \, dB)$$
$$= X(-\varphi Z \, dB) + Z2\widehat{B}(dB + \varphi \, dt) + (2\widehat{B} \, d\widehat{B})(-\varphi Z \, dB)$$

Using the expressions for $\widehat{B}$, and $d\widehat{B}$ (under $\mathbb{P}$) and rearranging gives

$$d[XZ] = Z[-\varphi X + 2\{B + \varphi t\}] \, dB$$

Then substituting the expressions for Z and X, and using the same approach as in (ii) and in footnote 2, shows that XZ is a martingale under $\mathbb{P}$.

---

[2] This can be done analytically, or numerically with Mathematica for arbitrary values of φ, t and u.

This completes the proof of the Girsanov transformation for the case when φ is a constant. It can also be shown to hold for the case when φ is a *non-random* function of time, and when φ is a *random* process. Then $\widehat{B}(t) \overset{\text{def}}{=} B(t) + \int_{u=0}^{t} \varphi(u)\,du$. In the random case there is an additional condition to ensure that Z is a martingale. A sufficient condition proposed by Novikov is that $\mathbb{E}_{\mathbb{P}}\big[\exp\big(\frac{1}{2}\int_{u=0}^{t}\varphi(u,\omega)^2\,du\big)\big]$ must be finite. Other conditions can be found in the technical literature. Their discussion is beyond the scope of this text. Clearly when φ is non-random this condition is satisfied. In the literature one also sees $-\varphi$ being used, then the $B(t)$ term in the exponent of Z changes sign. The Radon–Nikodym derivative used in the Girsanov transformation $\widehat{B}(t) = B(t) \pm \int_{u=0}^{t}\varphi(u,\omega)\,du$ can be easily 'remembered' as the solution to the SDE $dZ(t)/Z(t) = \mp\varphi(t,\omega)\,dB(t)$ with $Z(0) = 1$. When the time term in the transformation has a positive coefficient, the right hand side of the SDE has a negative coefficient. The Brownian paths that can exist under $\mathbb{P}$ and $\widehat{\mathbb{P}}$ are the same, the only difference is that they have different probabilities of occurring. This is the same as in Section 7.1, where the possible outcomes of the random variable are not changed.

Equivalent expression $B(t)^2 - t$ being a martingale implies that $[dB(t)]^2 = dt$, and vice versa. This can be summarized as $B(t)^2 - t \Leftrightarrow [dB(t)]^2 = dt$. Which of these expressions is the most convenient depends on the context. The above proof of the Girsanov transformation uses $\widehat{B}(t)^2 - t$. It is convenient to introduce the notation $X(t) \overset{\text{def}}{=} B(t)^2 - t$ which is a function of t and $B(t)$. Itō's formula gives

$$dX = \frac{\partial X}{\partial t}\,dt + \frac{\partial X}{\partial B}\,dB + \frac{1}{2}\frac{\partial^2 X}{\partial B^2}(dB)^2 = -dt + 2B\,dB + (dB)^2$$

Proof of $\Rightarrow$ If X is a martingale then the drift of dX must be zero, thus $(dB)^2$ must equal dt.
Proof of $\Leftarrow$ If $(dB)^2 = dt$ then $dX = 2B\,dB$ which is driftless so X is a martingale.

It is informative to get some practice in moving between Brownian motions B (under probability $\mathbb{P}$) and $\widehat{B}$ (under probability $\widehat{\mathbb{P}}$). The Radon–Nikodym derivative is always the ratio of the new probability over the original probability, expressed in terms of the random variable which is under the original probability.

First revisit the basic case of going from B to $\widehat{B}$ via $\widehat{B}(t) = B(t) + \varphi t$ with constant φ. New is $\widehat{B}$ under $\widehat{\mathbb{P}}$. The probability density of $\widehat{B}(t)$ at

$\widehat{B}(t) = y$ is

$$\frac{1}{\sqrt{t}\sqrt{2\pi}} \exp\left[-\frac{1}{2}\left(\frac{y}{\sqrt{t}}\right)^2\right]$$

The original is B under $\mathbb{P}$. The probability density of $B(t)$ at $B(t) = x$ is

$$\frac{1}{\sqrt{t}\sqrt{2\pi}} \exp\left[-\frac{1}{2}\left(\frac{x}{\sqrt{t}}\right)^2\right]$$

The ratio of densities is

$$\frac{d\widehat{\mathbb{P}}}{d\mathbb{P}} = \exp\left[-\frac{1}{2}\left(\frac{y}{\sqrt{t}}\right)^2\right] \Big/ \exp\left[-\frac{1}{2}\left(\frac{x}{\sqrt{t}}\right)^2\right]$$

This must be in terms of x. Substituting $y = x + \varphi t$ and rearrranging gives

$$\frac{d\widehat{\mathbb{P}}}{d\mathbb{P}} = \exp\left[-\varphi x - \tfrac{1}{2}\varphi^2 t\right] \qquad \text{at } B(t) = x$$

So the Radon–Nikodym derivative is

$$\frac{d\widehat{\mathbb{P}}}{d\mathbb{P}} = \exp\left[-\varphi B(t) - \tfrac{1}{2}\varphi^2 t\right]$$

Now the reverse, going from $\widehat{B}$ to B. New is B under $\mathbb{P}$. Original is $\widehat{B}$ under $\widehat{\mathbb{P}}$. The ratio of densities is

$$\frac{d\mathbb{P}}{d\widehat{\mathbb{P}}} = \exp\left[-\frac{1}{2}\left(\frac{x}{\sqrt{t}}\right)^2\right] \Big/ \exp\left[-\frac{1}{2}\left(\frac{y}{\sqrt{t}}\right)^2\right]$$

This must be in terms of y. Substituting $x = y - \varphi t$ and rearrranging gives

$$\frac{d\mathbb{P}}{d\widehat{\mathbb{P}}} = \exp\left[\varphi y - \tfrac{1}{2}\varphi^2 t\right] \qquad \text{at } \widehat{B}(t) = y$$

The Radon–Nikodym derivative is

$$\frac{d\mathbb{P}}{d\widehat{\mathbb{P}}} = \exp\left[\varphi \widehat{B}(t) - \tfrac{1}{2}\varphi^2 t\right]$$

Next analyse the relationship between $\frac{d\widehat{\mathbb{P}}}{d\mathbb{P}}$ and $\frac{d\mathbb{P}}{d\widehat{\mathbb{P}}}$. $\frac{d\widehat{\mathbb{P}}}{d\mathbb{P}}$ is a function of $B(t)$. Substituting $B(t) = \widehat{B}(t) - \varphi t$ gives

$$\frac{d\widehat{\mathbb{P}}}{d\mathbb{P}} = \exp\left[-\varphi\{\widehat{B}(t) - \varphi t\} - \tfrac{1}{2}\varphi^2 t\right] = \exp\left[-\varphi\widehat{B}(t) + \tfrac{1}{2}\varphi^2 t\right]$$

Then $1/d\widehat{\mathbb{P}}/d\mathbb{P} = \exp[\varphi\widehat{B}(t) - \tfrac{1}{2}\varphi^2 t]$. So $1/(d\widehat{\mathbb{P}}/d\mathbb{P}) = d\mathbb{P}/d\widehat{\mathbb{P}}$, and note that this is expressed in terms of what is the original variable from the perspective of using $d\mathbb{P}/d\widehat{\mathbb{P}}$, namely $\widehat{B}(t)$. Now the reverse. $d\mathbb{P}/d\widehat{\mathbb{P}}$ is a function of $\widehat{B}(t)$. Substituting $\widehat{B}(t) = B(t) + \varphi t$ gives

$$\frac{d\mathbb{P}}{d\widehat{\mathbb{P}}} = \exp\left[\varphi\{B(t) + \varphi t\} - \tfrac{1}{2}\varphi^2 t\right] = \exp\left[\varphi B(t) + \tfrac{1}{2}\varphi^2 t\right]$$

So

$$1\bigg/\frac{d\mathbb{P}}{d\widehat{\mathbb{P}}} = \exp\left[-\varphi B(t) - \tfrac{1}{2}\varphi^2 t\right] = \frac{d\widehat{\mathbb{P}}}{d\mathbb{P}}$$

This is expressed in terms of variable $B(t)$, the original when $d\widehat{\mathbb{P}}/d\mathbb{P}$ is used.

The case $\widehat{B}(t) = B(t) - \varphi t$ follows by changing the sign of φ in the expressions above.

The most general form of the Girsanov transformation is discussed in *Shreve II* Theorem 5.2.3, pp. 212–213, for a one-dimensional Brownian motion, and in Theorem 5.4.1 pp. 224–225 for the multi-dimensional case. Other references are *Kuo* Section 8.9, *Lin* Section 6.3, *Capasso/ Bakstein* Section 4.3. An interesting heuristic derivation of the Girsanov transformation is given in *Bond Pricing and Portfolio Analysis* by *de la Grandville* Section 16.7.2.

7.5   USE IN STOCK PRICE DYNAMICS – REVISITED

Let S^* denote the discounted stock price as used previously. Under the original probability $\mathbb{P}$

$$\frac{dS^*(t)}{S^*(t)} = (\mu - r)\,dt + \sigma\,dB(t) = \sigma\left[\left(\frac{\mu - r}{\sigma}\right)dt + dB(t)\right]$$

For option valuation $S^*(t)$ needs be a martingale. This is accomplished by the Girsanov transformation

$$\widehat{B}(t) \stackrel{\text{def}}{=} \left(\frac{\mu - r}{\sigma}\right) t + B(t)$$

so $[(\mu - r)/\sigma]\, dt + dB(t)$ is replaced by $d\widehat{B}(t)$, and

$$\frac{dS^*(t)}{S^*(t)} = \sigma\, d\widehat{B}(t) \qquad \text{under } \widehat{\mathbb{P}}$$

The Radon–Nikodym derivative that creates $\widehat{\mathbb{P}}$ is

$$Z(t) \stackrel{\text{def}}{=} \exp\left[-\left(\frac{\mu - r}{\sigma}\right) B(t) - \frac{1}{2}\left(\frac{\mu - r}{\sigma}\right)^2 t\right]$$

This is now illustrated numerially. The parameter values are: growth rate $\mu = 8\%$, interest rate $r = 5\%$, volatility $\sigma = 20\%$, $t = 1$, so $\varphi = 0.15$ and $Z = \exp[(-0.15)B(1) - \frac{1}{2}(0.15)^2 1]$ is a decreasing function of $B(1)$. On the horizontal axis of the chart are the values of $B(1)$. Multiplying the density of $B(1)$ by the decreasing Z has the effect of pulling up the original density to the left of the origin and pulling it down on the right. The original density pivots, as shown in Figure 7.1.

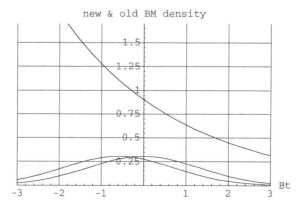

Figure 7.1 Pivoting of Brownian motion density

7.6   GENERAL DRIFT CHANGE

Let random process X be originally specified by

$$dX(t) = \mu_{\text{orig}}(.)\,dt + \sigma_{\text{orig}}(.)\,dB_{\text{orig}}(t)$$

where B_{orig} is a Brownian motion under the original probability distribution $\mathbb{P}_{\text{orig}}$, and $(.)$ indicates possible dependence on t and X. Now derive the dynamics under a new drift coefficient $\mu_{\text{new}}(.)$. To this end, rewrite the drift in terms of this new coefficient as

$$
\begin{aligned}
dX(t) &= \mu_{\text{new}}(.)\,dt - \mu_{\text{new}}(.)\,dt + \mu_{\text{orig}}(.)\,dt + \sigma_{\text{orig}}(.)\,dB_{\text{orig}}(t) \\
&= \mu_{\text{new}}(.)\,dt + [\mu_{\text{orig}}(.) - \mu_{\text{new}}(.)]\,dt + \sigma_{\text{orig}}(.)\,dB_{\text{orig}}(t) \\
&= \mu_{\text{new}}(.)\,dt + \sigma_{\text{orig}}(.)\left\{ \frac{\mu_{\text{orig}}(.) - \mu_{\text{new}}(.)}{\sigma_{\text{orig}}(.)}\,dt + dB_{\text{orig}}(t) \right\} \\
&= \mu_{\text{new}}(.)\,dt + \sigma_{\text{orig}}(.)[\varphi(.)\,dt + dB_{\text{orig}}(t)]
\end{aligned}
$$

where

$$\varphi(.) \stackrel{\text{def}}{=} \frac{\mu_{\text{orig}}(.) - \mu_{\text{new}}(.)}{\sigma_{\text{orig}}(.)}.$$

Assuming that the Novikov condition is satisfied, apply the Girsanov transformation

$$B_{\text{new}}(t) \stackrel{\text{def}}{=} \int_{s=0}^{t} \varphi(.)\,ds + B_{\text{orig}}(t)$$

or

$$dB_{\text{new}}(t) = \varphi(.)\,dt + dB_{\text{orig}}(t)$$

B_{new} is a Brownian motion under a new probability distribution $\mathbb{P}_{\text{new}}$ created from $\mathbb{P}_{\text{orig}}$ by the Radon–Nikodym derivative

$$
\begin{aligned}
\frac{d\mathbb{P}_{\text{new}}}{d\mathbb{P}_{\text{orig}}} = \exp\Bigg\{ &-\frac{1}{2}\int_{s=0}^{t} \left[\frac{\mu_{\text{orig}}(.) - \mu_{\text{new}}(.)}{\sigma_{\text{orig}}(.)} \right]^2 ds \\
&- \int_{s=0}^{t} \frac{\mu_{\text{orig}}(.) - \mu_{\text{new}}(.)}{\sigma(.)}\,dB_{\text{orig}}(s) \Bigg\}
\end{aligned}
$$

The new SDE for X is

$$dX(t) = \mu_{\text{new}}(.)\,dt + \sigma_{\text{orig}}(.)\,dB_{\text{new}}(t)$$

7.7   USE IN IMPORTANCE SAMPLING

Importance sampling can be an efficient method for increasing the accuracy of a simulation. It is based on a change of probability density. The method is illustrated for an option valuation for which the exact value is known, so as to have a benchmark for the impact of importance sampling. Consider the valuation of a European put option under the martingale method

$$p(0) = \mathbb{E}\{\exp(-rT)\max[K - S(T), 0]\}$$
$$= \exp(-rT)\int_{x=0}^{\infty}\max[K - x, 0]f(x)\,dx$$

where $f(x)$ denotes the lognormal probability density of $S(T)$ at $S(T) = x$. This integral can be approximated by simulating n values $x_1, \ldots, x_i, \ldots x_n$ of $S(T)$ according to density $f(x)$, computing the corresponding put payoffs $\max[K - x_i, 0]$, and taking the average

$$\frac{1}{n}\sum_{i=1}^{n}\max[K - x_i, 0]$$

An alternative way to formulate the integral is by introducing another probability density g and writing f as $(f/g)g$. The integration variable is now denoted y to emphasize the new formulation

$$\int_{y=0}^{\infty}\max[K - y, 0]\frac{f(y)}{g(y)}g(y)\,dy$$

As g is a probability density, this integral is the expected value of the random variable

$$\max[K - S(T), 0]\frac{f[S(T)]}{g[S(T)]}$$

where density g is used in taking the expectation. This expectation can be approximated by simulating n values $y_1, \ldots y_i, \ldots, y_n$ of $S(T)$ according to density $g(x)$, computing the corresponding quantities $\max[K - y_i, 0][f(y_i)/g(y_i)]$, and taking the average

$$\frac{1}{n}\sum_{i=1}^{n}\max[K - y_i, 0]\frac{f(y_i)}{g(y_i)}$$

By choosing g to be the density of $S(T)$ with stock price growth rate m lower than r, more of the simulated $S(T)$ values will be in the money ($<K$), so more positive put values are generated than when using r. The

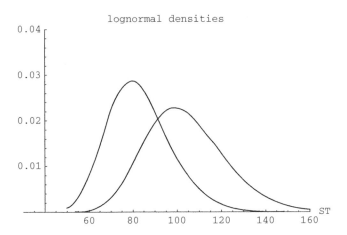

Figure 7.2   Shifted lognormal density

simulation now concentrates on the important values, and this method is therefore known as importance sampling. Density g is changed back to density f by multiplying it by the ratio f/g at y_i, which is a realization of the Radon–Nikodym derivative, the random variable $f[S(T)]/g[S(T)]$.

By a clever choice of m, the average from sampling under g has a much lower variance than ordinary sampling under f. In the martingale method $\mathbb{E}[S(T)] = S(0)e^{rT}$. Changing r to m, this becomes $\mathbb{E}_m[S(T)] = S(0)e^{mT}$. A good choice of m is the one for which this expected value equals K. That is $m = \ln[K/S(0)]/T$. The shift in the density and the numerical impact of importance sampling are shown in Figures 7.2 and 7.3 respectively.

Option parameter values			put value	call value
out of the money				
S(0)	100	exact	0.027522	0.022780
K	80 put	normal sampling	0.034216	0.027458
	130 call	importance sampling	0.027387	0.022571
T	0.25			
σ	20.00%	reduction factor		
r	5.00%	variance	183	264
		st error	13	16
		10000 simulation runs in Mathematica		

Figure 7.3   Importance sampling impact

A method for producing a family of probability distributions that can be used for importance sampling will now be described. Consider a continous random variable X with probability density $f(x)$ at $X = x$; refer to this as the original probability $\mathbb{P}_f$. The moment generating function of X, using parameter θ, is

$$m(\theta) \stackrel{\text{def}}{=} \mathbb{E}_f[e^{\theta X}] = \int_{x=-\infty}^{\infty} e^{\theta x} f(x)\, dx$$

assume that it is finite. Now create a new probability distribution, here denoted $\mathbb{P}_g$, from the original probability distribution $\mathbb{P}_f$, by the Radon–Nikodym derivative

$$\frac{d\mathbb{P}_g}{d\mathbb{P}_f} \stackrel{\text{def}}{=} \frac{e^{\theta X}}{m(\theta)}$$

The corresponding new density is $g(x) \stackrel{\text{def}}{=} \dfrac{e^{\theta x}}{m(\theta)} f(x).$[3] It is a density because $g(x)$ is non-negative and the area under g equals 1.

$$\int_{x=-\infty}^{\infty} g(x)\, dx = \int_{x=-\infty}^{\infty} \frac{e^{\theta x}}{m(\theta)} f(x)\, dx$$
$$= \frac{1}{m(\theta)} \int_{x=-\infty}^{\infty} e^{\theta x} f(x)\, dx = \frac{1}{m(\theta)} m(\theta) = 1$$

That is also the expected value of $d\mathbb{P}_g/d\mathbb{P}_f$ using density f. This transformation of probability is known as *exponential tilting*.

To illustrate, let X be standard normal. Then

$$m(\theta) = \mathbb{E}_f[e^{\theta X}] = e^{\mathbb{E}[X]+\frac{1}{2}\text{Var}[X]} = e^{\frac{1}{2}\theta^2}$$

and

$$\frac{d\mathbb{P}_g}{d\mathbb{P}_f} = e^{\theta X - \frac{1}{2}\theta^2}$$

The new density, at $X = y$, under $\mathbb{P}_g$ is

$$e^{\theta y - \frac{1}{2}\theta^2} \frac{1}{\sqrt{2\pi}} e^{-\frac{1}{2}y^2}$$

---

[3] An alternative formulation is as follows. Write $m(\theta) = e^{\ln[m(\theta)]}$. The ln of the mgf is known as the cumulant generating function, denoted $\psi(\theta)$ or $\kappa(\theta)$. Then $g(x) = e^{\theta x - \psi(\theta)} f(x)$.

The exponent rearranges to $-\frac{1}{2}(y-\theta)^2$. The transformation has increased the mean from 0 to θ. Under $\mathbb{P}_g$, $X \sim N(\theta, 1)$. Similarly, if $X \sim N(\mu, \sigma^2)$, then

$$m(\theta) = e^{\theta\mu+\frac{1}{2}\theta^2\sigma^2} \quad \text{and} \quad \frac{d\mathbb{P}_g}{d\mathbb{P}_f} = e^{\theta X - \theta\mu - \frac{1}{2}\theta^2\sigma^2}$$

The resulting new density is normal with mean $\mu + \theta\sigma^2$, an increase of $\theta\sigma^2$; the original variance σ^2 is unchanged.

Another example is for exponential random variable τ with density $\lambda e^{-\lambda t}$ at $\tau = t$. Here

$$m(\theta) = \int_{t=0}^{\infty} e^{\theta t} \lambda e^{-\lambda t} \, dt = \frac{\lambda}{\lambda - \theta}$$

provided $\theta < \lambda$, and

$$\frac{d\mathbb{P}_g}{d\mathbb{P}_f} = \exp\left(\theta t - \ln\left[\frac{\lambda}{\lambda-\theta}\right]\right) = \exp(\theta t)\exp\left(\ln\left[\frac{\lambda-\theta}{\lambda}\right]\right)$$

$$= \exp(\theta t)\frac{\lambda - \theta}{\lambda}$$

The new density at $\tau = t$ is

$$\frac{d\mathbb{P}_g}{d\mathbb{P}_f}\lambda e^{-\lambda t} = e^{\theta t}\frac{\lambda - \theta}{\lambda}\lambda e^{-\lambda t} = (\lambda - \theta)e^{-(\lambda-\theta)t}$$

which is again exponential with the parameter value reduced by θ.

Ordinary simulation computes an approximation of $\mathbb{E}_f[X]$. Under density g this becomes

$$\mathbb{E}_g\left[X\frac{d\mathbb{P}_f}{d\mathbb{P}_g}\right] \quad \text{where} \quad \frac{d\mathbb{P}_f}{d\mathbb{P}_g} = \frac{m(\theta)}{e^{\theta X}}.$$

Suggestions for suitable choices of the value of θ can be found in the references. Thus far, g and f were densities of the same type. That need not be the case. Density g can be a different one provided g and f assign probability zero to the same events. Exponential tilting can be applied to discrete distributions in the same way.

Applications include the simulation of expressions which depend on a rare event in risk management, such as a credit default by a company, or an excessive loss. Such events have a very low probability and ordinary simulation does not give enough accuracy.

A recommended first reference on importance sampling is *McLeish* Chapter 5. Extensive coverage is given in *Glasserman* and *Asmussen/ Glynn*.

7.8   USE IN DERIVING CONDITIONAL EXPECTATIONS

Conditional expectations of exponential expressions arise in options valuation.

Example 7.8.1. The value of a standard European call is computed as $\exp(-rT)\mathbb{E}_{\widehat{\mathbb{P}}}[\max[S(T) - K, 0]$ where $S(T) = S(0)\exp[(r - \frac{1}{2}\sigma^2)T + \sigma \widehat{B}(T)]$, $\widehat{B}$ is Brownian motion under probability $\widehat{\mathbb{P}}$, and $\mathbb{E}_{\widehat{\mathbb{P}}}$ is the corresponding expected value. This is now expressed using the indicator function 1 as

$$\exp(-rT)\mathbb{E}_{\widehat{\mathbb{P}}}[S(T)\,1_{\{S(T)>K\}}] - \exp(-rT)K\mathbb{E}_{\widehat{\mathbb{P}}}[1_{\{S(T)>K\}}]$$

It is known from elementary probability theory that the expected value of the indicator function of an event is the probability of that event. Thus the second term equals $\exp(-rT)K\widehat{\mathbb{P}}[S(T) > K]$, which can be readily evaluated. The first term cannot be handled the same way as $S(T)$ is a random variable and cannot be taken outside $\mathbb{E}_{\widehat{\mathbb{P}}}$. But the change of probability technique can help. The expected value of a random variable, say X, under a new probability can be written as the expected value under the original probability $\widehat{\mathbb{P}}$ multiplied by the Radon–Nikodym derivative that is used in creating the new probability,

$$\mathbb{E}_{\mathbb{P}_{\text{new}}}[X] = \mathbb{E}_{\widehat{\mathbb{P}}}\left[\frac{d\mathbb{P}_{\text{new}}}{d\widehat{\mathbb{P}}}X\right]$$

The idea now is to use the exponential expression in $S(T)$ to create a Radon–Nikodym derivative. Then the expected value under the original probability can be written as the expected value under a new probability, and then as a probability under the new probability, which can be readily evaluated. Rewrite $S(T)$ as $S(0)\exp(rT)\exp[-\frac{1}{2}\sigma^2 T + \sigma \widehat{B}(T)]$. The term $\exp[-\frac{1}{2}\sigma^2 T + \sigma \widehat{B}(T)]$ is the Radon–Nikodym derivative $\frac{d\mathbb{P}_{\text{new}}}{d\widehat{\mathbb{P}}}$ for the change of Brownian motion $B_{\text{new}}(T) \overset{\text{def}}{=} \widehat{B}(T) - \sigma T$. So

$$\exp(-rT)\mathbb{E}_{\widehat{\mathbb{P}}}[S(T)\,1_{\{S(T)>K\}}]$$
$$= \exp(-rT)S(0)\exp(rT)\mathbb{E}_{\widehat{\mathbb{P}}}\left\{\exp\left[-\tfrac{1}{2}\sigma^2 T + \sigma \widehat{B}(T)\right]1_{\{S(T)>K\}}\right\}$$
$$= S(0)\mathbb{E}_{\widehat{\mathbb{P}}}\left[\frac{d\mathbb{P}_{\text{new}}}{d\widehat{\mathbb{P}}}\,1_{\{S(T)>K\}}\right]$$
$$= S(0)\mathbb{E}_{\mathbb{P}_{\text{new}}}[1_{\{S_{\text{new}}(T)>K\}}]$$
$$= S(0)\mathbb{P}_{\text{new}}[S_{\text{new}}(T) > K]$$

Now $S_{\text{new}}(T)$ has to be expressed in terms of $B_{\text{new}}(T)$ by substituting $\widehat{B}(T) = B_{\text{new}}(T) + \sigma T$, giving

$$S_{\text{new}}(T) = S(0)\exp(rT)\exp\left[-\tfrac{1}{2}\sigma^2 T + \sigma\{B_{\text{new}}(T) + \sigma T\}\right]$$
$$= S(0)\exp(rT)\exp\left[\tfrac{1}{2}\sigma^2 T + \sigma B_{\text{new}}(T)\right]$$

The sign of $\frac{1}{2}\sigma^2 T$ has changed. It is now straightforward to evaluate the first term as

$$S(0)\mathbb{P}_{\text{new}}\left\{S(0)\exp(rT)\exp\left[\tfrac{1}{2}\sigma^2 T + \sigma B_{\text{new}}(T)\right] > K\right\}$$

Rearranging gives

$$S(0)\mathbb{P}_{\text{new}}\left\{\sigma B_{\text{new}}(T) > \ln\left[\frac{K}{S(0)}\right] - (r + \tfrac{1}{2}\sigma^2)T\right\}$$

As $B_{\text{new}}(T)$ is normal with mean 0 and standard deviation $\sqrt{T}$, the resulting probability equals $N(d_1)$ where

$$d_1 \overset{\text{def}}{=} \frac{\ln[\frac{S(0)}{K}] + (r + \tfrac{1}{2}\sigma^2)T}{\sigma\sqrt{T}}$$

so the first term equals $S(0)N(d_1)$. The first term could have been evaluated by elementary ordinary probability methods, but the purpose here was to illustrate a change of probability technique on a standard case from option valuation theory. The technique is particularly useful in the derivation of probability distributions which are used in the valuation of barrier options.

Example 7.8.2.   In the above example the Radon–Nikodym derivative was found by a simple decomposition of the stock price expression. Sometimes a bit more creativity is required, as when computing the expected value of $\exp[B(t)]$ for $B(T) > K$ where K is a constant. This can be written as $\mathbb{E}_{\mathbb{P}_{\mathrm{orig}}}[\exp[B(t)\,1_{\{B(t)>K\}}]]$. Here a Radon–Nikodym derivative can be constructed by rearranging the exponent.

$$\exp[B(t)] = \exp\left[\tfrac{1}{2}t - \tfrac{1}{2}t + B(t)\right] = \exp\left[\tfrac{1}{2}t\right]\exp\left[-\tfrac{1}{2}t + B(t)\right]$$
$$= \exp\left[\tfrac{1}{2}t\right]\frac{d\mathbb{P}_{\mathrm{new}}}{d\mathbb{P}_{\mathrm{orig}}}$$

The corresponding change of Brownian motion is $B_{\mathrm{new}}(t) \stackrel{\mathrm{def}}{=} B(t) - t$ so $B(t) = B_{\mathrm{new}}(t) + t$. Thus

$$\mathbb{E}_{\mathbb{P}_{\mathrm{orig}}}[\exp[B(t)\,1_{\{B(t)>K\}}]] = \exp\left[\tfrac{1}{2}t\right]\mathbb{E}_{\mathbb{P}_{\mathrm{orig}}}\left[\frac{d\mathbb{P}_{\mathrm{new}}}{d\mathbb{P}_{\mathrm{orig}}}1_{\{B(t)>K\}}\right]$$
$$= \exp\left[\tfrac{1}{2}t\right]\mathbb{E}_{\mathbb{P}_{\mathrm{new}}}\left[1_{\{B_{\mathrm{new}}(t)+t>K\}}\right]$$
$$= \exp\left[\tfrac{1}{2}t\right]\mathbb{P}_{\mathrm{new}}[B_{\mathrm{new}}(t) > K - t]$$

and

$$\mathbb{P}_{\mathrm{new}}[B_{\mathrm{new}}(t) > K - t] = \mathbb{P}_{\mathrm{new}}[\sqrt{t}Z > K - t]$$
$$= \mathbb{P}_{\mathrm{new}}[\sqrt{t}Z < t - K] = N\left(\frac{t-K}{\sqrt{t}}\right)$$

The final result is $\exp[\tfrac{1}{2}t]N[(t - K)/\sqrt{t}]$. For $K = -\infty$ this becomes $\exp[\tfrac{1}{2}t]$ which is the same as using the well-known formula for the normal random variable $B(t)$, $\exp\{\mathbb{E}[B(t)] + \tfrac{1}{2}\mathrm{Var}[B(t)]\}$.

Example 7.8.3. Here the probability distribution is derived of the running maximum of W, where $W(t) \stackrel{\mathrm{def}}{=} B(t) + \varphi t$, and B is Brownian motion under probability distribution $\mathbb{P}$. Let

$$M(t) \stackrel{\mathrm{def}}{=} \max[W(t)] \text{ on } [0, t]$$

denote the running maximum of W. The valuation of barrier options makes use of the joint probability distribution of W and M in the form $\mathbb{P}[W(t) \leq x, M(t) \geq y]$. This will now be derived and entails two successive Girsanov transformations. As W is not a Brownian motion under probability $\mathbb{P}$, the first step is to change to a probability $\widehat{\mathbb{P}}$ under which W is a Brownian motion. To this end, write the joint probability under $\mathbb{P}$ as the expected value of the indicator function as $\mathbb{E}_{\mathbb{P}}[1_{\{W(t)\leq x, M(t)\geq y\}}]$. For the transformation $W(t) \stackrel{\text{def}}{=} B(t) + \varphi t$ the Radon–Nikodym derivative is

$$\frac{d\widehat{\mathbb{P}}}{d\mathbb{P}} = \exp\left[-\tfrac{1}{2}\varphi^2 T - \varphi B(T)\right]$$

This produces a new probability $\widehat{\mathbb{P}}$ under which W is a Brownian motion. Converting from expectation under $\widehat{\mathbb{P}}$ to expectation under $\mathbb{P}$, requires $d\mathbb{P}/d\widehat{\mathbb{P}} = 1/(d\widehat{\mathbb{P}}/d\mathbb{P})$. This is $\exp[\tfrac{1}{2}\varphi^2 T + \varphi B(T)]$ in terms of $B(T)$, but as it will be used under $\widehat{\mathbb{P}}$, the random variable in $d\mathbb{P}/d\widehat{\mathbb{P}}$ must be Brownian motion under $\widehat{\mathbb{P}}$, which is W. Expressing B in terms of W, using $B(T) = W(T) - \varphi T$ gives

$$\frac{d\mathbb{P}}{d\widehat{\mathbb{P}}} = \exp\left[\tfrac{1}{2}\varphi^2 T + \varphi\{W(T) - \varphi T\}\right] = \exp\left[-\tfrac{1}{2}\varphi^2 T + \varphi W(T)\right]$$

$$\mathbb{E}_{\mathbb{P}}[1_{\{W(t)\leq x, M(t)\leq y\}}] = \mathbb{E}_{\widehat{\mathbb{P}}}\left[\frac{d\mathbb{P}}{d\widehat{\mathbb{P}}} 1\{W(t) \leq x, M(t) \leq y\}\right]$$

$$= \mathbb{E}_{\widehat{\mathbb{P}}}\left[\exp\left[-\tfrac{1}{2}\varphi^2 T + \varphi W(T)\right] 1_{\{W(t)\leq x, M(t)\leq y\}}\right]$$

Use will now be made of the so-called reflection principle for a Brownian motion.[4] This can be used here because W is a Brownian motion under the prevailing probability $\widehat{\mathbb{P}}$. Its application replaces W by its reflection against y, $2y - W$, giving

$$\mathbb{E}_{\widehat{\mathbb{P}}}\left[\exp\left[-\tfrac{1}{2}\varphi^2 T + \varphi\{2y - W(T)\}\right] 1_{\{2y-W(t)\leq x, M(t)\leq y\}}\right]$$

Write the indicator function as $1_{\{W(t)\geq 2y-x, M(t)\leq y\}}$. A further simplification is obtained by noting that $2y - x \geq y$, and that $W(t) \geq 2y - x$ implies $M(t) \geq y$ (because M is the max of W). The indicator function is then $1_{\{W(t)\geq 2y-x\}}$, leaving

$$\exp(2\varphi y)\mathbb{E}_{\widehat{\mathbb{P}}}\left[\exp\left[-\tfrac{1}{2}\varphi^2 T - \varphi W(T)\}\right] 1_{\{W(t)\geq 2y-x\}}\right]$$

As the first term inside $\mathbb{E}_{\widehat{\mathbb{P}}}$ is a Radon–Nikodym derivative, a further transformation to probability $\widehat{\widehat{\mathbb{P}}}$ can be made using

$$\frac{d\widehat{\widehat{\mathbb{P}}}}{d\widehat{\mathbb{P}}} = \exp\left[-\tfrac{1}{2}\varphi^2 T - \varphi W(T)\right]$$

---

[4] See Annex A, *Computations with Brownian Motion*.

The corresponding Brownian motion transformation is $\widehat{\widehat{W}}(t) \overset{def}{=} W(t) + \varphi t$. This gives

$$\exp\left(2\varphi y\right)\mathbb{E}_{\widehat{\mathbb{P}}}\left[\exp\left[-\tfrac{1}{2}\varphi^2 T - \varphi W(T)\right\}\right] 1_{\{W(t) \geq 2y - x\}}\right]$$

$$= \exp(2\varphi y)\mathbb{E}_{\widehat{\mathbb{P}}}\left[\frac{d\widehat{\widehat{\mathbb{P}}}}{d\widehat{\mathbb{P}}} 1_{\{W(t) \geq 2y - x\}}\right]$$

$$= \exp(2\varphi y)\mathbb{E}_{\widehat{\widehat{\mathbb{P}}}}\left[1_{\{W(t) \geq 2y - x\}}\right]$$

$$= \exp(2\varphi y)\widehat{\widehat{\mathbb{P}}}[W(t) \geq 2y - x]$$

To get this in terms of $\widehat{\widehat{W}}$, add φt , giving

$$\exp(2\varphi y)\widehat{\widehat{\mathbb{P}}}[W(t) + \varphi t \geq 2y - x + \varphi t]$$

$$= \exp(2\varphi y)\widehat{\widehat{\mathbb{P}}}[\widehat{\widehat{W}}(t) \geq 2y - x + \varphi t]$$

As $\widehat{\widehat{W}}$ is a Brownian motion under $\widehat{\widehat{\mathbb{P}}}$, this probability can be readily verified. Writing $\widehat{\widehat{W}}(t)$ as $Z\sqrt{t}$, Z standard normal, gives

$$\widehat{\widehat{\mathbb{P}}}[Z \geq (2y - x + \varphi t)/\sqrt{t}] = \widehat{\widehat{\mathbb{P}}}\left[\frac{Z \leq -(2y - x + \varphi t)}{\sqrt{t}}\right]$$

$$= N\left[\frac{(-2y + x - \varphi t)}{\sqrt{t}}\right]$$

The final result is

$$\mathbb{P}[W(t) \leq x, M(t) \geq y] = \exp(2\varphi y)N\left[\frac{(-2y + x - \varphi t)}{\sqrt{t}}\right]$$

The joint probability distribution in the standard form, $\mathbb{P}[W(t) \leq x, M(t) \leq y]$, can be derived from the above by noting that

$$\mathbb{P}[W(t) \leq x, M(t) \geq y] + \mathbb{P}[W(t) \leq x, M(t) \leq y] = \mathbb{P}[W(t) \leq x]$$

as it removes the conditioning on M. Then

$$\mathbb{P}[W(t) \leq x] = \mathbb{P}[B(t) + \varphi t \leq x]$$

$$= \mathbb{P}[B(t) \leq x - \varphi t]$$

$$= \mathbb{P}\left[\frac{Z \leq (x - \varphi t)}{\sqrt{t}}\right]$$

$$= N\left[\frac{x - \varphi t}{\sqrt{t}}\right]$$

gives

$$\mathbb{P}[W(t) \leq x, M(t) \leq y] = N\left[\frac{x - \varphi t}{\sqrt{t}}\right] - \exp(2\varphi y)N\left[\frac{-2y + x - \varphi t}{\sqrt{t}}\right]$$

The use of the change of probability in obtaining probability distributions for barrier options is given in *Epps* and *Lin*.

7.9   CONCEPT OF CHANGE OF PROBABILITY

The starting point is a random variable Z whose values are non-negative, and which has expected value 1. This expected value can always be created by scaling the values of an arbitrary non-negative random variable by its mean. If $f(z)$ denotes the density of Z at $Z = z$, then $\mathbb{E}[Z] = 1 = \int_{z=0}^{\infty} zf(z)\,dz$. Consider interval $A = [0, \alpha]$. Let 1_A denote the indicator variable on set A; it has value 1 on A and 0 elsewhere. Then $Z\,1_A$ is a random variable with values in the range $[0, \alpha]$. Its expected value is $\mathbb{E}[Z\,1_A] = \int_{z=0}^{\alpha} zf(z)\,dz$ which is a function of upper integration limit α. Denote it by $\widehat{\mathbb{P}}(\alpha)$. This function has the following four properties:

(i) If $\alpha = \infty$ then $\widehat{\mathbb{P}}(\infty) = \int\limits_{z=0}^{\infty} zf(z)\,dz = \mathbb{E}[Z] = 1$

(ii) If $\alpha = 0$ then $\widehat{\mathbb{P}}(0) = \int\limits_{z=0}^{0} zf(z)\,dz = 0$

(iii) For any $0 < \alpha < \infty$, $\int\limits_{z=0}^{\alpha} zf(z)\,dz$ is positive because z in the integrand is positive by the definition of Z as a strictly positive random variable, and $f(z)$ is positive because it is a probability density

(iv) Function $\widehat{\mathbb{P}}(\alpha)$ increases monotonically with α.

Thus $\widehat{\mathbb{P}}$ is a probability distribution function. The reason that $\mathbb{E}[Z]$ had to equal 1 is now clear. If $\mathbb{E}[Z] \neq 1$ then $\widehat{\mathbb{P}}(\infty)$ would not equal 1 and $\widehat{\mathbb{P}}$ would not be a probability distribution. The corresponding density, here denoted g, is obtained by differentiating the distribution function with respect to the upper integration limit α, $g(\alpha) = d\widehat{\mathbb{P}}(\alpha)/d\alpha = \alpha f(\alpha)$. So $\alpha = g(\alpha)/f(\alpha)$ is the ratio of densities at $Z = \alpha$. This can also be written as $g(\alpha)\,d\alpha = \alpha[f(\alpha)\,d\alpha]$ where $g(\alpha)\,d\alpha$ is a small amount of probability located at α which is often denoted as $d\widehat{\mathbb{P}}(\alpha)$, and similarly $f(\alpha)\,d\alpha$ is denoted $d\mathbb{P}(\alpha)$. In this notation $d\widehat{\mathbb{P}}(\alpha) = \alpha\,d\mathbb{P}(\alpha)$, or $\alpha = d\widehat{\mathbb{P}}(\alpha)/d\mathbb{P}(\alpha)$. As α is a value of random variable Z, this can be written as $Z = d\widehat{\mathbb{P}}(\alpha)/d\mathbb{P}(\alpha)$. It is the ratio of a small amount of new probability, $d\widehat{\mathbb{P}}$, to a small amount of original probability, $d\mathbb{P}$. In the literature this ratio is named after the mathematicians Radon and Nikodym who established a property in real analysis which was subsequently employed in probability theory. There are several names in use. The name Radon–Nikodym *derivative* comes from the appearance of the ratio as a derivative, and is used in this text. The name Radon–Nikodym *density* expresss how much probability mass there is in $d\widehat{\mathbb{P}}$ in terms of $d\mathbb{P}$. It is also known as the likelihood ratio. But detached from this mathematical origin, one might call it the *redistributor of probability mass*.

Because the Radon–Nikodym derivative is a ratio of probability masses, the probability of values that cannot occur must be the same under probability distributions $\widehat{\mathbb{P}}$ and $\mathbb{P}$, otherwise there could be a situation where a positive $d\mathbb{P}$ is divided by a zero value of $d\widehat{\mathbb{P}}$, or vice versa, $d\widehat{\mathbb{P}}$ is divided by a zero value of $d\mathbb{P}$.

In advanced probability books this is described in the following terminology:

Absolutely Continuous Probability: If for any event A, for which $\mathbb{P}[A] = 0$, it follows that $\widehat{\mathbb{P}}[A] = 0$, then probability $\widehat{\mathbb{P}}$ is said to be absolutely continuous with respect to probability $\mathbb{P}$. This is written as $\widehat{\mathbb{P}} << \mathbb{P}$ (double inequality sign).

Equivalent Probabilities: If both $\widehat{\mathbb{P}} << \mathbb{P}$ and $\mathbb{P} << \widehat{\mathbb{P}}$, that is if $\mathbb{P}[A] = 0$ implies $\widehat{\mathbb{P}}[A] = 0$, and $\widehat{\mathbb{P}}[A] = 0$ implies $\mathbb{P}[A] = 0$, then $\mathbb{P}$ and $\widehat{\mathbb{P}}$ are said to be equivalent probabilities. This is denoted as $\mathbb{P} \sim \widehat{\mathbb{P}}$. The corresponding probability spaces are $(\Omega, \Im, \mathbb{P})$ and $(\Omega, \Im, \widehat{\mathbb{P}})$.

Note that both probabilities are defined on the same σ-algebra of events $\Im$, that is, the events on which the probabilities are defined are the same, but there are two probabilities for each event. In more technical terminology: two probability measures $\mathbb{P}$ and $\widehat{\mathbb{P}}$ defined on the same event space $\Im$, are equivalent if they have the same null sets. From the work of Radon and Nikodym in real analysis it can be derived that if $\mathbb{P}$ and $\widehat{\mathbb{P}}$ are equivalent probability measures on the space $(\Omega, \Im)$, then there exists a unique random variable Z such that for all events $A \in \Im$ it holds that

$$\widehat{\mathbb{P}}[A] = \int_A Z\,d\mathbb{P} = \mathbb{E}_{\mathbb{P}}[Z\,1_A] \quad \text{and} \quad \mathbb{P}[A] = \int_A \frac{1}{Z}\,d\widehat{\mathbb{P}} = \mathbb{E}_{\widehat{\mathbb{P}}}\left[\frac{1}{Z}\,1_A\right]$$

7.9.1 Relationship between Expected Values under Equivalent Probabilities

For a function h of random variable X, the expectation under $\widehat{\mathbb{P}}$ can be expressed in terms of the expectation under $\mathbb{P}$ as

$$\mathbb{E}_{\widehat{\mathbb{P}}}[h(X)] = \int h(X)\,d\widehat{\mathbb{P}} = \int h(X)\frac{d\widehat{\mathbb{P}}}{d\mathbb{P}}\,d\mathbb{P} = \mathbb{E}_{\mathbb{P}}\left[h(X)\frac{d\widehat{\mathbb{P}}}{d\mathbb{P}}\right]$$
$$= \mathbb{E}_{\mathbb{P}}[h(X)Z(X)]$$

Note that $d\widehat{\mathbb{P}}/d\mathbb{P}$ is a function of the single random variable X, so the expression of which the expectation is taken involves only X. To get the expectation of a function of a random variable under probability $\widehat{\mathbb{P}}$, multiply the function by random variable $d\widehat{\mathbb{P}}/d\mathbb{P}$, and take the expected value under probability $\mathbb{P}$. *A change of probability does not affect the value of an expectation because the random variable of which the expectation is taken is adjusted.*

In the Girsanov transformation the Radon–Nikodym derivative is an exponential expression because it is the ratio of Brownian motion densities which are exponential. If the densities in the ratio are other than normal, then the Radon–Nikodym derivative is not necessarily exponential, as illustrated in exercises [7.10.1] and [7.10.7].

In the literature the original probability is usually labelled $\mathbb{P}$. It is the $\mathbb{P}$ in the familiar expression for the probability that a random variable X takes values in a specified interval, $\mathbb{P}[a \leq X \leq b]$. The new probability is often labelled $\mathbb{Q}$ in the literature and the new probability of this event is then written as $\mathbb{Q}[a \leq X \leq b]$. But in order to maintain the familiarity with the well-established $\mathbb{P}$ notation, notation such as $\mathbb{P}^*$, $\widetilde{\mathbb{P}}$, $\widehat{\mathbb{P}}$, is also widely used.

7.10 EXERCISES

[7.10.1] (from *Epps*) Random variable X has the exponential distribution with density $\exp(-x)$. Random variable Y has density $x\exp(-x)$. Determine the Radon–Nikodym derivative, and plot its graph.

[7.10.2] Random variable X is $N(0, 1)$ under probability $\mathbb{P}$. Let $Y \overset{\text{def}}{=} X + \mu$. Under probability $\widehat{\mathbb{P}}$, Y is $N(0, 1)$. Determine the Radon–Nikodym derivative.

[7.10.3] Random variable Y is $N(0, 1)$ under probability $\widehat{\mathbb{P}}$. Let $X \stackrel{\text{def}}{=} Y - \mu$. Under probability $\mathbb{P}$, X is $N(0, 1)$. Determine the Radon–Nikodym derivative.

[7.10.4] Over the time period $[0, T]$ the random terminal stock price $S(T)$ is related to the given initial stock price $S(0) = 1$ by $S(T) = \exp[(\mu - \frac{1}{2}\sigma^2)T] + \sigma B_{\mathbb{P}}(T)]$ under the original probability $\mathbb{P}$; μ is the true growth rate of the stock price. Under the risk-neutral probability $\widehat{\mathbb{P}}$ it is $S(T) = \exp[(r - \frac{1}{2}\sigma^2)T + \sigma B_{\widehat{\mathbb{P}}}(T)]$ where r is the risk-free interest rate.
   (a) Derive the expected value of $S(T)$ under $\mathbb{P}$.
   (b) Derive the Radon–Nikodym derivative $d\widehat{\mathbb{P}}/d\mathbb{P}$ by which the density of $B(t)$ can be transformed to the density of $B_{\widehat{\mathbb{P}}}(T)$.
   (c) Give the expected value of $S(T)$ under $\widehat{\mathbb{P}}$, and then transform this to the expected value of $S(T)$ under $\mathbb{P}$.

[7.10.5] Let random variable W be defined as $W(t) \stackrel{\text{def}}{=} B(t) + \varphi t$ where B is Brownian motion. Specify the Radon–Nikodym derivative Z that creates a new probability distribution $\widehat{\mathbb{P}}$ under which W is a Brownian motion. Then use Z to compute $\widehat{\mathbb{E}}[W(t)]$ from $\mathbb{E}[W(t)Z(t)]$.

[7.10.6] Revisit Chapter 6 and show the following:
   (a) $\mathbb{E}_P[Z(T)S^*(T)] = S(0)$
   (b) $\mathbb{E}_P[Z(T)V^*(T)] = V(0)$
   (c) $\mathbb{E}_P[Z(T)V^*(T)] = \mathbb{E}_{\widehat{P}}[V^*(T)]$

[7.10.7] *Redistribution of Poisson probability mass* Let random variable X have a Poisson distribution with parameter λ,

$$\mathbb{P}[X = k] = \frac{\lambda^k}{k!}e^{-\lambda} \quad \text{for} \quad k = 0, 1, 2, \ldots$$

Derive the Radon–Nikodym derivative for a change of probability from parameter λ_1 to λ_2. Construct a graph of the probability masses and the Radon–Nikodym derivative.

[7.10.8] *First passage of a barrier* Annex A, Section A.4, derives the probability density of the random time of first passage of a horizontal barrier. The object of this exercise is to derive the density when the barrier is straight with upwards slope μ, starting at level L. This can be done with a Girsanov transformation.

[7.10.9] *Bachelier type stock price dynamics* Let the SDE for stock price S be given by $dS(t) = \mu \, dt + \sigma \, dB(t)$, where μ and σ are constant. Derive the SDE for S under the money market account as the numeraire. Then turn this into a martingale by a suitable Girsanov transformation, and specify the expression for the Radon–Nikodym derivative. Thereafter derive the SDE for the undiscounted stock price and solve that SDE. Finally, using this latest stock price, derive the expected value of $\max[S(T) - K, 0]$ where K is a constant.

7.11   SUMMARY

What is generally called a change of probability measure in the literature, is a *redistribution of probability mass*. The best introduction to how this works is for a discrete probability distribution, because that has a distinct probability mass associated with each possible value of the discrete random variable. Section 7.1 explains in a simple example how the original probability masses p_1, p_2, p_3, are redistributed to new probability masses q_1, q_2, q_3. It then shows how an expected value of a function of the discrete random variable can be computed under the new probabilities q. The numerical expected value is the same because of the adjustment by the Radon–Nikodym derivative. If probabilities q and p are both known for all possible values of the discrete random variable, then the ratio q_i/p_i for each i is the factor by which the original probability mass p_i is changed; it is the value of the probability redistributor at i. Alternatively, if the values of the ratio are given, then the new probability distribution can be readily computed from the original one. This same recipe can be applied to continuous probability distributions. Let X be the random variable, f the original probability density at $X = x$, and g the new probability density. The probability that X takes a value in the interval $x + \Delta x$ is $f(x) \, \Delta x$ or $g(x) \, \Delta x$. The ratio of the new probability mass to the original is $\frac{g(x) \, \Delta x}{f(x) \, \Delta x}$, which is the ratio of the densities at $X = x$, $\frac{g(x)}{f(x)}$. Thus to use g instead of f, the density has to be multiplied by $g(x)/f(x)$. The redistribution method can be applied to any probability distribution, provided the range of outcomes is not modified. A redistribution of probability does not necessarily keep the type of density. However for some distributions it is possible to keep the type of distribution. This is shown in Section 7.2 for a normal distribution. Because the new density and the original are both of an exponential form, the ratio can only be of exponential form. Section 7.3 applies this to another normally distributed random variable, namely Brownian motion. The new random variable is connected to the original one via the constant φ. Section 7.4 extends this to where φ is a random process; that is the so-called Girsanov transformation. The Radon–Nikodym is a random process. Its values are positive, because otherwise it would create a new probability density that was negative.

Further information on the Radon–Nikodym derivative can be found on the Internet in Wikipedia.

8

Numeraire

In Chapter 6 the savings account was used as the numeraire in option valuation. In some applications it is advantageous to use a different numeraire. How this works is described here.

8.1 CHANGE OF NUMERAIRE

8.1.1 In Discrete Time

The method of option valuation on a one-period binomial tree, as described in Section 6.2, used the savings account as the numeraire. This produced a probability under which the price processes of all assets in the model were martingales when their values were expressed in terms of this numeraire. This so-called martingale probability appeared as a natural byproduct in the derivation. It turns out that there is a direct link between the numeraire and the martingale probability. If a different numeraire is chosen, then a different martingale probability results. To illustrate this, consider again the one-period binomial model, but now use the stock price as the numeraire. This is not a natural choice, but is used here to illustrate the change of numeraire concept. The values of the savings account at times 0 and T are now expressed in terms of the corresponding values of the stock price, as shown in Figure 8.1.

The martingale probability is determined from the condition that the savings account value process, measured in terms of the stock price as the numeraire, must be a martingale. The equation for finding this probability p is

$$p\left(\frac{1+r}{uS}\right) + (1-p)\left(\frac{1+r}{dS}\right) = \frac{1}{S}$$

The unique solution is

$$p = \frac{u(1+r) - ud}{(u-d)(1+r)} \qquad 1 - p = \frac{ud - d(1+r)}{(u-d)(1+r)}$$

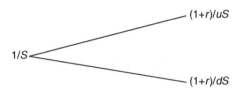

Figure 8.1   Savings account values under stock price numeraire

It is now shown that if the replicating portfolio value is expressed in terms of the stock price as the numeraire, then that process is also a martingale under probability p. That is the equivalent of what is shown in Section 6.2, but using the stock price as the numeraire instead of the savings account. At the beginning of the period, form a replicating portfolio of α shares of unit price S, and β amount of savings account. This initial replicating portfolio has the value $V(0) = \alpha S + \beta$. At the end of the period, this portfolio has the value $V(T)$. In the up-state, the stock price is uS and the savings account value has grown to $\beta(1 + r)$ so $V(T)_{\text{up}} = \alpha u S + \beta(1 + r)$. In the down-state, the stock price is dS and the savings account value has grown to $\beta(1 + r)$ so $V(T)_{\text{down}} = \alpha d S + \beta(1 + r)$. Now express the values of the replicating portfolio in terms of the numeraire.

initially $V(0)/S = [\alpha S + \beta]/S = \alpha + \beta[1/S]$

in up-state $V(T)_{\text{up}}/uS = [\alpha u S + \beta(1 + r)]/(uS)$
       $= \alpha + \beta[(1 + r)/(uS)]$

in down-state $V(T)_{\text{down}}/dS = [\alpha d S + \beta(1 + r)]/(dS)$
       $= \alpha + \beta[(1 + r)/(dS)]$

Note that in the above expressions the last $[\cdots]$ are the discounted values of the savings account. The dynamics are shown in Figure 8.2.

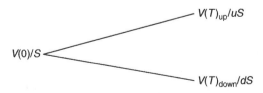

Figure 8.2   Replicating portfolio values under stock price numeraire

The expected terminal value of the discounted portfolio, using probability p, is

$$p\left[\alpha + \beta\left(\frac{1+r}{uS}\right)\right] + (1-p)\left[\alpha + \beta\left(\frac{1+r}{dS}\right)\right]$$

$$= \alpha + \beta\left[p\left(\frac{1+r}{uS}\right) + (1-p)\left(\frac{1+r}{dS}\right)\right]$$

$$= \alpha + \beta\left(\frac{1}{S}\right) = \frac{V(0)}{S}$$

The last line follows from the expression for p above, as the term in $[\cdots]$ equals $1/S$. Thus under probability p, the replicating portfolio value process under the stock price as numeraire is a martingale, just as it was a martingale under probability q when the savings account was used as the numeraire. Hence the option pricing expression under the stock price numeraire is

$$\frac{V(0)}{S} = p\left[\frac{V(T)_{\text{up}}}{uS}\right] + (1-p)\left[\frac{V(T)_{\text{down}}}{dS}\right]$$

The stock price process itself, expressed in terms of the stock price as numeraire, has the constant value 1, as its initial value is S/S, its up-value $uS/(uS)$, its down-value $dS/(dS)$, so it is also a martingale. All assets in this model are martingales under p, as soon as one of them is a martingale. That was also the case under q. The ratio of the probability masses is

for the up-movement $\quad \dfrac{p}{q} = \dfrac{u(1+r)-ud}{(u-d)(1+r)} \Big/ \dfrac{(1+r)-d}{(u-d)} = \dfrac{u}{1+r}$

for the down-movement $\quad \dfrac{1-p}{1-q} = \dfrac{ud-d(1+r)}{(u-d)(1+r)} \Big/ \dfrac{u-(1+r)}{(u-d)} = \dfrac{d}{1+r}$

Each equals the ratio of the numeraire, scaled by its initial value, that is, uS/S divided by $(1+r)/1$ and dS/S divided by $(1+r)/1$. The Radon–Nikodym derivative for expressing an expected under p as an expected value under q is a discrete random variable, say Z, with values

$$Z_{\text{up}} = \frac{p}{q} = \frac{u}{1+r} \qquad Z_{\text{down}} = \frac{1-p}{1-q} = \frac{d}{1+r}$$

If X is a discrete random variable with values X_{up} and X_{down}, then

$$
\begin{aligned}
\mathbb{E}_p[X] = \mathbb{E}_q[XZ] &= X_{\text{up}} Z_{\text{up}} q + X_{\text{down}} Z_{\text{down}} (1-q) \\
&= X_{\text{up}} \frac{p}{q} q + X_{\text{down}} \frac{1-p}{1-q} (1-q) \\
&= X_{\text{up}} p + X_{\text{down}} (1-p)
\end{aligned}
$$

8.1.2 In Continuous Time

Section 6.3 described the valuation of a European option by the martingale method. In that exposition the interest rate was treated as constant. Now consider the same valuation under a random interest rate. The initial value of the option is

$$
V(0) = \mathbb{E}_{\widehat{\mathbb{P}}}\left[\exp\left(-\int_{u=0}^{T} r(u)\,du \right) V(T) \right]
$$

The expected value $\mathbb{E}_{\widehat{\mathbb{P}}}$ is under martingale probability $\widehat{\mathbb{P}}$, which is the probability distribution under which asset prices are martingales when discounted by the savings account as numeraire. Generally the expression for the payoff at maturity also contains r, so when r is random the expression in $[\cdots]$ is the product of two random variables, the discount factor $\exp(-\int_{u=0}^{T} r(u)\,du)$ and the terminal payoff $V(T)$, which are not independent. Computing this expected value requires their joint probability distribution and is complicated. The valuation can be simplified by using another numeraire than the savings account. If the numeraire was the time-t value of a non-defaultable zero-coupon bond which matures at the same time T as the option, denoted $P(t, T)$, then there would be a corresponding probability distribution, known as the T-*forward probability measure,* here denoted $\mathbb{P}^T$, under which $V(t)/P(t, T)$ is a martingale, so $V(0)/P(0, T) = \mathbb{E}_{\mathbb{P}^T}[V(T)/P(T, T)]$ where $\mathbb{E}_{\mathbb{P}^T}$ denotes the expected value under $\mathbb{P}^T$, and $P(0, T)$ is the time-0 price of the bond which is observable. But as $P(T, T) = 1$, the valuation expression simplifies to $V(0) = P(0, T)\mathbb{E}_{\mathbb{P}^T}[V(T)]$. This simplification happened by design, namely by using a bond with maturity equal to the option maturity. $V(0)$ has the same form as the martingale expression using the savings account, but the expectation $\mathbb{E}_{\mathbb{P}^T}$ uses the different probability distribution $\mathbb{P}^T$. It remains to determine this martingale probability. There is *no joint probability distribution required.*

Another example is a so-called *exchange option* in the standard Black–Scholes framework. This is based on two stock prices, S_1 and S_2, which are driven by correlated Brownian motions. In the case of a European call, if $S_1(T) > S_2(T)$, the call holder can buy $S_1(T)$ by paying $S_2(T)$. Thus at maturity there is the option to exchange stocks. The dynamics of stock prices $S_i(t)$ under $\widehat{\mathbb{P}}$ are

$$\frac{dS_i(t)}{S_i(t)} = r\,dt + \sigma_i\,d\widehat{B}_i(t)$$

where r and σ_i are constants, and $d\widehat{B}_1(t)\,d\widehat{B}_2(t) = \rho\,dt$. The exchange option payoff is $c(T) \stackrel{\text{def}}{=} \max[S_1(T) - S_2(T), 0]$. Computing initial option value $c(0) = \exp^{-rT}\mathbb{E}_{\widehat{\mathbb{P}}}[c(T)]$ requires the joint probability distribution of S_1 and S_2 which is bi-variate lognormal. But the formulation can be simplified by factoring out $S_2(T)$. That makes S_2 the numeraire and $c(T)/S_2(T) = \max\{[S_1(T)/S_2(T)] - 1, 0\}$. Introducing random process F as the ratio of random processes S_1 and S_2, $F(t) \stackrel{\text{def}}{=} [S_1(t)/S_2(t)]$, formulates the option valuation with F as the underlying asset, as $c(T)/S_2(T) = \max[F(T) - 1, 0]$. For numeraire S_2 there is a unique probability distribution, say $\widetilde{\mathbb{P}}$, under which F is a martingale, and also the replicating portfolio value and hence the call value. When $\widetilde{\mathbb{P}}$ has been found, the initial value of the option, in terms of numeraire S_2, can be computed as

$$\frac{c(0)}{S_2(0)} = \mathbb{E}_{\widetilde{\mathbb{P}}}\{\max[F(T) - 1, 0]\}$$

where $\mathbb{E}_{\widetilde{\mathbb{P}}}$ is under $\widetilde{\mathbb{P}}$. To find the dynamics of F under $\widetilde{\mathbb{P}}$, apply Itō's formula to the ratio of the random processes $S_1(t)$ and $S_2(t)$, giving

$$\frac{dF(t)}{F(t)} = \left[\sigma_2^2 - \rho\sigma_1\sigma_2\right]dt + \sigma_1\,d\widehat{B}_1(t) - \sigma_2\,d\widehat{B}_2(t)$$

Using Section 4.6, the Brownian motion terms can be combined into a single new independent Brownian motion $\widehat{B}_3(t)$, under the same probability $\widehat{\mathbb{P}}$ hence the $\widehat{}$ notation, and

$$\frac{dF(t)}{F(t)} = \left[\sigma_2^2 - \rho\sigma_1\sigma_2\right]dt + \sigma_F\,d\widehat{B}_3(t)$$

where $\sigma_F^2 \stackrel{\text{def}}{=} \sigma_1^2 - 2\rho\sigma_1\sigma_2 + \sigma_2^2$. In order for F to be a martingale, its SDE has to be made driftless. This is achieved by the Girsanov transformation to a new Brownian motion $\widetilde{B}$ under probability $\widetilde{\mathbb{P}}$.

$$\sigma_F \, d\widetilde{B}(t) \overset{\text{def}}{=} \left[\sigma_2^2 - \rho\sigma_1\sigma_2\right] dt + \sigma_F \, d\widehat{B}_3(t)$$

Then $dF(t)/F(t) = \sigma_F \, d\widetilde{B}(t)$ under probability $\widetilde{\mathbb{P}}$.

8.2   FORWARD PRICE DYNAMICS

Again let $P(t, T)$ denote the price, at time t, of a non-defaultable zero-coupon bond maturing at T. The price of a traded asset, $X(t)$, expressed in terms of this bond price as numeraire is known as a *forward price*, denoted

$$F(t) \overset{\text{def}}{=} \frac{X(t)}{P(t, T)} \quad \text{for } 0 \leq t \leq T$$

The corresponding probability measure under which F is a martingale is now derived for two cases.

8.2.1   Dynamics of Forward Price of a Bond

The price of a bond maturing at T_2, expressed in terms of the price of a bond maturing at the earlier time T_1, is the ratio $P(t, T_2)/P(t, T_1)$. As the T_1-bond serves as the numeraire, it is the forward price for time T_1 of the T_2-bond. This ratio is here denoted with running clock time t as $F(t)$, where for greater readability references to the fixed maturity dates have been omitted. For further convenience $P(t, T_i)$ is abbreviated as P_i, so

$$F(t) \overset{\text{def}}{=} \frac{P(t, T_2)}{P(t, T_1)} = \frac{P_2}{P_1}$$

The bond dynamics under $\widehat{\mathbb{P}}$ are given as $dP_i/P_i = r(t)\,dt + \sigma_i(t)\,d\widehat{B}_P(t)$, so P_i is a martingale when discounted by the savings account. Note that the bond prices are driven by the same $\widehat{B}_P(t)$, only their time dependent non-random volatilities $\sigma_i(t)$ are different. By Itô's formula

$$\frac{dF}{F} = -\frac{dP_1}{P_1} + \frac{dP_2}{P_2} + \left(\frac{dP_1}{P_1}\right)^2 - \frac{dP_1}{P_1}\frac{dP_2}{P_2}$$

Using the bond dynamics dP_i/P_i then gives

$$\frac{dF}{F} = -[r(t)\,dt + \sigma_1(t)\,d\widehat{B}_P] + [r(t)\,dt + \sigma_2(t)\,d\widehat{B}_P] + \sigma_1(t)^2\,dt$$
$$- \sigma_1(t)\sigma_2(t)\,dt$$
$$= [\sigma_1(t)^2 - \sigma_1(t)\sigma_2(t)]\,dt + [\sigma_2(t) - \sigma_1(t)]\,d\widehat{B}_P$$
$$= [\sigma_2(t) - \sigma_1(t)][-\sigma_1(t)\,dt + d\widehat{B}_P]$$

Applying the Girsanov transformation $dB \stackrel{\text{def}}{=} -\sigma_1(t)\,dt + d\widehat{B}_P$ gives

$$\frac{dF}{F} = [\sigma_2(t) - \sigma_1(t)]\,dB$$

With the notation $\sigma_F(t) \stackrel{\text{def}}{=} \sigma_2(t) - \sigma_1(t)$,

$$\frac{dF(t)}{F(t)} = \sigma_F(t)\,dB(t)$$

This is driftless, so F is a martingale under the probability distribution of Brownian motion B. The SDE is the familiar geometric Brownian motion, now with a volatility that is time dependent, and solution

$$F(t) = F(0)\exp\left[-\frac{1}{2}\int_{s=0}^{t}\sigma_F(s)^2\,ds + \int_{s=0}^{t}\sigma_F(s)\,dB(s)\right]$$

8.2.2 Dynamics of Forward Price of any Traded Asset

Let the dynamics of traded asset X under $\widehat{\mathbb{P}}$ be specified as

$$\frac{dX(t)}{X(t)} = r(t)\,dt + \sigma_X(t)\,d\widehat{B}_X(t)$$

and the bond price process as

$$\frac{dP(t, T)}{P(t, T)} = r(t)\,dt + \sigma_P(t)\,d\widehat{B}_P(t)$$

Both processes have a time-dependent non-random volatility. When discounted by the savings account, $\exp[\int_{s=0}^{t} r(s)\,ds]$, both have a driftless SDE. This drift specification entails no loss of generality as risk-neutrality can always be achieved by a Girsanov transformation. In general, the above Brownian motions $\widehat{B}_X$ and $\widehat{B}_P$ can be correlated with coefficient ρ, so $d\widehat{B}_X(t)\,d\widehat{B}_P(t) = \rho\,dt$. As the forward price $F = X(t)/P(t, T)$ is a function of the two variables X and P, Itō's formula gives

$$\frac{dF}{F} = \frac{dX}{X} - \frac{dP}{P} + \left(\frac{dP}{P}\right)^2 - \frac{dX}{X}\frac{dP}{P}$$

Substituting the dynamics of X and P gives

$$\frac{dF}{F} = [r(t)\,dt + \sigma_X(t)\,d\widehat{B}_X] - [r(t)\,dt + \sigma_P(t)\,d\widehat{B}_P] +$$

$$+ \sigma_P(t)^2\,dt - \rho\sigma_X(t)\sigma_P(t)\,dt$$

$$\frac{dF(t)}{F(t)} = [\sigma_P(t)^2 - \rho\sigma_X(t)\sigma_P(t)]\,dt + \sigma_X(t)\,d\widehat{B}_X(t)$$

$$- \sigma_P(t)\,d\widehat{B}_P(t) \qquad\qquad (^*)$$

The two correlated Brownian motion terms will be first combined into $\sigma(t)\,dB_1(t)$ where

$\sigma(t)^2 \overset{\text{def}}{=} \sigma_X(t)^2 - 2\rho\sigma_X(t)\sigma_P(t) + \sigma_P(t)^2$, according to Section 4.6. Then

$$\frac{dF(t)}{F(t)} = [\sigma_P(t)^2 - \rho\sigma_X(t)\sigma_P(t)]\,dt + \sigma(t)\,dB_1(t)$$

The Girsanov transformation $\sigma(t)\,dB_2(t) \overset{\text{def}}{=} [\sigma_P(t)^2 - \rho\sigma_X(t)\sigma_P(t)]\,dt + \sigma(t)\,dB_1(t)$ now gives the driftless SDE

$$\frac{dF(t)}{F(t)} = \sigma(t)\,dB_2(t)$$

So F is a martingale under the probability distribution of B_2.

Alternatively, in expression (*) the drift term can be removed by a Girsanov transformation which uses either $\widehat{B}_P$ or $\widehat{B}_X$. Taking $\widehat{B}_P$, a new independent Brownian motion B_3 is defined by

$$-\sigma_P(t)\,dB_3(t) \overset{\text{def}}{=} [\sigma_P(t)^2 - \rho\sigma_X(t)\sigma_P(t)]\,dt - \sigma_P(t)\,d\widehat{B}_P(t)$$

giving the driftless SDE

$$\frac{dF(t)}{F(t)} = \sigma_X(t)\,d\widehat{B}_X(t) - \sigma_P(t)\,dB_3(t)$$

This is a geometric SDE with two driving Brownian motions which can be solved by using $\ln[F]$ as the trial solution. Itō's formula gives $d\{\ln[F]\} = \frac{dF}{F} - \frac{1}{2}(\frac{dF}{F})^2$. As $\widehat{B}_X$ and B_3 are independent

$$\left(\frac{dF}{F}\right)^2 = \sigma_X(t)^2\,dt + \sigma_P(t)^2\,dt$$

and

$$d\{\ln[F]\} = -\frac{1}{2}[\sigma_X(t)^2 + \sigma_P(t)^2]\,dt + \sigma_X(t)\,d\widehat{B}_X(t) - \sigma_P(t)\,dB_3(t)$$

For option valuation it is convenient to have a single driving Brownian motion. This can be obtained by combining $\sigma_X(t)\,d\widehat{B}_X(t) - \sigma_P(t)\,dB_3(t)$ into $\sigma(t)\,dB_4(t)$ where $\sigma(t)^2 \overset{\text{def}}{=} \sigma_X(t)^2 - 2\rho\sigma_X(t)\sigma_P(t) + \sigma_P(t)^2$. Thus F can be written as $\frac{dF(t)}{F(t)} = \sigma(t)\,dB_4(t)$.

This may seem a bit long, but it is intended as practice in creating martingale dynamics for use in option valuation.

8.3   OPTION VALUATION UNDER MOST SUITABLE NUMERAIRE

In Section 6.5 it was shown that if, under a probability $\widehat{\mathbb{P}}$, a random process S^* is a martingale with SDE $dS^*(t)/S^*(t) = \sigma\,d\widehat{B}(t)$, then for a standard European call

$$\mathbb{E}_{\widehat{\mathbb{P}}}\{\max[S^*(T) - K^*, 0]\} = S^*(0)N(d_1) - K^*N(d_2)$$

where

$$d_1 \overset{\text{def}}{=} \frac{\ln[S^*(0)/K^*] + \frac{1}{2}\sigma^2 T}{\sigma\sqrt{T}} \qquad d_2 \overset{\text{def}}{=} d_1 - \sigma\sqrt{T}$$

This will now be used to write down the value of various options based on a martingale F with SDE of the form $dF/F = \sigma_F(t)\,dB(t)$, without the need for rederivation from basics.

8.3.1   Exchange Option

$$\frac{c(0)}{S_2(0)} = \mathbb{E}_{\widehat{\mathbb{P}}}\{\max[F(T) - 1, 0]\}$$

Here $S^*(T)$ becomes $F(T) = S_1(T)/S_2(T)$, K^* becomes 1, $\sigma^2 T$ becomes $\sigma_F^2 T$ where $\sigma_F^2 = \sigma_1^2 + \sigma_2^2 - 2\sigma_1\sigma_2\rho$. The expected value expression becomes

$$\frac{S_1(0)}{S_2(0)}N(\tilde{d}_1) - N(\tilde{d}_2) \qquad \text{where} \quad \tilde{d}_1 \overset{\text{def}}{=} \frac{\ln\left[\frac{S_1(0)}{S_2(0)}\right] + \frac{1}{2}\sigma_F^2 T}{\sigma_F\sqrt{T}}$$

and

$$\widetilde{d}_2 \stackrel{\text{def}}{=} \widetilde{d}_1 - \sigma_F \sqrt{T}$$

Thus

$$c(0) = S_2(0) \left\{ \frac{S_1(0)}{S_2(0)} N(\widetilde{d}_1) - N(\widetilde{d}_2) \right\}$$

$$= S_1(0)N(\widetilde{d}_1) - S_2(0)N(\widetilde{d}_2)$$

Both stock prices grow at the same rate r, thus their ratio has no growth, and the option price is independent of r.

8.3.2   Option on Bond

$$\frac{c(0)}{P(0, T_1)} = \mathbb{E}_{\widehat{\mathbb{P}}}\{\max[F(T) - K, 0]\}$$

Here $S^*(T)$ becomes $F(T_1) = P(T_1, T_2)/P(T_1, T_1)$, K^* becomes K, $\sigma^2 T$ becomes $\Sigma^2 = \int_{t=0}^{T} \sigma_F(t)^2 \, dt$ where $\sigma_F(t) \stackrel{\text{def}}{=} \sigma_2(t) - \sigma_1(t)$. The expected value expression becomes

$$\frac{P(0, T_2)}{P(0, T_1)} N\left(\widetilde{d}_1\right) - N\left(\widetilde{d}_2\right)$$

where $\widetilde{d}_1 \stackrel{\text{def}}{=} \dfrac{\ln\left[\frac{P(0,T_2)}{P(0,T_1)K}\right] + \frac{1}{2}\Sigma^2}{\Sigma}$ and $\widetilde{d}_2 \stackrel{\text{def}}{=} \widetilde{d}_1 - \Sigma$.

Thus

$$c(0) = P(0, T_1) \left\{ \frac{P(0, T_2)}{P(0, T_1)} N\left(\widetilde{d}_1\right) - K N\left(\widetilde{d}_2\right) \right\}$$

$$= P(0, T_2)N\left(\widetilde{d}_1\right) - P(0, T_1)K N\left(\widetilde{d}_2\right)$$

Compared to the pricing expression of a standard European call on S, the equivalent of $S(0)$ is $P(0, T_2)$, the initial value of the underlying bond price. The discount factor on K is $P(0, T_1)$, the random version of $\exp(-rT_1)$.

8.3.3   European Call under Stochastic Interest Rate

The SDE for stock price $S(t)$ under $\widehat{P}$ is

$$\frac{dS(t)}{S(t)} = r(t) \, dt + \sigma_S \, d\widehat{B}_S(t)$$

The analysis will use $P(t, T)$ as numeraire, where

$$\frac{dP(t, T)}{P(t, T)} = r(t)\,dt + \sigma_P(t)\,d\widehat{B}_P(t)$$

Assume that the driving Brownian motions are independent, $d\widehat{B}_S(t)\,d\widehat{B}_P(t) = 0$. The forward price of S for delivery at T is $F(t) \overset{\text{def}}{=} S(t)/P(t, T)$. As pointed out earlier, the bond used in the forward price must mature at the same time as the option. The SDE for F follows from the routine application of Itō's formula to the ratio of random processes S and P, as

$$\frac{dF}{F} = \sigma_p(t)^2\,dt + \sigma_S\,d\widehat{B}_S - \sigma_p(t)\,d\widehat{B}_P$$
$$= \sigma_p(t)[\sigma_p(t)\,dt - d\widehat{B}_P] + \sigma_S\,d\widehat{B}_S$$

Applying the Girsanov transformation $-dB(t) \overset{\text{def}}{=} \sigma_p(t)\,dt - d\widehat{B}_P(t)$ produces the driftless SDE

$$\frac{dF}{F} = -\sigma_p(t)\,dB + \sigma_S\,d\widehat{B}_S$$

The linear combination of the two independent Brownian motions can be replaced by a single Brownian motion W with volatility $\sigma_F(t)$, where $\sigma_F(t)^2 \overset{\text{def}}{=} \sigma_p(t)^2 + \sigma_S^2$, giving

$$\frac{dF(t)}{F(t)} = \sigma_F(t)\,dW(t)$$

Use as numeraire the bond that matures at the call maturity date, $P(t, T)$. Then $c(t)/P(t, T)$ is a martingale and

$$\frac{c(0)}{P(0, T)} = \mathbb{E}_{\mathbb{P}^T}\left\{ \frac{\max[S(T) - K, 0]}{P(T, T)} \right\}$$
$$= \mathbb{E}_{\mathbb{P}^T}\left\{ \max\left[\frac{S(T)}{P(T, T)} - K, 0 \right] \right\}$$
$$= \mathbb{E}_{\mathbb{P}^T}\{\max[F(T) - K, 0]\}$$

Here $S^*(t)$ becomes $F(t) = S(t)/P(t, T)$, K^* becomes K, $\sigma^2 T$ becomes

$$\sum{}^2 \overset{\text{def}}{=} \int_{t=0}^{T} \sigma_F(t)^2\,dt = \int_{t=0}^{T} [\sigma_p(t)^2 + \sigma_S^2]\,dt = \int_{t=0}^{T} \sigma_p(t)^2\,dt + \sigma_S^2 T$$

The expected value expression becomes

$$\frac{S(0)}{P(0, T)} N\left(\tilde{d}_1\right) - N\left(\tilde{d}_2\right) \qquad \text{where} \quad \tilde{d}_1 \overset{\text{def}}{=} \frac{\ln\left[\frac{S(0)}{P(0,T)K}\right] + \frac{1}{2}\Sigma^2}{\Sigma}$$

and $\quad \tilde{d}_2 \overset{\text{def}}{=} \tilde{d}_1 - \sigma_F \sqrt{T}$.

Thus

$$c(0) = P(0, T)\left\{\frac{S(0)}{P(0, T)} N\left(\tilde{d}_1\right) - K N\left(\tilde{d}_2\right)\right\}$$
$$= S(0) N\left(\tilde{d}_1\right) - P(0, T) K N\left(\tilde{d}_2\right)$$

To go any further, the expression for the bond price volatility $\sigma_p(t)$ must be known, and this depends on the underlying interest rate model that is used. By way of check, if the interest rate is constant, $\sigma_p(t) = 0$, $\sigma_F(t) = \sigma_S$, $\Sigma^2 = \sigma_S^2 T$, $P(0, T) = \exp(-rT)$, and the standard Black–Scholes pricing expression is recovered.

8.4   RELATING CHANGE OF NUMERAIRE TO CHANGE OF PROBABILITY

In the first part of Section 8.1 it was shown in the binomial framework that each choice of numeraire induces its own unique probability distribution. The expectations under these probability distributions are connected via the Radon–Nikodym derivative, as discussed in Chapter 7. This Radon–Nikodym derivative can be expressed as a ratio of the numeraires, as will now be shown.

Let $N(t)$ denote the value of a numeraire at time t. The value of the terminal option payoff $V(T)$ expressed in units of this numeraire is $V(T)/N(T)$. For this numeraire there is a probability distribution under which the discounted $V(T)$ is a martingale. Now consider two numeraires, N_{orig} and N_{new}, with respective martingale probabilities $\mathbb{P}_{\text{orig}}$ and $\mathbb{P}_{\text{new}}$. For greater readability the conditioning is suppressed in what follows.

$$\mathbb{E}_{\text{orig}}\left[\frac{V(T)}{N_{\text{orig}}(T)}\right] = \frac{V(0)}{N_{\text{orig}}(0)} \quad \text{or} \quad V(0) = N_{\text{orig}}(0)\mathbb{E}_{\text{orig}}\left[\frac{V(T)}{N_{\text{orig}}(T)}\right]$$

$$\mathbb{E}_{\text{new}}\left[\frac{V(T)}{N_{\text{new}}(T)}\right] = \frac{V(0)}{N_{\text{new}}(0)} \quad \text{or} \quad V(0) = N_{\text{new}}(0)\mathbb{E}_{\text{new}}\left[\frac{V(T)}{N_{\text{new}}(T)}\right]$$

where $\mathbb{E}_{orig}$ and $\mathbb{E}_{new}$ denote expectations under the respective probabilities. Equating the expressions for $V(0)$ gives

$$N_{new}(0)\mathbb{E}_{new}\left[\frac{V(T)}{N_{new}(T)}\right] = N_{orig}(0)\mathbb{E}_{orig}\left[\frac{V(T)}{N_{orig}(T)}\right]$$

$$\mathbb{E}_{new}\left[\frac{V(T)}{N_{new}(T)}\right] = \frac{N_{orig}(0)}{N_{new}(0)}\mathbb{E}_{orig}\left[\frac{V(T)}{N_{orig}(T)}\right]$$

Writing $\frac{V(T)}{N_{orig}(T)}$ as $\frac{V(T)}{N_{new}(T)}\frac{N_{new}(T)}{N_{orig}(T)}$

$$\mathbb{E}_{new}\left[\frac{V(T)}{N_{new}(T)}\right] = \mathbb{E}_{orig}\left[\frac{V(T)}{N_{new}(T)}\frac{N_{new}(T)/N_{new}(0)}{N_{orig}(T)/N_{orig}(0)}\right]$$

The left is the expected value of $V(T)$ based on the new numeraire N_{new}, computed under the new probability $\mathbb{P}_{new}$. The right is the expected value of the same random variable $V(T)/N_{new}(T)$ multiplied by the random variable $\frac{N_{new}(T)/N_{new}(0)}{N_{orig}(T)/N_{orig}(0)}$, computed under the original probability $\mathbb{P}_{orig}$. This random variable is the Radon–Nikodym derivative, say $Z(T)$, and gives the relationship between expectations under different numeraires as

$$\mathbb{E}_{new}\left[\frac{V(T)}{N_{new}(T)}\right] = \mathbb{E}_{orig}\left[\frac{V(T)}{N_{new}(T)}Z(T)\right]$$

$$\int\frac{V(T)}{N_{new}(T)}\,d\mathbb{P}_{new} = \int\frac{V(T)}{N_{new}(T)}\frac{d\mathbb{P}_{new}}{d\mathbb{P}_{orig}}\,d\mathbb{P}_{orig}$$

Equating this to

$$\int\frac{V(T)}{N_{new}(T)}\frac{N_{new}(T)/N_{new}(0)}{N_{orig}(T)/N_{orig}(0)}\,d\mathbb{P}_{orig}$$

gives

$$\frac{d\mathbb{P}_{new}}{d\mathbb{P}_{orig}} = \frac{N_{new}(T)/N_{new}(0)}{N_{orig}(T)/N_{orig}(0)}$$

for going from $\mathbb{P}_{orig}$ to $\mathbb{P}_{new}$. The numeraire values at time T are scaled by their initial values. A discrete example was shown in the first part of Section 8.1.

Example 8.4.1 Let X be the price of an asset under the money market account M as numeraire; recall that $dM(t)/M(t) = r(t)\,dt$, $M(0) = 1$.

As before, let $P(t, T)$ denote the time-t price of a non-defaultable zero coupon bond maturing at T, which evolves according to

$$\frac{dP(t, T)}{P(t, T)} = r(t) \, dt + \sigma_p(t) \, d\widehat{B}(t)$$

where Brownian motion $\widehat{B}(t)$ is under probability $\widehat{\mathbb{P}}$, the probability under which the bond price expressed in terms of M as numeraire is a martingale. The probability which corresponds to the use of the T-bond as numeraire is called the T-forward probability measure, and is denoted $\mathbb{P}^T$.

$$\frac{d\mathbb{P}^T}{d\widehat{\mathbb{P}}} = \frac{P(t, T)/P(0, T)}{M(t)/M(0)} = \frac{P(t, T)/P(0, T)}{M(t)}$$

The expressions for $P(t, T)$ and $M(t)$ are

$$P(t, T) = P(0, T) \exp\left\{ \int_{s=0}^{t} \left[r(s) - \tfrac{1}{2}\sigma_p(s)^2 \right] ds + \int_{s=0}^{t} \sigma_p(s) \, d\widehat{B}(s) \right\}$$

$$M(t) = \exp\left[\int_{s=0}^{t} r(s) \, ds \right]$$

So

$$\frac{d\mathbb{P}^T}{d\widehat{\mathbb{P}}} = \exp\left\{ -\tfrac{1}{2} \int_{s=0}^{t} \sigma_p(s)^2 \, ds + \int_{s=0}^{t} \sigma_p(s) \, d\widehat{B}(s) \right\}$$

8.5   CHANGE OF NUMERAIRE FOR GEOMETRIC BROWNIAN MOTION

Let N_1 and N_2 be numeraires specified by

$$\frac{dN_1(t)}{N_1(t)} = \mu_1 \, dt + \sigma_1 \, dB_1(t) \qquad N_1(0) \text{ known}$$

$$\frac{dN_2(t)}{N_2(t)} = \mu_2 \, dt + \sigma_2 \, dB_2(t) \qquad N_2(0) \text{ known}$$

where B_1 and B_2 have correlation coefficient ρ, so $dB_1 \, dB_2(t) = \rho \, dt$, and $\mu_1, \sigma_1, \mu_2, \sigma_2$, are constants. Numeraires have to be strictly positive, and here they are as the SDEs are geometric Brownian motion. The corresponding probabilities under which asset prices based on these numeraires are martingales, are denoted $\mathbb{P}_1$ and $\mathbb{P}_2$. The Radon–Nikodym derivative for constructing $\mathbb{P}_2$ from $\mathbb{P}_1$ is denoted $Z(t) \overset{\text{def}}{=} d\mathbb{P}_2(t)/d\mathbb{P}_1(t)$.

It is the solution to the driftless SDE for Z, $dZ(t)/Z(t) = -\varphi \, dB(t)$; $Z(t) = \exp[-\frac{1}{2}\varphi^2 t - \varphi B(t)$. It is also the ratio of the numeraires scaled by their initial values, $Z(t) = \frac{N_2(t)/N_2(0)}{N_1(t)/N_1(0)}$. Thus Z can be found by

(i) deriving a driftless SDE for Z of the form $dZ(t)/Z(t) = -\varphi \, dB(t)$
(ii) using the expressions for N_1 and N_2 and taking their ratio.

The derivations are as follows.

Method (i)   SDE for Z   Itō's formula applied to Z as the ratio of N_2 and N_1 gives

$$\frac{dZ}{Z} = -\frac{dN_1}{N_1} + \frac{dN_2}{N_2} + \left(\frac{dN_1}{N_1}\right)^2 - \frac{dN_1}{N_1}\frac{dN_2}{N_2}$$

Substituting the SDEs for N gives

$$\frac{dZ}{Z} = -[\mu_1 \, dt + \sigma_1 \, dB_1(t)] + [\mu_2 \, dt + \sigma_2 \, dB_2(t)] +$$

$$+ \sigma_1{}^2 \, dt - \rho\sigma_1\sigma_2 \, dt$$
$$= [-\mu_1 + \mu_2 + \sigma_1^2 - \rho\sigma_1\sigma_2] \, dt - \sigma_1 \, dB_1(t) + \sigma_2 \, dB_2(t)$$
$$= m \, dt - \sigma_1 \, dB_1(t) + \sigma_2 \, dB_2(t)$$

where m is shorthand for the drift coefficient. To deal with the correlation, write $B_2(t) \overset{\text{def}}{=} \rho B_1(t) + \sqrt{1-\rho^2} B_3(t)$, where B_3 is another independent Brownian motion. Then the random terms become

$$-\sigma_1 \, dB_1(t) + \sigma_2[\rho \, dB_1(t) + \sqrt{1-\rho^2} \, dB_3(t)]$$
$$= [\rho\sigma_2 - \sigma_1] \, dB_1(t) + \sigma_2\sqrt{1-\rho^2} \, dB_3(t)$$

This is a linear combination of two independent Brownian motions which can be replaced by $\sigma B_4(t)$ where

$$\sigma^2 \overset{\text{def}}{=} [\rho\sigma_2 - \sigma_1]^2 + [\sigma_2\sqrt{1-\rho^2}]^2 = \sigma_1^2 - 2\rho\sigma_1\sigma_2 + \sigma_2^2$$

so

$$\frac{dZ}{Z} = m \, dt + \sigma \, dB_4(t) = \sigma\left[\frac{m}{\sigma} \, dt + dB_4(t)\right]$$

The Girsanov transformation $dB_5(t) \overset{\text{def}}{=} (m/\sigma) \, dt + dB_4(t)$ turns this into the driftless SDE $dZ(t)/Z(t) = \sigma \, dB_5(t)$. Thus $\varphi = -\sigma$.

Method (ii) Ratio   Solving the SDEs gives

$$\frac{N_1(t)}{N_1(0)} = e^{(\mu_1 - \frac{1}{2}\sigma_1^2)t + \sigma_1 B_1(t)}$$

$$\frac{N_2(t)}{N_2(0)} = e^{(\mu_2 - \frac{1}{2}\sigma_2^2)t + \sigma_2 B_2(t)}$$

so

$$Z = \frac{e^{(\mu_2 - \frac{1}{2}\sigma_2^2)t + \sigma_2 B_2(t)}}{e^{(\mu_1 - \frac{1}{2}\sigma_1^2)t + \sigma_1 B_1(t)}}$$

$$= e^{[(\mu_2 - \mu_1) - \frac{1}{2}(\sigma_2^2 - \sigma_1^2)]t + \sigma_2 B_2(t) - \sigma_1 B_1(t)}$$

In the exponent, replace $\sigma_2 B_2(t) - \sigma_1 B_1(t)$ by $\sigma B_3(t)$, as in Method (i), giving

$$e^{[(\mu_2 - \mu_1) - \frac{1}{2}(\sigma_2^2 - \sigma_1^2)]t + \sigma B_3(t)}$$

But this is not yet is the standard form $d\mathbb{P}_2(t)/d\mathbb{P}_1(t)$. The first term in the exponent needs to be changed to $-\frac{1}{2}\sigma^2 t$. This can be done by the Girsanov transformation

$$-\frac{1}{2}\sigma^2 t + \sigma B_4(t) \stackrel{\text{def}}{=} \left[(\mu_2 - \mu_1) - \frac{1}{2}\left(\sigma_2^2 - \sigma_1^2\right)\right]t + \sigma B_3(t)$$

or

$$B_4(t) = \left[\frac{(\mu_2 - \mu_1) - \frac{1}{2}\left(\sigma_2^2 - \sigma_1^2\right) + \frac{1}{2}\varphi^2 t}{\sigma}\right]t + \sigma B_3(t)$$

Then $d\mathbb{P}_2(t)/d\mathbb{P}_1(t) = \exp\left[-\frac{1}{2}\sigma^2 t + \sigma B_4(t)\right]$ as before.

8.6   CHANGE OF NUMERAIRE IN LIBOR MARKET MODEL

This section illustrates the use of the ratio property of the Radon–Nikodym derivative in the interest rate model known as the LIBOR market model. LIBOR is a forward interest rate quoted at present time t for the future period $[T_i, T_i + \delta]$, and is denoted here as $L(t, T_i)$. It is defined from the prices of non-defaultable zero-coupon bonds as $P(t, T_i)/P(t, T_{i+1}) = 1 + \delta L(t, T_i)$ where $P(t, T_i)$ denotes the time-t price of a bond which matures at T_i. In this model there are pairs of driftless SDEs by which the LIBORs for successive periods are related. For the purpose of this discussion these are given, and there is no need to know how they were derived. The dynamics for the L that is in effect in the period $[T_N, T_{N+1}]$ is $dL(t, T_N)/L(t, T_N) = \lambda_N \, dW_{N+1}$ where W denotes Brownian motion. The L for the preceding period $[T_{N-1}, T_N]$ is specified under a different Brownian motion by $dL(t, T_{N-1})/L(t, T_{N-1}) = \lambda_{N-1} \, dW_N$. Each L is a martingale under its own probability distribution. Each W is indexed by the end date of the period during which it is in force. The aim is to express the SDE for the earlier L in terms of the W_{N+1} of the later L (the N index is used backwards). The link is the Girsanov transformation

$$dW_N(t) \stackrel{\text{def}}{=} dW_{N+1}(t) - \varphi(t) \, dt$$

The corresponding Radon–Nikodym derivative is thus

$$Z(t) = \exp\left[\int_{s=0}^{t} \varphi(s) \, dW_{N+1}(s) - \tfrac{1}{2} \int_{s=0}^{t} \varphi(s)^2 \, ds \right]$$

where random process φ has to be determined. This can be done by equating two expressions for Z. One expression is the SDE

$$\frac{dZ(t)}{Z(t)} = \varphi(t) \, dW_{N+1}(t)$$

The other expression follows from Z being the ratio of numeraires

$$Z(t) = \frac{P(t, T_N)/P(0, T_N)}{P(t, T_{N+1})/P(0, T_{N+1})} = \frac{P(0, T_{N+1})}{P(0, T_N)} \frac{P(t, T_N)}{P(t, T_{N+1})}$$

Expressing $P(t, T_N)/P(t, T_{N+1})$ in terms of L via its definition $L(t, T_N) = \tfrac{1}{\delta}[P(t, T_N)/P(t, T_{N+1}) - 1]$ gives

$$\frac{P(t, T_N)}{P(t, T_{N+1})} = 1 + \delta L(t, T_N)$$

Then

$$Z(t) = \frac{P(0, T_{N+1})}{P(0, T_N)}[1 + \delta L(t, T_N)]$$

where $P(0, T_{N+1})/P(0, T_N)$ is an observable constant. So

$$dZ(t) = \frac{P(0, T_{N+1})}{P(0, T_N)} \delta \, dL(t, T_N)$$

To express this in the form dZ/Z, $P(0, T_{N+1})/P(0, T_N)$ is expressed in terms of Z

$$\frac{P(0, T_{N+1})}{P(0, T_N)} = \frac{Z(t)}{1 + \delta L(t, T_N)}$$

Substituting this, and using the SDE for $L(t, T_N)$ gives

$$dZ(t) = \frac{Z(t)}{1 + \delta L(t, T_N)} \delta \lambda_N L(t, T_N) \, dW_{N+1}(t)$$

$$\frac{dZ(t)}{Z(t)} = \frac{\delta L(t, T_N)}{1 + \delta L(t, T_N)} \lambda_N \, dW_{N+1}(t)$$

Equating the two expressions for $dZ(t)/Z(t)$ then gives

$$\varphi(t) = \frac{\delta L(t, T_N)}{1 + \delta L(t, T_N)} \lambda_N$$

which is a *random process*. In *Björk*, Chapter 25, it is argued that this satisfies the Novikov condition that was mentioned in Section 7.4. So

$$dW_N(t) = dW_{N+1}(t) - \frac{\delta L(t, T_N)}{1 + \delta L(t, T_N)} \lambda_N \, dt$$

Similarly, by changing N to $N - 1$

$$dW_{N-1}(t) = dW_N(t) - \frac{\delta L(t, T_{N-1})}{1 + \delta L(t, T_{N-1})} \lambda_{N-1} \, dt$$

The following sequence of transformations then results in the dynamics of all LIBORs being expressed in terms of one driving Brownian motion, namely the terminal W_{N+1}.

Start with

$$\frac{dL(t, T_N)}{L(t, T_N)} = \lambda_N \, dW_{N+1}$$

Expressing

$$\frac{dL(t, T_{N-1})}{L(t, T_{N-1})} = \lambda_{N-1} \, dW_N$$

in terms of dW_{N+1} by substituting

$$dW_N = dW_{N+1} - \frac{\delta L(t, T_N)}{1 + \delta L(t, T_N)} \lambda_N \, dt$$

gives

$$\frac{dL(t, T_{N-1})}{L(t, T_{N-1})} = \lambda_{N-1} \left[dW_{N+1} - \frac{\delta L(t, T_N)}{1 + \delta L(t, T_N)} \lambda_N \, dt \right]$$

$$= -\frac{\delta L(t, T_N)}{1 + \delta L(t, T_N)} \lambda_N \lambda_{N-1} \, dt + \lambda_{N-1} \, dW_{N+1}$$

Now expressing the preceding

$$\frac{dL(t, T_{N-2})}{L(t, T_{N-2})} = \lambda_{N-2} \, dW_{N-1}$$

in terms of dW_{N+1} by substituting

$$dW_{N-1} = dW_N - \frac{\delta L(t, T_{N-1})}{1 + \delta L(t, T_{N-1})} \lambda_{N-1} \, dt$$

and

$$dW_N = dW_{N+1} - \frac{\delta L(t, T_N)}{1 + \delta L(t, T_N)} \lambda_N \, dt$$

gives

$$dW_{N-1} = dW_{N+1} - \frac{\delta L(t, T_N)}{1 + \delta L(t, T_N)} \lambda_N \, dt - \frac{\delta L(t, T_{N-1})}{1 + \delta L(t, T_{N-1})} \lambda_{N-1} \, dt$$

$$\frac{dL(t, T_{N-2})}{L(t, T_{N-2})} = \lambda_{N-2} \left[dW_{N+1} - \frac{\delta L(t, T_N)}{1 + \delta L(t, T_N)} \lambda_N \, dt \right.$$

$$\left. - \frac{\delta L(t, T_{N-1})}{1 + \delta L(t, T_{N-1})} \lambda_{N-1} \, dt \right]$$

$$= \lambda_{N-2} \left[-\frac{\delta L(t, T_{N-1})}{1 + \delta L(t, T_{N-1})} \lambda_{N-1} \right.$$

$$\left. - \frac{\delta L(t, T_N)}{1 + \delta L(t, T_N)} \lambda_N \right] dt + \lambda_{N-2} \, dW_{N+1}$$

$$= -\sum_{i=N-1}^{N} \frac{\delta L(t, T_i)}{1 + \delta L(t, T_i)} \lambda_i \lambda_{N-2} \, dt + \lambda_{N-2} \, dW_{N+1}$$

The general expression is the system of coupled SDEs

$$\frac{dL(t, T_j)}{L(t, T_j)} = -\lambda_j \sum_{i=j+1}^{N} \frac{\delta L(t, T_i)}{1 + \delta L(t, T_i)} \lambda_i \, dt + \lambda_j \, dW_{N+1}$$

It is not possible to find an analytical expression for L. Their values can only be obtained by simulation. Key references for this are *Efficient Methods for Valuing Interest Rate Derivatives* by *Pelsser* Chapter 8, and *Glasserman*.

8.7 APPLICATION IN CREDIT RISK MODELLING

In credit risk modelling, λ is the rate with which defaults on payment obligations occur. Empirical evidence suggests that λ and interest rate r are not independent. Deriving various characterizations of the default situation then requires a change to the forward probability. Let the interest rate process r and the default intensity process λ be specified as

$$dr(t) = k_r[\bar{r} - r(t)] \, dt + \sigma_r \, dW_r(t)$$
$$d\lambda(t) = k_\lambda[\bar{\lambda} - \lambda(t)] \, dt + \sigma_\lambda \, dW_\lambda(t)$$

where k_r, $\bar{r}$, σ_r, k_λ, $\bar{\lambda}$, σ_λ, are given constants. The driving Brownian motions $W_r(t)$ and $W_\lambda(t)$ have correlation coefficient ρ, and are under a probability measure $\widehat{\mathbb{P}}$ that corresponds to using the money market account M as numeraire. The dynamics of M are given by $dM(t) = r(t)M(t) \, dt$, $M(0) = 1$, $M(t) = \exp[\int_{s=0}^{t} r(s) \, ds]$. Now take the bond price $P(t, T)$ as the numeraire. As specified earlier, the bond price dynamics under M as numeraire has drift term $r(t) \, dt$, and its driving Brownian motion is the same as the one for r, so it is under the same $\widehat{\mathbb{P}}$. Corresponding to numeraire $P(t, T)$ is a probability measure, known as the T-forward probability measure, here denoted $\mathbb{P}^T$. What will now be derived is

(i) the expression for the Radon–Nikodym that is needed to create $\mathbb{P}^T$
(ii) the SDE for r and for λ under $\mathbb{P}^T$.

(i) The Radon–Nikodym derivative, here denoted Z, is the ratio of the numeraires scaled by their respective initial values,

$$Z(t) = \frac{P(t, T)/P(0, T)}{M(t)/M(0)}$$

Using

$$\frac{P(t, T)}{P(0, T)} = \exp\left\{ \int_{s=0}^{t} [r(s) - \tfrac{1}{2}\sigma_p(s)^2] \, ds + \int_{s=0}^{t} \sigma_p(s) \, dW_r(t) \right\}$$

and

$$\frac{M(t)}{M(0)} = \exp\left[\int_{s=0}^{t} r(s)\,ds\right]$$

gives

$$Z(t) = \exp\left[-\frac{1}{2}\int_{s=0}^{t} \sigma_p(s)^2\,ds + \int_{s=0}^{t} \sigma_p(s)\,dW_r(t)\right]$$

This is the Radon–Nikodym derivative for the Girsanov transformation $dW^T(t) \overset{\text{def}}{=} dW_r(t) - \sigma_p(s)\,dt$ where $W^T(t)$ denotes a Brownian motion under $\mathbb{P}^T$.

An alternative way to find this is to derive a driftless SDE for Z. Write

$$Z = \frac{P(t,T)}{M(t)}\frac{1}{P(0,T)}$$

where the last term is observable in the market at time 0, hence constant, say c. Letting $F(t) \overset{\text{def}}{=} P(t,T)/M(t)$, and applying Itō's formula gives, after simplification,

$$\frac{dF}{F} = \frac{dP}{P} - \frac{dM}{M} = r(t)\,dt + \sigma_p(t)\,dW_r(t) - r(t)\,dt$$
$$= \sigma_p(t)\,dW_r(t)$$

and

$$\frac{dZ}{Z} = \frac{d(Fc)}{Fc} = \frac{d(F)c}{Fc} = \frac{dF}{F} = \sigma_p(t)\,dW_r(t)$$

For $dW^T(t) \overset{\text{def}}{=} dW_r(t) + \varphi\,dt$, the corresponding dZ/Z is $-\varphi\,dW_r(t)$, thus here $\varphi = -\sigma_p(t)$.

(ii) To get the SDE of r under $\mathbb{P}^T$, substitute $dW_r(t) = dW^T(t) + \sigma_p(t)\,dt$, giving

$$dr(t) = k_r[\bar{r} - r(t)]\,dt + \sigma_r[dW^T(t) + \sigma_p(t)\,dt]$$
$$= \{k_r[\bar{r} - r(t)] + \sigma_r\sigma_p(t)\}\,dt + \sigma_r\,dW^T(t)$$

The drift coefficient has increased by $\sigma_r\sigma_p(t)$.

For the SDE of λ, first deal with the correlation by writing

$$W_\lambda(t) \overset{\text{def}}{=} \rho W_r(t) + \sqrt{1-\rho^2}\,W(t)$$

where W is a new independent Brownian motion under $\widehat{\mathbb{P}}$. Substituting this gives the dynamics under $\widehat{\mathbb{P}}$ as

$$d\lambda(t) = k_\lambda[\bar{\lambda} - \lambda(t)] \, dt + \sigma_\lambda[\rho \, dW_r(t) + \sqrt{1 - \rho^2} \, dW(t)]$$
$$= k_\lambda[\bar{\lambda} - \lambda(t)] \, dt + \sigma_\lambda \rho \, dW_r(t) + \sigma_\lambda \sqrt{1 - \rho^2} \, dW(t)$$

Now substitute $dW_r(t) \stackrel{\text{def}}{=} dW^T(t) + \sigma_p(t) \, dt$, giving

$$d\lambda(t) = k_\lambda[\bar{\lambda} - \lambda(t)] \, dt + \sigma_\lambda \rho [dW^T(t) + \sigma_p(t) \, dt] +$$
$$+ \sigma_\lambda \sqrt{1 - \rho^2} \, dW(t)$$
$$= \{k_\lambda[\bar{\lambda} - \lambda(t)] + \rho\sigma_\lambda\sigma_p(t)\} \, dt + \sigma_\lambda \rho \, dW^T(t) +$$
$$+ \sigma_\lambda \sqrt{1 - \rho^2} \, dW(t)$$

The two Brownian motions W^T and W are independent and can be combined into a new independent Brownian motion W^* by $\sigma \, dW^*(t) \stackrel{\text{def}}{=} \sigma_\lambda \rho \, dW^T(t) + \sigma_\lambda \sqrt{1 - \rho^2} \, dW(t)$ where

$$\sigma^2 = (\sigma_\lambda \rho)^2 + (\sigma_\lambda \sqrt{1 - \rho^2})^2$$
$$= \sigma_\lambda^2 \rho^2 + \sigma_\lambda^2(1 - \rho^2)$$
$$= \sigma_\lambda^2$$

so

$$d\lambda(t) = \{k_\lambda[\bar{\lambda} - \lambda(t)] + \rho\sigma_\lambda\sigma_p(t)\} \, dt + \sigma_\lambda \, dW^*(t)$$

As $W^*(t)$ has the same probability density as $W^T(t)$, both new SDEs can be written under the new Brownian motion W^T as

$$dr(t) = \{k_r[\bar{r} - r(t)] + \sigma_r\sigma_p(s)\} \, dt + \sigma_r \, dW^T(t)$$
$$d\lambda(t) = \{k_\lambda[\bar{\lambda} - \lambda(t)] + \rho\sigma_\lambda\sigma_p(t)\} \, dt + \sigma_\lambda \, dW^T(t)$$

8.8   EXERCISES

[8.8.1] For the exchange option in Section 8.3, verify $c(0)$ by evaluating $\widetilde{\mathbb{E}}$ as in Section 6.5.

[8.8.2] For the option on a bond in Section 8.3, verify $c(0)$ by evaluating $\widetilde{\mathbb{E}}$ as in Section 6.5.

[8.8.3] For the European call under a stochastic interest rate in Section 8.3, verify $c(0)$ by evaluating $\widetilde{\mathbb{E}}$ as in Section 6.5.

8.9   SUMMARY

This chapter showed how options can be valued by the use of a convenient numeraire that is different from the traditional money market account. Each numeraire induces a corresponding probability distribution, under which asset values are martingales. It was shown that the Radon–Nikodym derivative for changing from one probability distribution to another can be expressed as the ratio of the respective numeraires. Option values V can generally be expressed under numeraire N in the form

$$\frac{V(t)}{N(t)} = \mathbb{E}_{\mathbb{P}^N}\left[\frac{V(T)}{N(T)}\right]$$

where $\mathbb{E}_{\mathbb{P}^N}$ denotes expectation under the probability distribution under which the process $V(t)/N(t)$ is a martingale. Evaluation of this expectation can be readily done by analogy to the standard Black–Scholes formula. Further examples of the use of a change of numeraire are given in the following journal articles:

- Sundaram: 'Equivalent Martingale Measures and Risk-Neutral Pricing: An Expository Note', *The Journal of Derivatives* 1997, Fall, pp. 85–98.
- Benninga/Björk/Wiener: 'On the Use of Numeraires in Option Pricing', *The Journal of Derivatives* 2002, Winter, pp. 85–98.

Annexes

—— Annex A ——
Computations with
Brownian Motion ——

A.1 MOMENT GENERATING FUNCTION AND MOMENTS OF BROWNIAN MOTION

The moment generating function (mgf) of a random variable X is defined as the expected value of an exponential function of X, $\mathbb{E}[\exp(\theta X)]$, where θ (or any other letter) is a dummy parameter. This expectation is a function of θ which transforms the probability distribution of the random variable into a function of θ. It can be used to find the moments of the random variable, $\mathbb{E}[X^k]$, for $k = 1, 2, \ldots$ as will be shown below. But what makes it really useful is the property that the mgf of a random variable is *unique*. So if in an analysis one has a particular random variable for which the probability distribution is not known, then that distribution can be sometimes determined by deriving its mgf and comparing it to the mgf of known distributions. This is used in Chapters 2 and 4. That property is also the key in showing the convergence in distribution of a sequence of independent identically distributed random variables.

The moment property is obtained by using the power series expansion

$$\exp(y) = 1 + y + \frac{1}{2!}y^2 + \cdots + \frac{1}{k!}y^k + \cdots$$

giving

$$\mathbb{E}[\exp(\theta X)] = \mathbb{E}\left[1 + \theta X + \frac{1}{2!}(\theta X)^2 + \cdots + \frac{1}{k!}(\theta X)^k + \cdots\right]$$

$$= 1 + \theta\mathbb{E}[X] + \frac{1}{2!}\theta^2\mathbb{E}[X^2] + \cdots + \frac{1}{k!}\theta^k\mathbb{E}[X^k] + \cdots$$

The terms have the moments of random variable X. The kth moment can be singled out by differentiating k times with respect to θ, and interchanging the differentiation and $\mathbb{E}$ (which is an integration)

$$\frac{d^k}{d\theta^k}\left\{\frac{1}{k!}\theta^k\mathbb{E}[X^k]\right\} = \frac{1}{k!}\theta^k\frac{d^k}{d\theta^k}\mathbb{E}[X^k]$$

That reduces the first k terms to 0, and leaves a power series starting with

$$\frac{1}{k!}k(k-1)(k-2)\cdots(2)1\mathbb{E}[X^k] = \frac{1}{k!}k!\mathbb{E}[X^k] = \mathbb{E}[X^k]$$

All higher terms have coefficients which are powers of θ. By setting $\theta = 0$, these all disappear, and the result is that the kth moment of X is the kth derivative of the mgf of X, at $\theta = 0$

$$\mathbb{E}[X^k] = \frac{d^k \text{ mgf of } X}{d\theta^k}\Big|_{\theta=0}$$

For a vector of random variables $X = (X_1, \ldots, X_n)$ there is the *multivariate mgf* which uses the vector $\theta = (\theta_1, \ldots, \theta_n)$

$$\mathbb{E}[\exp(\theta_1 X_1 + \cdots + \theta_n X_n)]$$

Mgf of Brownian motion

The mgf of the Brownian motion position at time t is

$$\mathbb{E}[\exp(\theta B(t))] = \int_{x=-\infty}^{\infty} \exp(\theta x)\frac{1}{\sqrt{t}\sqrt{2\pi}}\exp\left[-\frac{1}{2}\left(\frac{x}{\sqrt{t}}\right)^2 dx\right]$$

where x is a value of $B(t)$

Change of variable $z \overset{\text{def}}{=} x/\sqrt{t}$ gives

$$\mathbb{E}[\exp(\theta B(t))] = \int_{z=-\infty}^{\infty} \exp(\theta\sqrt{t}z)\frac{1}{\sqrt{2\pi}}\exp\left(-\frac{1}{2}z^2\right) dz$$

Rearranging exponent $\theta\sqrt{t}z - \frac{1}{2}z^2$ to $-\frac{1}{2}(z - \theta\sqrt{t})^2 + \frac{1}{2}\theta^2 t$ gives

$$\exp\left(\frac{1}{2}\theta^2 t\right)\int_{z=-\infty}^{\infty}\frac{1}{\sqrt{2\pi}}\exp\left(-\frac{1}{2}(z - \theta\sqrt{t})^2\right) dz$$
where $\int$ = area under a normal density = 1

Thus the mgf of $B(t)$ is

$$\mathbb{E}[\exp(\theta B(t))] = \exp\frac{1}{2}\theta^2 t$$

Moments of Brownian Motion

The *first moment* of Brownian motion is

$$\mathbb{E}[B(t)] = \frac{d}{d\theta}\exp\left(\frac{1}{2}\theta^2 t\right) = \exp\left(\frac{1}{2}\theta^2 t\right)\theta t \text{ at } \theta = 0,$$

which is 0, as already known. The *second moment* of Brownian motion is

$$\mathbb{E}[B(t)^2] = \frac{d}{d\theta} \text{ (first moment) at } \theta = 0$$

$$\frac{d}{d\theta}(\text{first moment}) = \exp\left(\frac{1}{2}\theta^{2t}\right) t + \exp\left(\frac{1}{2}\theta^{2t}\right)\theta t \theta t$$

$$= \exp\left(\frac{1}{2}\theta^{2t}\right)[t + (\theta t)^2]$$

Substituting $\theta = 0$ gives $\mathbb{E}[B(t)^2] = t$, as already known. The *third moment* of Brownian motion is

$$\mathbb{E}[B(t)^3] = \frac{d}{d\theta}(\text{second moment}) \text{ at } \theta = 0$$

$$\frac{d}{d\theta}(\text{second moment}) = \exp\left(\frac{1}{2}\theta^2 t\right)\theta t[t + (\theta t)^2] + \exp\left(\frac{1}{2}\theta^2 t\right)2\theta t^2$$

$$= \exp\left(\frac{1}{2}\theta^2 t\right)[3\theta t^2 + (\theta t)^2]$$

Substituting $\theta = 0$ gives $\mathbb{E}[B(t)^3] = 0$. The *fourth moment* of Brownian motion is

$$\mathbb{E}[B(t)^4] = 3t^2 = \frac{d}{d\theta}(\text{third moment}) \text{ at } \theta = 0.$$

$$\frac{d}{d\theta}(\text{third moment}) = \exp\left(\frac{1}{2}\theta^2 t\right)\theta t\left[3\theta t^2 + (\theta t)^2\right]$$

$$+ \exp\left(\frac{1}{2}\theta^2 t\right)\left[3t^2 + 2\theta t^2\right]$$

Substituting $\theta = 0$ gives $\mathbb{E}[B(t)^4] = 3t^2$.
The kth *moment* of $B(t)$ is

$$\mathbb{E}[B(t)^k] = \int_{x=-\infty}^{\infty} x^k \frac{1}{\sqrt{t}\sqrt{2\pi}} \exp\left[-\frac{1}{2}\left(\frac{x}{\sqrt{t}}\right)^2\right] dx$$

For odd k this is zero, as positive values x^k are cancelled against the corresponding negative values x^k in the summation. Moments for even k can be expressed in terms of second moments. Consequently the mean and variance completely characterize the probability distribution of a

Brownian motion (as was already known since this is the case for any normal distribution).

A.2   PROBABILITY OF BROWNIAN MOTION POSITION

The position of a Brownian motion at time t is described in probabilistic terms by

$$\mathbb{P}[B(t) \leq a] = \int_{x=-\infty}^{a} \frac{1}{\sqrt{t}\sqrt{2\pi}} \exp\left[-\frac{1}{2}\left(\frac{x}{\sqrt{t}}\right)^{2}\right] dx$$

There exists no closed form expression for this integral but manual computations can be readily carried out by transforming to the probability distribution of a standard normal, for which numerical tables are included in elementary books on probability. This is achieved by the change of variable $z \overset{\text{def}}{=} x/\sqrt{t}$ so $dz = dx/\sqrt{t}$. Lower integration limit $x = -\infty$ becomes $z = -\infty$, upper integration limit $x = a$ becomes $z = a/\sqrt{t}$, and

$$\mathbb{P}[B(t) \leq a] = \int_{x=-\infty}^{a/\sqrt{t}} \frac{1}{\sqrt{2\pi}} \exp[-\tfrac{1}{2}z^{2}] dz$$

As an example compute the probability that at time $t = 0.81$ the position of the Brownian motion is no greater than 0.25. It is not relevant whether the motion has been above this level at a prior time. Upper integration limit $a/\sqrt{t} = 0.25/\sqrt{0.81} \cong 0.28$. Looking up tables gives $\mathbb{P}[B(t) \leq 0.25] \cong 0.6103$. To get greater precision requires interpolation on the table. There is an easier way. Spreadsheets have a built-in normal probability distribution for which the mean and the standard deviation can be specified. The answer is 0.609408. Computations of this kind only require elementary probability theory. There exists fast numerical algorithms for approximating the normal probability distribution which can be used when the normal distribution is needed in a computer program.

A.3   BROWNIAN MOTION REFLECTED AT THE ORIGIN

This random process is the absolute value of a standard Brownian motion, $Z(t) \overset{\text{def}}{=} |B(t)|$. Whenever the Brownian motion position is negative, $Z(t)$ has a positive value of the same magnitude. The value of $Z(t)$ is

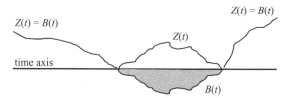

Figure A.1   Reflected Brownian motion

the reflection of $B(t)$ about the time axis whenever $B(t)$ is negative, (see Figure A.1).

For a new random process one usually wants to determine its probability distribution, its probability density, the value its takes on average, and the degree of variability about this average. To analyse a newly defined process, give it a name, say $Z(t)$, so here $Z(t) \overset{def}{=} |B(t)|$. Then determine what values the new process can take. Here $B(t)$ can take any positive or negative values, but Z can only take non-negative values. To determine the probability distribution of $Z(t)$ at time t, the expression $\mathbb{P}[Z(t) \leq z]$ has to be evaluated. The first step is to substitute the expression for the new process, $\mathbb{P}[Z(t) \leq z] = \mathbb{P}[|B(t)| \leq z]$. As the probability distribution of $B(t)$ is known, the next step is to rearrange the above expression in terms of this known underlying probability distribution. Once the probability distribution of $Z(t)$ is known, the corresponding probability density follows by differentiation.

Probability Distribution

The event $|B(t)| \leq z$ means that the values of $B(t)$ lie in the interval from $-z$ to z. This is the same as the event of lying to the left of z, and the event of not lying to the left of $-z$, as illustrated in Figure A.2.

Thus $\mathbb{P}[|B(t)| \leq z] = \mathbb{P}[B(t) \leq z] - \mathbb{P}[B(t) \leq -z]$. As the density of $B(t)$ is symmetric about 0, $\mathbb{P}[B(t) \leq -z] = \mathbb{P}[B(t) \geq z]$, which equals $1 - \mathbb{P}[B(t) \leq z]$. So

$$\mathbb{P}[|B(t)| \leq z] = \mathbb{P}[B(t) \leq z] - \{1 - \mathbb{P}[B(t) \leq z]\} = 2\mathbb{P}[B(t) \leq z] - 1$$

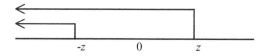

Figure A.2   Event diagram

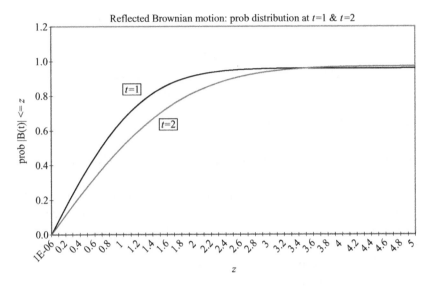

Figure A.3   Probability distribution of reflected Brownian motion

The end result (see Figure A.3) is

$$\mathbb{P}[Z(t) \leq z] = \mathbb{P}[|B(t)| \leq z] = 2\int_{x=-\infty}^{z} \frac{1}{\sqrt{t}\sqrt{2\pi}} \exp\left[-\frac{1}{2}\left(\frac{x}{\sqrt{t}}\right)^2\right] dx - 1$$

Probability Density

To derive the probability density of $Z(t)$, differentiate its probability distribution with respect to z. This gives the density of $|B(t)|$ as

$$2\frac{1}{\sqrt{t}\sqrt{2\pi}} \exp\left[-\frac{1}{2}\left(\frac{z}{\sqrt{t}}\right)^2\right]$$

which is double the density of $B(t)$ (see Figure A.4).

Expected Value

$$\mathbb{E}[|B(t)|] = \int_{z=0}^{\infty} z2\frac{1}{\sqrt{t}\sqrt{2\pi}} \exp\left[-\frac{1}{2}\left(\frac{z}{\sqrt{t}}\right)^2\right] dz$$

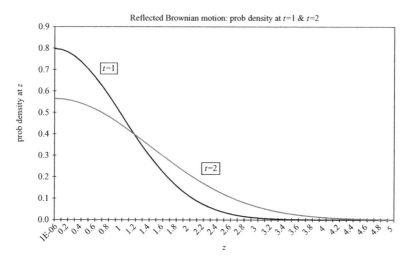

Figure A.4 Probability density of reflected Brownian motion

Note that the lower integration limit is 0 since $|B(t)|$ is non-negative. Change to the new variable $y \overset{\text{def}}{=} z/\sqrt{t}$. Then $z = \sqrt{t}\,y$ and $dz = \sqrt{t}\,dy$. The corresponding integration limits are: $z = 0$ gives $y = 0$ and $z = \infty$ gives $y = \infty$. Thus

$$\mathbb{E}[|B(t)|] = \int_{y=0}^{\infty} \sqrt{t}\,y 2 \frac{1}{\sqrt{2\pi}} \exp\left[-\tfrac{1}{2}y^2\right] dy$$

$$= 2\sqrt{t} \int_{y=0}^{\infty} y \frac{1}{\sqrt{2\pi}} \exp\left[-\tfrac{1}{2}y^2\right] dy$$

Knowing that $\int \exp(w)\,dw = \exp(w)$, use the y in the integrand to create an integrator which is the same as the exponent, namely $-\tfrac{1}{2}y^2$. That gives

$$-2\sqrt{t} \int_{y=0}^{\infty} \frac{1}{\sqrt{2\pi}} \exp\left[-\frac{1}{2}y^2\right] d\left(-\frac{1}{2}y^2\right)$$

which can be integrated, giving

$$-2\sqrt{t}\frac{1}{\sqrt{2\pi}} \exp\left[-\frac{1}{2}y^2\right]\Bigg|_{y=0}^{\infty} = -2\sqrt{t}\frac{1}{\sqrt{2\pi}}(0 - 1).$$

Thus,

$$\mathbb{E}[|B(t)|] = \sqrt{\frac{2t}{\pi}} \approx 0.8\sqrt{t}$$

compared to $\mathbb{E}[B(t)] = 0$. Symbolic integration in Mathematica confirms this result.

Variance

$$\mathbb{V}ar[Z(t)] = \mathbb{E}[Z(t)^2] - \{\mathbb{E}[Z(t)]\}^2$$

$$= \mathbb{E}[Z(t)^2] - \frac{2t}{\pi}$$

The first term

$$\mathbb{E}[Z(t)^2] = \int_{z=0}^{\infty} z^2 2 \frac{1}{\sqrt{t}\sqrt{2\pi}} \exp\left[-\frac{1}{2}\left(\frac{z}{\sqrt{t}}\right)^2\right] dz.$$

Change of variable $y = z/\sqrt{t}$, $z = \sqrt{t}y$, $z^2 = ty^2$, $dz = \sqrt{t}\,dy$, gives

$$\mathbb{E}[Z(t)^2] = \int_{y=0}^{\infty} ty^2 2 \frac{1}{\sqrt{2\pi}} \exp\left[-\frac{1}{2}y^2\right] dy$$

$$= 2t \int_{y=0}^{\infty} y^2 \frac{1}{\sqrt{2\pi}} \exp\left[-\frac{1}{2}y^2\right] dy$$

$$= -2t \int_{y=0}^{\infty} y \frac{1}{\sqrt{2\pi}} d\left(\exp\left[-\frac{1}{2}y^2\right]\right)$$

This integral is of the form $\int_{y=0}^{\infty} g(y)\,df(y)$, where $g(y) = 1/\sqrt{2\pi}$ and $f(y) = \exp[-\frac{1}{2}y^2]$. Partial integration[1] gives

$$\int_{y=0}^{\infty} g(y)\,df(y) = g(y)f(y)|_{y=0}^{\infty} - \int_{y=0}^{\infty} f(y)\,dg(y)$$

with

$$dg(y) = d\left(y\frac{1}{\sqrt{2\pi}}\right) = \frac{1}{\sqrt{2\pi}}dy$$

Substituting $f(y)$, $g(y)$ and $dg(y)$ then gives

$$\mathbb{E}[Z(t)^2] = -2ty\frac{1}{\sqrt{2\pi}} \exp[-\tfrac{1}{2}y^2]|_{y=0}^{\infty} + 2t \int_{y=0}^{\infty} \frac{1}{\sqrt{2\pi}} \exp[-\tfrac{1}{2}y^2]\,dy$$

---

[1] See Annex B, *Ordinary Integration*.

The integral has value 0.5 since it covers half the area under the standard normal density. So

$$\mathbb{E}[Z(t)^2] = 0 + 2t(0.5) = t$$

The final result is

$$\mathbb{V}ar[|B(t)|] = \mathbb{E}[Z(t)^2] - \frac{2t}{\pi} = t - \frac{2t}{\pi} = t(1 - \frac{2}{\pi}) \approx 0.36t$$

compared to $\mathbb{V}ar[B(t)] = t$. Symbolic integration in Mathematica confirms this. Note that the second moment of the reflected Brownian motion, $\mathbb{E}[Z(t)^2]$, is the same as that of a standard Brownian motion, because $Z(t)^2$ is the same as $B(t)^2$. But the mean increases non-linearly with t. This results in the variance of the reflected Brownian motion only being just over one-third of the variance of a standard Brownian motion.

Simulation of Reflected Brownian Motion

A sample path is shown in Figure A.5.

A batch of 1000 simulations of a reflected Brownian motion over the time period [0, 1] gave the statistics shown in Figure A.6 for the position at time 1.

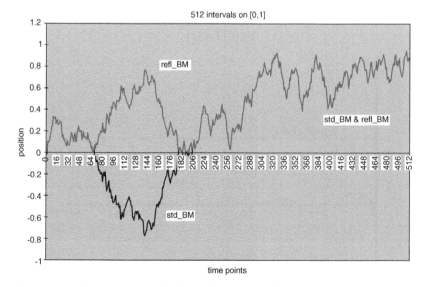

Figure A.5   Simulated path of Reflected Brownian motion

reflected Brownian motion 1000 runs		
	sample	exact
mean	0.827050	0.7978846
variance	0.340505	0.363380

Figure A.6   Simulation statistics

A.4   FIRST PASSAGE OF A BARRIER

Suppose there is a barrier situated at a positive level L. The time at which a Brownian motion first reaches this barrier is a random variable, which will be denoted T_L. This time is called the first passage time or hitting time. The event $T_L \leq t$ says that the barrier (Figure A.7) was reached by no later than time t. Its probability distribution $\mathbb{P}[T_L \leq t]$ will now be derived.

In order to make use of the known probability distribution of $B(t)$, the probability distribution of T_L is computed by considering the probability distribution of $B(t) \geq L$, which is the event that the position of the Brownian motion at time t is above L, and conditioning on whether or not $T_L \leq t$. The events $T_L \leq t$ and $T_L > t$ are mutually exclusive, and there is no other possibility, so $\mathbb{P}[B(t) \geq L]$ can be written as

$$\mathbb{P}[B(t) \geq L] = \mathbb{P}[B(t) \geq L | T_L \leq t]\mathbb{P}[T_L \leq t] + \mathbb{P}[B(t) \geq L | T_L > t]\mathbb{P}[T_L > t]$$

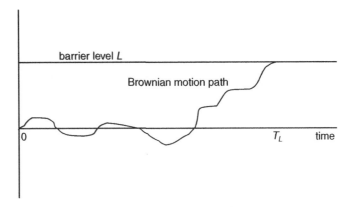

Figure A.7   Barrier diagram

where the condition is specified behind the vertical divider. But the event $B(t) \geq L$ that at time t the motion is above the barrier can only have happened if $T_L \leq t$. Thus the second term is zero. So $\mathbb{P}[B(t) \geq L] = \mathbb{P}[B(t) \geq L|T_L \leq t]\mathbb{P}[T_L \leq t]$. If the Brownian motion has reached the barrier at some time prior to t, then where is it at time t? It can be either above the barrier, which is the event $B(t) \geq L$, or below the barrier, which is the event $B(t) < L$. The key observation now is that *both events are equally likely*. This symmetry gives $\mathbb{P}[B(t) \geq L|T_L \leq t] = \frac{1}{2}$. Thus $\mathbb{P}[B(t) \geq L] = \frac{1}{2}\mathbb{P}[T_L \leq t]$ and the probability of interest, $\mathbb{P}[T_L \leq t]$, equals $2\mathbb{P}[B(t) \geq L]$. As $B(t)$ is normally distributed, the probability on the right can be readily written down as

$$\mathbb{P}[B(t) \geq L] = \int_{x=L}^{\infty} \frac{1}{\sqrt{t}\sqrt{2\pi}} \exp\left[-\frac{1}{2}\left(\frac{x}{\sqrt{t}}\right)^2\right] dx$$

There exists no closed form expression for this integral. Write it in terms of the standard normal density by the transformation $z = x/\sqrt{t}$. This gives

$$\int_{z=L/\sqrt{t}}^{\infty} \frac{1}{\sqrt{2\pi}} \exp[-\frac{1}{2}z^2] dz$$

The end result is

$$\mathbb{P}[T_L \leq t] = 2\mathbb{P}[B(t) \geq L] = 2\int_{z=L/\sqrt{t}}^{\infty} \frac{1}{\sqrt{2\pi}} \exp[-\frac{1}{2}z^2] dz$$

$$= 2\left[1 - N\left(\frac{L}{\sqrt{t}}\right)\right]$$

The value for given t and L can be readily found on a spreadsheet.

To get the probability density of random variable T_L, differentiate the above distribution function with respect to t, giving

$$-2\frac{1}{\sqrt{2\pi}} \exp\left[-\frac{1}{2}\left(\frac{L}{\sqrt{t}}\right)^2\right] \frac{d}{dt}\left(\frac{L}{\sqrt{t}}\right) = \frac{L}{\sqrt{2\pi}\sqrt{t^3}} \exp\left[-\frac{1}{2}\left(\frac{L}{\sqrt{t}}\right)^2\right]$$

Simulation of First Passage of a Barrier

With a barrier situated at $L = 0.4$, the probability of a first hit by no later than time 1 is

$$\mathbb{P}[T_L \leq 1] = 2\left[1 - N\left(\frac{0.4}{\sqrt{1}}\right)\right] = 2[1 - 0.6554] = 0.6892$$

From a batch of 500 simulations of a standard Brownian motion, run over the time period [0, 1], 348 hit the barrier by time 1. That is a sample proportion of 0.6960 compared to the exact probability of 0.6892. The average time of first hit in the batch was 0.278163 compared to the exact expected time of first hit of 0.267502.

A.5   ALTERNATIVE BROWNIAN MOTION SPECIFICATION

In some expositions, Brownian motion is introduced by the joint probability distribution that is shown in the beginning of Section 1.7. What follows here is an explanation of how that links to the original specification of the Brownian motion density. Consider the path of a Brownian motion which is at level x at time t, and at level y at time $(t + u)$, see Figure A.8.

The increment over this time interval has the probability distribution

$$\mathbb{P}[B(t + u) - B(t) \leq a] \overset{\text{def}}{=} \int_{w=-\infty}^{a} \frac{1}{\sqrt{u}\sqrt{2\pi}} \exp\left[-\frac{1}{2}\left(\frac{w}{\sqrt{u}}\right)^2\right] dw \tag{1}$$

Convenient notation for the above density is

$$f(u, w) \overset{\text{def}}{=} \frac{1}{\sqrt{u}\sqrt{2\pi}} \exp\left[-\frac{1}{2}\left(\frac{w}{\sqrt{u}}\right)^2\right]$$

where u denotes the time interval and w the corresponding increment. Applying expression (1) to time period $[0, t]$, and using $B(0) = 0$ gives

$$\mathbb{P}[B(t) - B(0) \leq a] = \int_{x=-\infty}^{a} \frac{1}{\sqrt{t}\sqrt{2\pi}} \exp\left[-\frac{1}{2}\left(\frac{x}{\sqrt{t}}\right)^2\right] dx \tag{2}$$

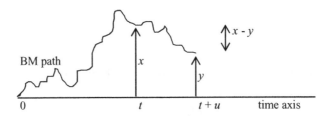

Figure A.8   Successive Brownian motion movements

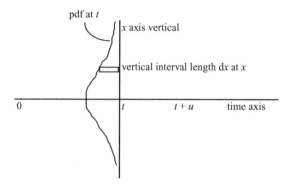

Figure A.9 Probability density function

Similarly for the time period $[0, t + u]$

$$\mathbb{P}[B(t + u) - B(0) \leq a] = \int_{y=-\infty}^{a} \frac{1}{\sqrt{t+u}\sqrt{2\pi}} \exp\left[-\frac{1}{2}\left(\frac{y}{\sqrt{t+u}}\right)^2\right] dy$$

(3)

Note that (2) and (3) are expressions for the position of the motion at different times, whereas (1) is an expression for the increment between these times. Looking at these expressions it is not immediately obvious how (2) and (3) relate to (1). This will now be shown. The expression for the probability density (Figure A.9) is

$$f(t, x) = \text{pdf of (position } x \text{ at time } t \mid \text{if starting position 0 at time 0)}$$

$$= \frac{1}{\sqrt{t}\sqrt{2\pi}} \exp\left[-\frac{1}{2}\left(\frac{x}{\sqrt{t}}\right)^2\right]$$

The expression for the probability of being in a specified interval is

$$\mathbb{P}[\text{position in interval } [x, x + dx] \text{ at time } t \mid \text{if starting position 0}$$
$$\text{at time 0]}$$

$$= f(t, x)dx = \frac{1}{\sqrt{t}\sqrt{2\pi}} \exp\left[-\frac{1}{2}\left(\frac{x}{\sqrt{t}}\right)^2\right] dx$$

$$f(u, y - x) = \text{pdf of (position is } y \text{ at time } (t + u) \mid \text{if starting}$$
$$\text{position } x \text{ at time } t)$$

$$= \frac{1}{\sqrt{u}\sqrt{2\pi}} \exp\left[-\frac{1}{2}\left(\frac{y - x}{\sqrt{u}}\right)^2\right]$$

as the increment is $(y - x)$ over time interval of length u.

$\mathbb{P}$[position in interval $[y, y + dy]$ at time $(t + u)$ | if starting position x at time t]

$$= f(u, y - x)dy = \frac{1}{\sqrt{u}\sqrt{2\pi}} \exp[-\tfrac{1}{2} \left(\tfrac{y-x}{\sqrt{u}}\right)^2] dy$$

As event $B(t + u) - B(t) \leq a$ is determined by the two random variables $B(t + u)$ and $B(t)$, computing $\mathbb{P}[B(t + u) - B(t) \leq a]$ requires the joint pdf of these two random variables. Note that the random variables $B(t + u)$ and $B(t)$, which each give the position of the motion, are not independent. But the random variables which represent the increments, $[B(t) - B(0)]$ and $[B(t + u) - B(t)]$ are independent. Recall that for any two random variables X and Y, $f(y \mid x) = f(x, y)/f_X(x)$ where

$f(y \mid x)$ is the conditional pdf given random variable X has occurred

$f(x, y)$ is the joint pdf of X and Y

$f_X(x)$ is the pdf of X (subscripted); this is a marginal density

The joint pdf can thus be expressed as $f(x, y) = f(y \mid x)f_X(x)$. If X and Y were independent then $f(y \mid x)$ would be the same as the pdf of Y, $f_Y(y)$, and the joint density $f(x, y)$ would be the product of the two densities, $f_Y(y)f_X(x)$. That, however is not the case here. $B(t)$ plays the role of X, $B(t + u)$ the role of Y, and $B(t + u) - B(t) \leq a$ corresponds to $y - x \leq a$ so

$$\mathbb{P}[B(t + u) - B(t) \leq a] = \int_{x=-\infty}^{\infty} \int_{y-x=-\infty}^{a} f(t, x)f(u, y-x)\, dx\, dy$$

$$= \int_{x=-\infty}^{\infty} \int_{y-x=-\infty}^{a} \left\{ \frac{1}{\sqrt{t}\sqrt{2\pi}} \exp\left[-\frac{1}{2}\left(\frac{x}{\sqrt{t}}\right)^2 \right] \right\}$$
$$\times \left\{ \frac{1}{\sqrt{u}\sqrt{2\pi}} \exp\left[-\frac{1}{2}\left(\frac{y - x}{\sqrt{u}}\right)^2 \right] \right\} dx\, dy$$

The two terms in the integrand both involve x. At first sight this looks like a complicated double integral, but it turns out that it can be simplified as follows. The integration region for y is found with the help of Figure A.10.

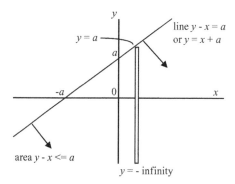

Figure A.10 Integration region

The double integral can then be rewritten as

$$\int_{x=-\infty}^{\infty} \int_{y=-\infty}^{x+a} \left\{ \frac{1}{\sqrt{t}\sqrt{2\pi}} \exp\left[-\frac{1}{2} \left(\frac{x}{\sqrt{t}} \right)^2 \right] \right\}$$
$$\times \left\{ \frac{1}{\sqrt{u}\sqrt{2\pi}} \exp\left[-\frac{1}{2} \left(\frac{y-x}{\sqrt{u}} \right)^2 \right] \right\} dx\, dy$$

The exponent in the second term is simplified by changing to variable $w \stackrel{\text{def}}{=} y - x$. Integration limit $y = x + a$ becomes $w = a$, and $y = -\infty$ becomes $w = -\infty$, $dw = dy$, giving

$$\int_{x=-\infty}^{\infty} \int_{w=-\infty}^{a} \frac{1}{\sqrt{t}\sqrt{2\pi}} \exp\left[-\frac{1}{2} \left(x/\sqrt{t} \right)^2 \right]$$
$$\times \frac{1}{\sqrt{u}\sqrt{2\pi}} \exp\left[-\frac{1}{2} \left(w/\sqrt{u} \right)^2 \right] dx\, dw$$

The w integral does not involve x and can thus be treated as a constant in the x integral, and taken outside that integral, giving

$$\left\{ \int_{x=-\infty}^{\infty} \frac{1}{\sqrt{t}\sqrt{2\pi}} \exp\left[-\frac{1}{2} \left(\frac{x}{\sqrt{t}} \right)^2 \right] dx \right\}$$
$$\times \left\{ \int_{w=-\infty}^{a} \frac{1}{\sqrt{u}\sqrt{2\pi}} \exp\left[-\frac{1}{2} \left(\frac{w}{\sqrt{u}} \right)^2 \right] dw \right\}$$

The x integral is the area under a normal density, so equals 1, giving

$$\mathbb{P}[B(t+u) - B(t) \leq a] = \int_{w=-\infty}^{a} \frac{1}{\sqrt{u}\sqrt{2\pi}} \exp\left[-\frac{1}{2}\left(\frac{w}{\sqrt{u}}\right)^2\right] dw$$

as was to be shown.

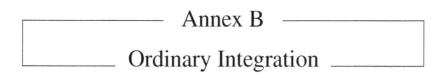

Annex B

Ordinary Integration

B.1 RIEMANN INTEGRAL

The ordinary integral from calculus is known as a Riemann integral (after the German mathematician Riemann, middle of 19th century). It is the area (Figure B.1) under a function $f(x)$.

There are two approaches to constructing this integral. They both start from a discretization of the x-axis and approximate the area under the function using step-functions. The simplest setting is where the function is bounded and smooth. These methods will now be reviewed on the example of the area under the function $f(x) = x$.

Consider, without loss of generality, the interval $[0, 1]$ on the x-axis. Divide this into n subintervals of equal length $\Delta x = 1/n$. Equal length is convenient but not essential for the arguments that follow. The endpoints of the subintervals are multiples of $\frac{1}{n}$:

$$0, \ldots, k\frac{1}{n}, (k+1)\frac{1}{n}, \ldots, 1.$$

This collection of endpoints is called a partition. The word partition is often used loosely to refer to either the set of endpoints or the collection of subintervals.

B.1.1 Darboux Construction

The method devised by the French mathematician Darboux proceeds as follows. Consider the area under f in the subinterval $[k/n, (k+1)/n]$ – see Figure B.2. Approximate this area in the following manner. Find the maximum value of the function $f(x)$. In the example this is at the right endpoint of this subinterval, $x = (k+1)/n$, and equals $f[(k+1)/n] = (k+1)/n$. Form a rectangle with the maximum value of the function as the height, and the length of the subinterval, $1/n$, as the base. The area of this rectangle equals $[(k+1)/n]/n$ and is greater than the actual area under f. It is an upper bound for the actual area (Figure B.3).

Similarly find the minimum value of f on this subinterval. In the example this is at the left endpoint of the subinterval and equals k/n.

0   ---> x   1   **Figure B.1**   Area

Form a rectangle with this minimum value. The area of this rectangle equals $(k/n)/n$ and is less than the actual area under f. It is a lower bound for the actual area (Figure B.4).

On this subinterval the difference between the upper rectangle and the lower rectangle equals $1/n^2$. Now sum over all subintervals. That gives a so-called upper Darboux sum, denoted $U(n)$, and a lower Darboux sum, denoted $L(n)$, where n is a reminder that they are based on n subintervals. In the example

$$U(n) \overset{\text{def}}{=} \sum_{k=0}^{n-1} \frac{k+1}{n} \frac{1}{n} = \frac{1}{n^2} \sum_{k=0}^{n-1}(k+1) = \frac{1}{2}\left(1 + \frac{1}{n}\right)$$

$$L(n) \overset{\text{def}}{=} \sum_{k=0}^{n-1} \frac{k}{n} \frac{1}{n} = \frac{1}{n^2} \sum_{k=0}^{n-1} k = \frac{1}{2}\left(1 - \frac{1}{n}\right)$$

(both sums were computed in Mathematica). Summed over all subintervals the difference between the upper sum and the lower sum is $U(n) - L(n) = 1/n$. It is the shaded area shown in Figure B.5.

Now the partition is refined. This is done by splitting subintervals, for example by halving all the existing subintervals. It creates additional

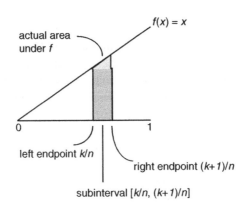

Figure B.2   Subinterval

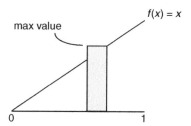

Figure B.3   Upper approximation

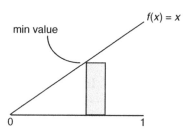

Figure B.4   Lower approximation

endpoints and keeps the existing partition points (see Figure B.6). Re-
fining the partition makes the upper sum decrease, and the lower sum
increase, both get closer to the actual area under f. This can be seen
directly from the example by substituting $2n$ for n in the expressions for
$U(n)$ and $L(n)$, as well as from the following close up of the subinterval
$[k/n, (k + 1)/n]$.

Repeated refinement of the partition keeps decreasing the value of
$U(n)$ and increasing the value of $L(n)$. The minimum value of $U(n)$ over
all possible partitions is called the upper Darboux integral, and is denoted
$U(f)$. Similarly the maximum value of $L(n)$ over all possible partitions
is called the lower Darboux integral, denoted $L(f)$. A lower sum is never

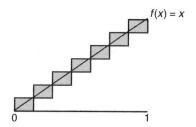

Figure B.5   Upper and lower approximation

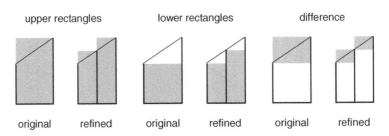

upper rectangles   lower rectangles   difference

original   refined   original   refined   original   refined

Figure B.6 Refinements

greater than an upper sum, $U(f) \geq L(f)$. (To be mathematically correct, the word 'minimum' should be replaced by 'greatest lower bound' and 'maximum' by 'least upper bound'[1]). If $U(f) = L(f)$ then that value is defined as the value of the integral. In the example, $U(f)$ and $L(f)$ can be readily found as the limit for $n \to \infty$ of $U(n)$ and $L(n)$.

$U(n) = \frac{1}{2}(1 + 1/n)$ tends to the value $\frac{1}{2}$ but 'never gets quite that low'.

Similarly

$L(n) = \frac{1}{2}(1 - 1/n)$ tends to the value $\frac{1}{2}$ but 'never gets quite that high'.

With $U(f) = L(f) = \frac{1}{2}$ the value of the integral is $\frac{1}{2}$. How both $U(n)$ and $L(n)$ tend to the value $\frac{1}{2}$ is illustrated in Figure B.7.

Thus it has been shown, by using the Darboux construction of the integral, that $\int_{y=0}^{x} x \, dx = \frac{1}{2}$. This is correct as the area is half of a rectangle with sides of length 1. The maxima of f on the respective subintervals form a step function which approximates the function f from above, the minima form a step-function which approximates the function f from below.

B.1.2 Riemann Construction

The method originally devised by Riemann takes the value of the function at an arbitrary point ξ_k in the subinterval, $k/n \leq \xi_k \leq (k+1)/n$ (note that the endpoints are included). In the above example this gives $f(\xi_k)\Delta x = f(\xi_k)/n$ as the approximation of the area. Then, as before, sum over all subintervals. That gives the so-called Riemann sum, denoted $R(n)$. In the example, $R(n) \stackrel{\text{def}}{=} \sum_{k=0}^{n-1} f(\xi_k)/n$. Now successively

---

[1] See below in section "Sup and Max".

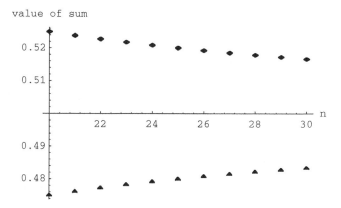

Figure B.7   Convergence

refine the partition. If $R(n)$ converges to a limit as $n \to \infty$, then that limit is defined as the value of the integral $\int_{x=0}^{1} f(x)\,dx$. To illustrate on the example, take ξ_k as the midpoint (see Figure B.8) of the subinterval $[k/n, (k+1)/n]$ so

$$\xi_k = \frac{k/n + [(k+1)/n]}{2} \quad \text{with} \quad f(\xi_k) = \xi_k.$$

Then, according to Mathematica, the Riemann sum equals

$$\sum_{k=0}^{n-1} \xi_k\,\Delta x = \sum_{k=0}^{n-1} \frac{k/n + [(k+1)/n]}{2}\,\frac{1}{n} = \frac{1}{2}$$

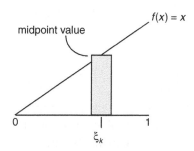

Figure B.8   Midpoint approximation

In this special example the value of the Riemann sum does not depend on n. The limit equals $\frac{1}{2}$, which is the value of the integral, as above. As for any ξ_k the value of the function $f(\xi_k)$ lies between the 'max' and the 'min' of f, the value of a Riemann sum lies between those of the upper and lower Darboux sums, so both methods give the same result. *The value of the integral does not depend on the position of ξ_k in the subinterval.*

B.2   RIEMANN–STIELTJES INTEGRAL

Integration in the familiar sense is computing an area under a curve. In a wider sense it is an averaging process where values of a function f are weighted by the length of subintervals. More generally, the weighting associated with each subinterval needs not be its length. The 'importance' of a subinterval can depend on where on the x-axis the subinterval is located. In general the weighting can be given by another function $g(x)$. Here the integral can also be constructed by both the Darboux method and the Riemann method. To illustrate how the integral arises, consider a horizontal metal bar of unit length with a variable cross-section. Its volume up to a cross-section at distance x from the left is given by $g(x)$, an increasing function of x. Suppose the (physical) density of the metal is not constant but given by $f(x)$ at position x. To compute the total weight of this bar, conceptually create a partition of n subintervals as described above. The volume of the metal bar over the subinterval

$$\left[\frac{k}{n}, \frac{k+1}{n}\right] \quad \text{is} \quad g\left(\frac{k+1}{n}\right) - g\left(\frac{k}{n}\right).$$

Following Darboux, the weight of this section of the bar lies between

$$\max\left[f\left(\frac{k+1}{n}\right)\right]\left[g\left(\frac{k+1}{n}\right) - g\left(\frac{k}{n}\right)\right].$$

and

$$\min\left[f\left(\frac{k+1}{n}\right)\right]\left[g\left(\frac{k+1}{n}\right) - g\left(\frac{k}{n}\right)\right].$$

As before, sum over all subintervals, and find the lowest and highest value of these sums over all possible partitions. If these are equal then that value is defined as the Riemann–Stieltjes integral, co-named after the Dutch mathematician Stieltjes, and is denoted $\int_{x=0}^{1} f(x)\, dg(x)$.

Being constructed from the path x, this construction is known as path-wise integration. The Riemann–Stieltjes integral is a generalization of the Riemann–integral. In the case where $g(x) = x$ it reduces to a Riemann integral. To numerically illustrate the Riemann–Stieltjes integral, consider $\int_{x=0}^{1} x \, d(x^2)$. The integrator which provides the weighting is the function $g(x) = x^2$. Constructing the value of the integral from its definition follows the same steps as for the Riemann integral above. The only difference is that the max and the min of the integrand $f(x) = x$ are now multiplied by the change in x^2 over a subinterval instead of by the change in x. Using the same notation as above

$$\mathsf{U}(n) \overset{\text{def}}{=} \sum_{k=0}^{n-1} \frac{k+1}{n} \left[\left(\frac{k+1}{n} \right)^2 - \left(\frac{k}{n} \right)^2 \right] = \frac{2}{3} - \frac{1}{6n^2} + \frac{1}{2n}$$

$$\mathsf{L}(n) \overset{\text{def}}{=} \sum_{k=0}^{n-1} \frac{k}{n} \left[\left(\frac{k+1}{n} \right)^2 - \left(\frac{k}{n} \right)^2 \right] = \frac{2}{3} - \frac{1}{6n^2} - \frac{1}{2n}$$

(both according to Mathematica). The difference between the upper sum and the lower sum is $\mathsf{U}(n) - \mathsf{L}(n) = 1/n$. Again $\mathsf{U}(f)$ and $\mathsf{L}(f)$ can be found here by letting $n \to \infty$, giving $\mathsf{U}(f) = \mathsf{L}(f) = \frac{2}{3}$, so this is the value of the integral (Figure B.9). Thus it was shown by using the definition of the Riemann–Stieltjes integral that $\int_{x=0}^{1} x \, d(x^2) = \frac{2}{3}$. This example is illustrated in Figure B.10 for $n = 50$, designed in Excel.

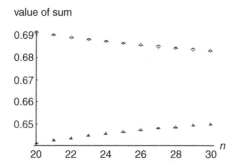

Figure B.9 Convergence

Riemann-Stieltjes: Integrator g, Upper sum U, Lower sum L

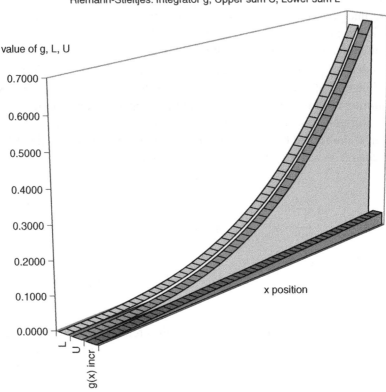

Figure B.10   Riemann–Stieltjes schematic

This example was convenient because it can also be written as a Riemann integral since $d(x^2) = 2x\,dx$ and can thus serve as a check.

$$\int_{x=0}^{1} x\,d(x^2) = \int_{x=0}^{1} x2x\,dx = 2\int_{x=0}^{1} x^2\,dx = \frac{2}{3}.$$

For the Riemann–Stieltjes integral $\int_{x=0}^{1} f(x)\,dg(x)$ to exist, the *integrator* $g(x)$ *must be of bounded variation* on [0, 1]. This is discussed in the next section.

B.2.1   Bounded Variation Condition

This section discusses why the integrator in a Riemann–Stieltjes integral must be of bounded variation. Using a partition of $0 \le x \le 1$ into n

equal subintervals, and the formulation using the intermediate points $k/n \leq \xi_k \leq (k+1)/n$, the sum

$$S(n) \overset{\text{def}}{=} \sum_{k=0}^{n-1} f(\xi_k) \left[g\left(\frac{k+1}{n}\right) - g\left(\frac{k}{n}\right) \right]$$

is an approximation to the total weight of the bar in the above example. Repeated refinement of the partition produces a sequence of such sums. If that sequence converges to a limit, then the Riemann–Stieltjes integral is defined as that limit, and is denoted $\int_{x=0}^{1} f(x)\, dg(x)$. For the limit to exist, a condition on the variability of the function $g(x)$ must be satisfied, as will now be explained. Refine the n-partition by halving all subintervals. This gives a partition with $2n$ subintervals. All points of the original partition have been kept, and n additional endpoints have been created. For convergence to hold it must be true that the absolute difference of the sums corresponding to these partitions, $|S(n) - S(2n)|$, can be made arbitrarily small for sufficiently large n. The sums $S(n)$ and $S(2n)$ can both be expressed in terms of the refined partition,[2] that is, with k running from 0 to $(2n - 1)$.

$$S(n) = \sum_{k=0}^{2n-1} f(\xi_k) \left[g\left(\frac{k+1}{2n} - g\frac{k}{2n}\right) \right]$$

$$S(2n) = \sum_{k=0}^{2n-1} f(\eta_k) \left[g\left(\frac{k+1}{2n} - g\frac{k}{2n}\right) \right]$$

where η_k is in a subinterval of the refined partition. The absolute difference is

$$|S(n) - S(2n)| = \left| \sum_{k=0}^{2n-1} \left[f(\xi_k) - f(\eta_k) \right] \left[g\left(\frac{k+1}{2n}\right) - g\left(\frac{k}{2n}\right) \right] \right|$$

Using the well known inequality $|a + b| \leq |a| + |b|$ gives

$$|S(n) - S(2n)| \leq \sum_{k=0}^{2n-1} |f(\xi_k) - f(\eta_k)| \left| g\left(\frac{k+1}{2n}\right) - g\left(\frac{k}{2n}\right) \right|$$

---

[2] This is a technical point that is explained in the *Bartle* reference mentioned at the end of this Annex.

If $f(x)$ is a continuous function, then $|f(\xi_k) - f(\eta_k)|$ can be made arbitrarily small ($\leq \epsilon$) so

$$|S(n) - S(2n)| \leq \epsilon \sum_{k=0}^{2n-1} \left| g\left(\frac{k+1}{2n}\right) - g\left(\frac{k}{2n}\right) \right|$$

Then $|S(n) - S(2n)|$ can only be arbitrarily small if the sum $\sum_{k=0}^{2n-1} |g[(k+1)/2n] - g(k/2n)|$ is a finite number, for any partition, that is, the sum must be bounded. The largest of these sums over all possible partitions is known as the variation of the function $g(x)$; more precisely, it is the least upper bound of these sums, that is defined as the variation. Thus for this Riemann–Stieltjes integral to exist the integrator $g(x)$ must be of bounded variation on $0 \leq x \leq 1$. In the case of $g(x) = x$, the Riemann–Stieltjes integral becomes an ordinary Riemann integral. The sum becomes

$$\sum_{k=0}^{2n-1} \left| \frac{k+1}{2n} - \frac{k}{2n} \right| = \sum_{k=0}^{2n-1} \frac{1}{2n} = 2n\frac{1}{2n} = 1.$$

So for an ordinary Riemann integral the condition is always satisfied.

B.2.2   Random integrand

The above is now used to define $\int_{t=0}^{T} f(t, \omega)\, dt$ where $f(t, \omega)$ is a random process whose paths are continuous. For a fixed path ω, let

$$S(n) \stackrel{\text{def}}{=} \sum_{k=0}^{n-1} f(\tau_k, \omega)(t_{k+1} - t_k)$$

where $t_k = k(T/n)$ and $t_k \leq \tau_k \leq t_{k+1}$. Because the time interval $[0, T]$ includes the end points, each path $f(t, \omega)$ is uniformly continuous. That

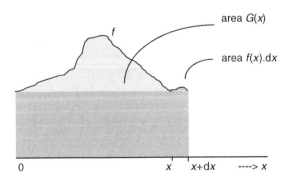

Figure B.11   Area increment

means that for any small $\epsilon > 0$ there exists a small $\delta(\epsilon)$ which depends only on ϵ, such that for $|t_2 - t_1| \le \delta(\epsilon)$

$$|f(t_2, \omega) - f(t_1, \omega)| \le \epsilon$$

This $\delta(\epsilon)$ is the same for all choices of t_2 and t_1. Then

$$|S(n) - S(2n)| \le \varepsilon T$$

which can be made arbitrarily small by choice of ε. On any path ω, $S(n)$ is a Cauchy sequence and converges in probability, according to section E.4. The limit is denoted $\int_{t=0}^{T} f(t, \omega) \, dt$.

B.3   OTHER USEFUL PROPERTIES

B.3.1   Fundamental Theorem of Calculus

Consider the ordinary integral of a function $f(t)$ from 0 to x, $\int_{t=0}^{x} f(t) \, dt$. The value of this integral depends on the upper integration limit x, and this can be captured by writing its value as $G(x) \overset{\text{def}}{=} \int_{t=0}^{x} f(t) \, dt$. Differentiating $G(x)$ with respect to x gives $dG(x)/dx = f(x)$, the value of the function $f(t)$ at the upper integration limit $t = x$. The fact that $dG(x)/dx = f(x)$ comes about because $f(x)$ is the rate at which this area is increasing. This can be demonstrated as follows (Figure B.11). Suppose for purpose of illustration that f is a positive function. $G(x)$ is the area under the curve from $t = 0$ to $t = x$. $G(x + \Delta x)$ is the area under the curve from $t = 0$ to $t = x + \Delta x$. Then $G(x + \Delta x) - G(x) = f(x) \Delta x$. Dividing by Δx and letting Δx go to zero gives the definition of $dG(x)/dx$ on the left-hand side, and leaves $f(x)$ on the right hand, so $dG(x)/dx = f(x)$.

B.3.2   Use in Deriving Probability Density Function from Probability Distribution

For a continuous random variable, the probability density function (pdf), here denoted f, is found by differentiating its probability distribution $F(x) \overset{\text{def}}{=} \int_{y=-\infty}^{x} f(y) \, dy$ with respect to upper integration limit x, $f(x) = dF(x)/dx$. This is because the pdf is defined as that function f which makes it possible to express $F(x)$ as the integral of f. That is, the probability distribution F is the starting point. In more complicated cases, the upper integration limit is not x itself, but a function of x, say

$a(x)$, so $F(x) \overset{\text{def}}{=} \int_{y=-\infty}^{a(x)} f(y)\,dy$. To get the first derivative of $F(x)$ with respect to x, the chain rule of differentiation then has to be applied. First differentiate with respect to upper limit $a(x)$ treated as a variable, say call it $b = a(x)$. That gives $dF(x)/db = f(b) = f[a(x)]$; the variable y of the integrand $f(y)$ is replaced by upper integration limit $a(x)$. This is then multiplied by the derivative of $a(x)$ with respect to x, provided $a(x)$ is an increasing function of x, as is often the case. The final result is

$$\frac{dF(x)}{dx} = f[a(x)]\frac{da(x)}{dx}.$$

In the case where $a(x)$ is a decreasing function of x, use $|da(x)/dx|$.

B.3.3   Integration by Parts

Let $h(x)$ be a function which has a continuous first derivative $dh(x)/dx$ on the interval $a \leq x \leq b$. Consider the integral with respect to x of this first derivative, $\int_{x=a}^{b} [dh(x)/dx]\,dx$. The Fundamental Theorem of Calculus says that the value of this integral is

$$h(x)\,\big|_{x=a}^{x=b} = h(b) - h(a).$$

Now let $h(x)$ be the product of the functions $f(x)$ and $g(x)$ which have continuous derivatives $df(x)/dx$ and $dg(x)/dx$. By the chain rule of differentiation

$$\frac{dh(x)}{dx} = \frac{df(x)}{dx}g(x) + f(x)\frac{dg(x)}{dx}.$$

Then

$$\int_{x=a}^{b} \frac{dh(x)}{dx}\,dx = \int_{x=a}^{b} \left[\frac{df(x)}{dx}g(x) + f(x)\frac{dg(x)}{dx}\right] dx = f(x)g(x)\,\big|_{x=a}^{x=b}$$

Rearranging gives

$$\int_{x=a}^{b} \frac{df(x)}{dx}g(x)\,dx = f(x)g(x)\,\big|_{x=a}^{x=b} - \int_{x=a}^{b} f(x)\frac{dg(x)}{dx}\,dx$$

Writing $[df(x)/dx]\,dx$ as $df(x)$ and $[dg(x)/dx]\,dx$ as $dg(x)$ gives

$$\int_{x=a}^{b} g(x)\,df(x) = f(x)g(x)\,\big|_{x=a}^{x=b} - \int_{x=a}^{b} f(x)\,dg(x).$$

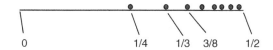

Figure B.12   Towards upper bound

Omitting the arguments, x gives a compact way of writing the expression for integration by parts which can be easily remembered

$$\int_{x=a}^{b} g \ df = fg \left|_{x=a}^{x=b}\right. - \int_{x=a}^{b} f \ dg$$

B.3.4   Interchanging the Order of Integration (Fubini)

Let f be a continuous function of variables $x \in [x_1, x_2]$ and $y \in [y_1, y_2]$. Consider the double integral $\int_{y=y_1}^{y_2} [\int_{x=x_1}^{x_2} f(x, y) \, dx] \, dy$. The first integration is the inner one with respect to x, and the result is a function of y. This is then integrated with respect to y by the outer integral. A theorem by *Fubini* says that the order of integration and the integration limits can be interchanged to $\int_{x=x_1}^{x_2} [\int_{y=y_1}^{y_2} f(x, y) \, dy] \, dx$ for smooth enough functions that are typically encountered in applications. This property is used in the main text when expected value operator $\mathbb{E}$, which is an integration, is moved inside an integral. One says that this is justified 'by Fubini'.

B.3.5   Sup and Max

In the review of integration, the function $\mathbb{L}(n) = \frac{1}{2}(1 - \frac{1}{n})$ appears, where n is a positive integer. For $n = 1, 2, 3, 4, \ldots$ this has the values $0, \frac{1}{4}, 1/3, 3/8, \ldots$ (see Figure B.12). As n increases, the values of $\mathbb{L}(n)$ get closer and closer to $\frac{1}{2}$. But however large n is, $\mathbb{L}(n)$ will never be exactly equal to $\frac{1}{2}$.

What is the maximum value of $\mathbb{L}(n)$? Suppose that it is x. As $\mathbb{L}(n)$ never equals $\frac{1}{2}$, its maximum can only be strictly less than $\frac{1}{2}$. So it is possible to find another integer value $n^* > n$ such that $\mathbb{L}(n^*) = x^*$. Thus x cannot be the maximum. This reasoning shows that the set of values $\mathbb{L}(n)$ has no maximum. What it does have is upper bounds. All numbers $\geq \frac{1}{2}$ are upper bounds. Since $\mathbb{L}(n)$ will never be greater than $\frac{1}{2}$, the upper bounds that are greater than $\frac{1}{2}$ are of no interest. The only useful upper bound is the smallest of all possible upper bounds. That is called the *least upper bound* or *supremum*, abbreviated *lub* or *sup*.

Similarly the function $U(n) = \frac{1}{2}(1 + 1/n)$ does not have a minimum. But it is bounded below by lots of numbers $\leq \frac{1}{2}$, but the ones that are $< \frac{1}{2}$ are are all too small. The only lower bound that is of interest is the one that 'just fits'. That is called the *greatest lower bound* or *infimum*, and abbreviated *glb* or *inf*.

The least upper bound is less than every other possible upper bound. The greatest lower bound is greater than every other possible lower bound: sup $\geq$ inf. The greatest lower bound and the least upper bound of a set always exist, whereas the minimum and the maximum may not exist. That is why in Mathematics sup and inf are used in situations that involve greatest and smallest. For everyday use one can think of sup as a maximum and inf as a minimum.

B.3.6   Differential

The first derivative (slope) of a function $f(x)$ is denoted as $df(x)/dx$, or as $f'(x)$, or as dy/dx when the function value is denoted as $y = f(x)$. It is tempting to think of the latter as the ratio of two quantities dy and dx which can be used separately; but that is not correct; dy/dx is simply another notation for f'. If Δx denotes a change in x, then the differential is *defined* as $\Delta y \stackrel{def}{=} f'(x)\, \Delta x$. This is the change on the vertical axis when moving along the slope $f'(x)$. It is *not* the change in the value of the function f itself, unless f happens to be the $45°$ line, so $f'(x) = 1$. The use of the term differential in stochastic calculus is inspired by this, but now the symbol d in the left-hand side of an equation denotes the change in the value of a function resulting from changes in d in the right-hand side.

B.4   REFERENCES

Riemann integration is explained in numerous books on calculus and real analysis. The exposition in *Brennan* Chapter 7 is splendid. Also recommended is *Basic Elements of Real Analysis* by *Protter*. The most comprehensive coverage of the RS integral is given in the mathematical classic *The Elements of Real Analysis* by *Bartle*, Chapter VI. A useful discussion on the limitations of the RS integral is given in *Kuo* Chapter 1. *Klebaner* Chapter 1 recaps various concepts from calculus which are of direct relevance to stochastic calculus.

Brownian Motion Variability

C.1 QUADRATIC VARIATION

Partition the time period $[0, T]$ into n intervals of equal length $\Delta t = T/n$. The time points in the partition are $t_k \overset{\text{def}}{=} k\,\Delta t$. The sum of the squared Brownian motion increments over these intervals is $\sum_{k=0}^{n-1}[B(t_{k+1}) - B(t_k)]^2$. This is a random variable. Over the interval $[t_k, t_{k+1}]$, $\mathbb{E}\{[B(t_{k+1}) - B(t_k)]^2\} = \Delta t$ which is the same for all intervals. That suggests that the sum of $[B(t_{k+1}) - B(t_k)]^2$ over all intervals may converge to $\sum_{k=0}^{n-1} \Delta t$ in mean-square, that is, that $\sum_{k=0}^{n-1}[B(t_{k+1}) - B(t_k)]^2 \to T$ as $n \to \infty$, so $\Delta t \to 0$. This will now be shown in two ways, first heuristically.

C.1.1 Heuristic Explanation

As the sum of the squared increments is a random variable, it is natural to analyze its expected value and its variance.

$$
\mathbb{E}\left\{\sum_{k=0}^{n-1}[B(t_{k+1}) - B(t_k)]^2\right\} = \sum_{k=0}^{n-1} \mathbb{E}\left\{[B(t_{k+1}) - B(t_k)]^2\right\}
$$

$$
= \sum_{k=0}^{n-1}[t_{k+1} - t_k] = T
$$

$\mathbb{Var}\{\sum_{k=0}^{n-1}[B(t_{k+1}) - B(t_k)]^2\}$ can be written as the sum of the variances of the terms, as the Brownian motion increments over non-overlapping time intervals are independent. Writing $\Delta B(t_k)$ for $B(t_{k+1}) - B(t_k)$, and using the second and the fourth Brownian motion moment

$$
\mathbb{Var}\{[\Delta B(t_k)]^2\} = \mathbb{E}\{[\Delta B(t_k)]^4\} - [\mathbb{E}\{[\Delta B(t_k)]^2\}]^2
$$
$$
= 3(\Delta t)^2 - (\Delta t)^2 = 2(\Delta t)^2
$$

$$
\sum_{k=0}^{n-1} \mathbb{Var}\{[B(t_{k+1}) - B(t_k)]^2\} = \sum_{k=0}^{n-1} 2(\Delta t)^2 = \sum_{k=0}^{n-1} 2n(\Delta t)^2 = 2T\,\Delta t
$$

As $\Delta t \to 0$, $\mathbb{V}\text{ar}\left\{\sum_{k=0}^{n-1}[B(t_{k+1}) - B(t_k)]^2\right\} \to 0$ and $\sum_{k=0}^{n-1}[B(t_{k+1}) - B(t_k)]^2$ homes in on the fixed value T.

C.1.2   Heuristic Explanation of $[dB(t)]^2 = dt$

The quadratic variation property is also the explanation for the notation $[dB(t)]^2 = dt$ which is used in the application of Itō's formula. Consider $\Delta B(t_k)$ over the finite interval Δt

$$\mathbb{E}\{[\Delta B(t_k)]^2\} = \Delta t \qquad \mathbb{V}\text{ar}\{[\Delta B(t_k)]^2 = 2(\Delta t)^2$$

For very small Δt, the order of magnitude of $(\Delta t)^2$ is negligible compared to the order of magnitude of Δt. Thus the variance of $\{\Delta B(t)\}^2$ is much smaller than the expected value of $[\Delta B(t)]^2$. As the time interval becomes infinitely small, $[\Delta B(t)]^2$ is written as $[dB(t)]^2$ and Δt as dt. So the random quantity $[dB(t)]^2$ approaches the non-random dt. This is commonly written with an equality sign, but *must be understood in the mean-square sense*.

C.1.3   Derivation of Quadratic Variation Property

The heuristic explanation above gives some feel for the quadratic variation property of Brownian motion. Now follows the complete derivation of the convergence in mean square of $\sum_{k=0}^{n-1}[B(t_{k+1}) - B(t_k)]^2$ to $\sum_{k=0}^{n-1}\Delta t$, that is, to T as $n \to \infty$. To recap what this means, take the difference between the left and the right, $\sum_{k=0}^{n-1}[B(t_{k+1}) - B(t_k)]^2 - T$. Square this difference, to get $[\sum_{k=0}^{n-1}[B(t_{k+1}) - B(t_k)]^2 - T]^2$. Then take the expected value of this expression

$$\mathbb{E}\left\{\left[\sum_{k=0}^{n-1}[B(t_{k+1}) - B(t_k)]^2 - T\right]^2\right\}$$

If this $\to 0$ as $n \to \infty$, there is convergence in mean-square. This will now be shown. The notation is somewhat cumbersome but the derivation is not conceptually complicated.

Step 1   Replace T by $\sum_{k=0}^{n-1} \Delta t$. That brings all terms under the summation sign $\sum_{k=0}^{n-1}$, and changes $\mathbb{E}\{...\}$ to

$$\mathbb{E}\left\{\left[\sum_{k=0}^{n-1}[B(t_{k+1}) - B(t_k)]^2 - \Delta t\right]^2\right\}$$

Carefully observe the expression of which the expected value is to be taken. Inside [...] is a sum of n terms, and [...]^2 is the square of this sum of n terms.

Step 2   Expand the square and apply $\mathbb{E}$. The resulting expression consists of full square terms $([B(t_{k+1}) - B(t_k)]^2 - \Delta t)^2$ and cross terms $([B(t_{k+1}) - B(t_k)]^2 - \Delta t)([B(t_{m+1}) - B(t_m)]^2 - \Delta t)$ where $m \neq k$. The two parts of a cross term are independent random variables, so the expected value of a cross term can be written as the product of the expected value of each of the parts. As these are each zero, what remains is the expected value of the full square terms. So the expected value of the square of this sum of n terms can be written as the expected value of the sum of n squares:

$$\mathbb{E}\left\{\left[\sum_{k=0}^{n-1}([B(t_{k+1}) - B(t_k)]^2 - \Delta t)\right]^2\right\}$$

$$= \mathbb{E}\left\{\sum_{k=0}^{n-1}([B(t_{k+1}) - B(t_k)]^2 - \Delta t)^2\right\}$$

Taking $\mathbb{E}$ inside $\sum_{k=0}^{n-1}$, the right-hand side is $\sum_{k=0}^{n-1}\mathbb{E}\{([B(t_{k+1}) - B(t_k)]^2 - \Delta t)^2\}$.

Step 3   Expand the square in the expression on the right, and apply $\mathbb{E}$. Expanding the term in the sum gives

$$\mathbb{E}\{[B(t_{k+1}) - B(t_k)]^4\} - 2\mathbb{E}\{[B(t_{k+1}) - B(t_k)]^2\}\Delta t + (\Delta t)^2$$

For the first term, use $\mathbb{E}[B(t)^4] = 3t^2$ so $\mathbb{E}\{[B(t_{k+1}) - B(t_k)]^4\} = 3(\Delta t)^2$. In the second term, $\mathbb{E}\{[B(t_{k+1}) - B(t_k)]^2\} = \Delta t$. Putting it together gives $3(\Delta t)^2 - 2\Delta t\,\Delta t + (\Delta t)^2 = 2(\Delta t)^2 = 2(T/n)^2$. Summing over all terms gives $\sum_{k=0}^{n-1}2(T/n)^2 = n2(T/n)^2 = 2T^2/n$ which tends to zero as $n \to \infty$. So it has been shown that the sum of the squared increments convergences in mean square to T.

C.2   FIRST VARIATION

Another measure of the variability of a Brownian motion path is the sum of the absolute values of the increments, $\sum_{k=0}^{n-1} |B(t_{k+1}) - B(t_k)|$. It is the length of a Brownian motion path. The analysis that follows is aimed at finding a bound on this. The method of analysis is not particularly intuitive. The first step is to analyse the quadratic variation, then use the resulting property to analyse the total increments.

Step 1   Find a bound on the term $|B(t_{k+1}) - B(t_k)|^2$ in the quadratic variation. Find the largest of the absolute increments, denote it $b \stackrel{\text{def}}{=} \max |B(t_{k+1} - B(t_k)|$ where max is over all k. Then $|B(t_{k+1}) - B(t_k)|^2 = |B(t_{k+1}) - B(t_k)||B(t_{k+1}) - B(t_k)|$ is no greater than $b|B(t_{k+1}) - B(t_k)|$.

Step 2   Sum over all n terms.

$$\sum_{k=0}^{n-1} |B(t_{k+1}) - B(t_k)|^2 \leq b \sum_{k=0}^{n-1} |B(t_{k+1}) - B(t_k)|$$

since b is not a function of k. The sum on the right is the first variation. Rearrange in terms of this first variation

$$\sum_{k=0}^{n-1} |B(t_{k+1}) - B(t_k)| \geq \sum_{k=0}^{n-1} |B(t_{k+1}) - B(t_k)|^2/b$$

The right hand side is a *lower bound* for the first variation on the left.

Step 3   Analyse what happens to this lower bound as $n \to \infty$. The denominator b goes to zero because as Brownian motion is continuous, $B(t_{k+1})$ will be arbitrarily close to $B(t_k)$ for sufficiently large n. The numerator $\sum_{k=0}^{n-1} |B(t_{k+1}) - B(t_k)|^2$ is the quadratic variation of the Brownian motion which tends to T. Therefore the right-hand side becomes unbounded and since this is a lower bound on the variation on the left, it has been demonstrated that *Brownian motion has unbounded variation.*

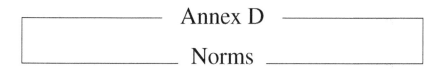

Annex D
Norms

The construction of the stochastic integral uses approximations of random variables and random processes. This Annex explains how the closeness of these approximations can be expressed. The notion of *distance* between objects is derived from the fundamental concept of a *norm* which is the generalization of the concept of length, and expresses a magnitude.

D.1 DISTANCE BETWEEN POINTS

D.1.1 One Dimension

Consider an arbitrary point x on the real line. The *magnitude* of x is denoted $|x|$, and is called modulus x (Figure D.1). It is a function of x defined as

$$|x| \overset{\text{def}}{=} \begin{cases} x & \text{if } x \geq 0 \\ -x & \text{otherwise} \end{cases}$$

so $|x|$ is always ≥ 0.

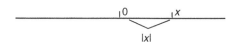

Figure D.1 Modulus x

Given the notion of magnitude, the *distance* between two points x and y, denoted $d(x, y)$, is defined as

$$d(x, y) \overset{\text{def}}{=} |x - y|$$

It is a function of x and y. For example, $d(3, -2) = |3 - (-2)| = |5| = 5$, and $d(-3, -7) = |-3 - (-7)| = |4| = 4$.

D.1.2 Two Dimensions

Consider the point x in two dimensional space (Figure D.2), with coordinates x_1 and x_2, $x = (x_1, x_2)$.

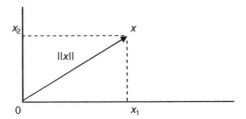

Figure D.2   Point in two dimensions

The straight line from the origin to point x is known as a vector. For the triangle under the arrow, Pythagoras proved that the square of the length of the vector equals the sum of the squares of the two sides adjacent to the 90° angle. Based on this, the length of vector x is defined as $\sqrt{x_1^2 + x_2^2}$. This is a measure of the magnitude of the vector. Once the length has been defined, the concept of distance follows. Let y be another point whose coordinates are y_1 and y_2. The distance between points x and y can then be defined as

$$d(x, y) \overset{\text{def}}{=} \sqrt{(x_1 - y_1)^2 + (x_2 - y_2)^2}$$

That is the distance light travels between x and y. For example, if $x = (1, 1)$ and $y = (4, 3)$ then $d(x, y)^2 = (4 - 1)^2 + (3 - 1)^2$ so $d(x, y) = 3.6$. There are several other ways in which length can be defined.

A second way to measure length is $|x_1| + |x_2|$. The distance between x and y is then

$$d(x, y) \overset{\text{def}}{=} |x_1 - y_1| + |x_2 - y_2|$$

This is the distance a car travels from x to y in Manhattan. Now $d(x, y) = |4 - 1| + |3 - 1| = 5$.

A third way to measure length is $\max(|x_1|, |x_2|)$. The distance between x and y is then

$$d(x, y) \overset{\text{def}}{=} \max\{|x_1 - y_1|, |x_2 - y_2|\}$$

Here $d(x, y) = \max\{3, 2\} = 3$.

D.1.3 n Dimensions

The above can be extended to n-dimensional space where an object x is called a point or vector, specified by its n coordinates $x_1, ..., x_n$. The length (magnitude) of this vector can be defined as a generalization of Pythagoras, by the square root of the sum of the squares of the coordinates of x, $\sqrt{\sum_{i=1}^{n} x_i^2}$. The distance between vector x and y can then be measured by

$$d(x, y) = \sqrt{\sum_{i=1}^{n}(x_i - y_i)^2}$$

Similarly the n-dimensional equivalents of the other distances can be used. What is most suitable depends on the application at hand.

D.1.4 Properties

Although the distances corresponding to these different definitions are different (3.6; 5; 3), it can be shown that they all have the following properties:

(1) $d(x, y) \geq 0$
(2) $d(x, y) = 0$ if and only if the coordinates of x equal those of y
(3) $d(x, y) = d(y, x)$, the distance measured from a to b is the same as measured from b to a
(4) $d(z, y) \leq d(z, x) + d(x, y)$, in the case of three vectors x, y, z; this property is called the *triangle inequality* and can be seen in Figure D.3.

   The length as described above measures the magnitude of an object. In general the term *norm* is used for this, and is denoted by $||.||$. Once a norm of an object has been defined, the distance between objects follows, so a norm is the fundamental concept, and distance (or metric) is the derived

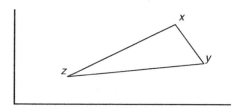

Figure D.3   Distance triangle

concept. The norm of object x is a function $||.||$ which is defined by the following properties:

(1) $||x|| \geq 0$.
(2) $||x|| = 0$ if and only if all its coordinates are zero.
(3) For any number λ, λx is the object whose coordinates are a multiple λ of the coordinates of x. Its norm is $||\lambda x||$ and equals $|\lambda| \, ||x||$, the value of the norm of x multiplied by the absolute value of λ. Thus $||-x|| = ||-1x|| = |-1| \, ||x|| = ||x||$.
(4) For two objects x and y, $x + y$ is the object whose coordinates are the respective sums of the coordinates of x and y. For this, $||x + y|| \leq ||x|| + ||y||$

The properties of a norm induce the properties of a distance that is derived from it. For the examples given earlier, the norms in n-dimensions are labelled and defined as

$$||x||_2 = \sqrt{x_1^2 + \cdots + x_n^2}$$
$$||x||_1 = |x_1| + \cdots + |x_n|$$
$$||x||_\infty = \max\{|x_1|, \cdots, |x_n|\}$$

D.2   NORM OF A FUNCTION

D.2.1   One Dimension

Consider a function f which is defined at discrete points $x_i, i = 1 \ldots n$. A norm can be defined for f analogous to the length of vector x above, as $||f|| \stackrel{\text{def}}{=} \sqrt{\sum_{i=1}^n f(x_i)^2}$. If f can take values for all x in an interval $[a, b]$ then the discrete sum is replaced by an integral, and that norm is $||f|| \stackrel{\text{def}}{=} \sqrt{\int_{x=a}^b f(x)^2 \, dx}$. The difference between functions f and g can then be expressed by

$$||f - g|| = \sqrt{\int_{x=a}^b [f(x) - g(x)]^2 \, dx}$$

This measures the average difference between the two functions. In this definition, the value of the difference at any particular x is not the decisive consideration. All positions are equally important.

There are other ways in which the norm of a function can be defined (Figure D.4). For example, the maximum absolute value of f on $[a, b]$

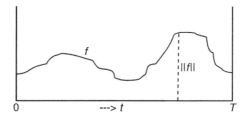

Figure D.4   Max norm of function

can be viewed as a measure of the magnitude of this function. This defines the norm as $||f|| \stackrel{\text{def}}{=} \max_{a \leq x \leq b} |f(x)|$.

The difference between two functions f and g is then

$$||f - g|| = \max_{a \leq x \leq b} |f(x) - g(x)|$$

It is where they are farthest apart, regardless of which function is the largest at the point x where this maximum difference occurs (Figure D.5).

So just as in the case of points, once a norm has been defined, a difference can be constructed.

D.2.2   Two Dimensions

Let f be a function of variables x_1 and x_2. It f can take values for $a \leq x_1 \leq b$ and $c \leq x_2 \leq d$, then a norm can be defined as

$$||f|| \stackrel{\text{def}}{=} \sqrt{\int_{x_2=c}^{d} \int_{x_1=a}^{b} f(x_1, x_2)^2 \, dx_1 \, dx_2}$$

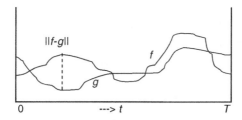

Figure D.5   Max norm of difference of functions

D.3   NORM OF A RANDOM VARIABLE

In probability theory, a random variable X is simply a different name for a function that takes the outcome ω of an 'experiment' and assigns it a numerical value $X(\omega)$. If X is a *discrete* random variable which can take numerical values x_i, with discrete probabilities p_i, $i = 1 \ldots . n$, then its norm is defined by using the above concept of the norm of a function, but now with each possible value weighted by the likelihood that it occurs

$$||X|| \overset{\text{def}}{=} \sqrt{\sum_{i=1}^{n} x_i{}^2 p_i} = \sqrt{\mathbb{E}(X^2)}$$

It is a measure of the magnitude of the random variable. As $\mathbb{E}(X^2)$ captures the variability of X, the definition of the norm says that the greater the variability of the random variable, the greater its magnitude. If X is a *continuous* random variable then the summation sign is replaced by an integral and the norm is

$$||X|| \overset{\text{def}}{=} \sqrt{\int_{x=-\infty}^{\infty} x^2 \varphi(x)\,dx}$$

where x denotes a value of random variable X and φ denotes the probability density of X at $X = x$. With this definition, the difference between two random variables X and Y, which is itself a random variable, can then be expressed by $||X - Y|| = \sqrt{\mathbb{E}[(X - Y)^2]}$. Actually computing the value of $||X - Y||$ requires knowledge of the joint probability distribution of X and Y. The concept of *convergence in mean-square* of a sequence of random variables X_n to a limit random variable X, is defined as $\mathbb{E}[(X_n - X)^2] \to 0$ as $n \to \infty$. This can be equivalently expressed in terms of norms as $||X_n - X|| \to 0$ when $n \to \infty$.

D.4   NORM OF A RANDOM PROCESS

A random process has two dimensions, time t and outcome ω. Its norm is defined using the above concept of the norm of a function in two dimensions, with each outcome weighted by its likelihood $d\mathbb{P}(\omega)$ (as for the norm of a random variable).

$$||f|| \overset{\text{def}}{=} \sqrt{\int_{t=0}^{T} \int_{\text{all }\omega} f(t, \omega)^2\, d\mathbb{P}(\omega)\, dt}$$

The inner integral is the summation over all possible values a random variable can take at time t, weighted by their probabilities; the second integral is the summation over all times, that is over the entire collection of random variables that make up the random process. As the inner integral is an expected value, the norm is commonly written as

$$||f|| \overset{\text{def}}{=} \sqrt{\int_{t=0}^{T} \mathbb{E}[f(t, \omega)^2] \, dt} \quad \text{or} \quad ||f||^2 \overset{\text{def}}{=} \int_{t=0}^{T} \mathbb{E}[f(t, \omega)^2] \, dt$$

It can be shown that this definition satisfies the specification of a norm mentioned previously. For fixed t^*, $f(t^*, \omega)$ is a random variable with squared norm $\mathbb{E}[f(t^*, \omega)^2]$. The squared norm of random process f on $[0, T]$ is the continuous sum (integral) of all these. Divided by T it is the average of the squared norms of random variables $f(t^*, \omega)$.

Example 1   In the discussion of stochastic integration it says that a general random integrand f can be approximated with any desired degree of accuracy by a random step-function $f^{(n)}$ on a partition of n intervals, by taking n large enough. The closeness of the approximation $f^{(n)}$ to the actual function f is measured by this norm. It is convenient to use its square

$$\left|\left| f^{(n)} - f \right|\right|^2 = \int_{t=0}^{T} \mathbb{E}\left\{ \left[(f^{(n)}(t, \omega) - f(t, \omega) \right]^2 \right\} \, dt$$

For $f^{(n)}$ to approximate f with any desired degree of accuracy, the above expression should go to zero as n becomes very large.

Example 2   For any fixed t, $B(t)$ is a random variable with squared norm $||B(t)||^2 = \mathbb{E}[B(t)^2|| = t$. Brownian motion as a random process has squared norm

$$\int_{t=0}^{T} \mathbb{E}[B(t)]^2 \, dt = \int_{t=0}^{T} t \, dt = \tfrac{1}{2} T^2$$

This is the average of the squared norm of random variables $B(t)$. A non-anticipating step-function approximation of the Brownian motion

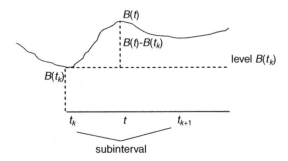

Figure D.6   Approximation of Brownian motion path

path (Figure D.6) has squared norm

$$\int_{t=0}^{1\,\Delta t} \mathbb{E}[B(0\,\Delta t)]^2 \, dt + \cdots + \int_{t=k\,\Delta t}^{(k+1)\,\Delta t} \mathbb{E}[B(k\,\Delta t)]^2 \, dt + \cdots$$

$$+ \int_{t=(n-1)\,\Delta t}^{n\,\Delta t} \mathbb{E}[B((n-1)\,\Delta t)]^2 \, dt$$

Using

$$\int_{t=k\,\Delta t}^{(k+1)\,\Delta t} \mathbb{E}[B(k\,\Delta t)]^2 \, dt = \int_{t=k\,\Delta t}^{(k+1)\,\Delta t} k\,\Delta t \, dt = k(\Delta t)^2$$

gives

$$[0 + 1 + \cdots + (n-1)](\Delta t)^2 = \tfrac{1}{2}(n-1)n \left(\tfrac{T}{n}\right)^2 = \tfrac{1}{2}\left(1 - \tfrac{1}{n}\right)T^2$$

As $n \to \infty$, the squared norm of the approximation converges to the squared norm of the process it approximates, $\tfrac{1}{2}T^2$.

D.5   REFERENCE

An excellent systematic compact exposition of norms with complete proofs is given in *Lectures on Real Analysis* by *Yeh* Chapter 4: Section I introduces the concept, Section II applies it to vectors in n-dimensions, and Section III applies it to continuous functions. This is highly recommended mathematical reading.

Consider an experiment which is repeated under identical conditions. Let the numerical outcomes of these trials be recorded by the sequence of random variables $X_1 \cdots X_i \cdots X_n \cdots$. These random variables are independent and identically distributed. A key question is whether there exists another random variable X which is the limit of this sequence *in some sense*. Recall that a random variable is a function defined on a sample space Ω of points ω. So $X_n(\omega)$ and $X(\omega)$ are the values of random variables X_n and X when outcome ω occurs, that is, $X_n(\omega)$ and $X(\omega)$ are the values of functions X_n and X when the function argument is ω. This is the same as in ordinary calculus where a function f of argument y produces the value $f(y)$. Convergence is about comparing $X_n(\omega)$ and $X(\omega)$ for various ω. This can be done in several ways.

E.1 CENTRAL LIMIT THEOREM

The Central Limit Theorem (CLT) is one of the most remarkable results in probability theory. It says that the sum of a large number of independent identically distributed random variables has a probability distribution that is approximately normal. The probability distribution of these X_i need not not specified; only the mean $\mathbb{E}[X_i] = \mu$, and the variance $\mathbb{V}\mathrm{ar}[X_i] = \sigma^2$ are needed. Let random variable S_n record the sum of the first n outcomes, $S_n \overset{\text{def}}{=} X_1 + \cdots + X_i + \cdots + X_n$. Then its mean is $\mathbb{E}[S_n] = \mathbb{E}[X_1 + \cdots + X_n] = \mathbb{E}[X_1] + \cdots + \mathbb{E}[X_n] = n\mu$. Its variance is $\mathbb{V}\mathrm{ar}[S_n] = \mathbb{V}\mathrm{ar}[X_1 + \cdots + X_n]$ which equals $\mathbb{V}\mathrm{ar}[X_1] + \cdots + \mathbb{V}\mathrm{ar}[X_n]$ because the random variables in the sum are independent. As all $\mathbb{V}\mathrm{ar}[X_i] = \sigma^2$, $\mathbb{V}\mathrm{ar}[S_n] = n\sigma^2$. The standard deviation of S_n is thus $\sigma\sqrt{n}$. Note that if the number of terms n in the sum S_n is increased, both the expected value and the variance of S_n increase linearly with n. That suggests expressing S_n on a scale which corresponds to its rate of growth. To this end, deduct from S_n its expected value, and divide that difference by its standard deviation. That gives the so-called standardized random variable Z

$$Z_n \stackrel{\text{def}}{=} \frac{S_n - \mathbb{E}[S_n]}{\sqrt{\mathbb{V}\text{ar}[S_n]}} = \frac{S_n - n\mu}{\sigma\sqrt{n}}$$

It is called standardized because its expected value is 0 and its standard deviation is 1, as can be seen as follows:

$$\mathbb{E}[Z_n] = \mathbb{E}\left[\frac{S_n - n\mu}{\sigma\sqrt{n}}\right] = \frac{1}{\sigma\sqrt{n}}\{\mathbb{E}[S_n] - n\mu\} = 0$$

$$\mathbb{V}\text{ar}[Z_n] = \mathbb{V}\text{ar}\left[\frac{S_n - n\mu}{\sigma\sqrt{n}}\right] = \left(\frac{1}{\sigma\sqrt{n}}\right)^2 \mathbb{V}\text{ar}[S_n] = \frac{1}{\sigma^2 n}n\sigma^2 = 1$$

The expected value and variance of Z do not change as n changes.

Conducting an experiment n times generates a particular *sample* of values for the random variables X_i, and a corresponding value of Z_n. If the experiment is conducted another n times, another sample of n values is produced, and hence another value of Z_n. Repeating this produces a large number of sample values Z_n which can be arranged into a cumulative histogram. The Central Limit Theorem says that for a 'large n', this histogram resembles the shape of the standard normal distribution function. The probability distribution of Z_n converges to the standard normal distribution as $n \to \infty$. This property comes as no surprise if the individual X_i have a normal distribution, as the scaled sum then also has a normal distribution. But even if the individual distributions do not resemble a normal distribution at all, the scaled sum is still approximately normal. The CLT serves as a simple device for making probabilistic calculations about sums without having to know much about the probabilistic nature of the individual random variables. The CLT can also be expressed in terms of the average X values, $\bar{X}_n$

$$Z_n = \frac{S_n - n\mu}{\sigma\sqrt{n}} = \frac{S_n/n - \mu}{\sigma\sqrt{n}/n} = \frac{S_n/n - \mu}{\sigma/\sqrt{n}} = \frac{\bar{X}_n - \mu}{\sigma/\sqrt{n}}$$

Described here is the standard CLT. Its proof is found in the probability books given in the references. More technical versions of the CLT exist for when the X_i are not identically distributed.

E.2   MEAN-SQUARE CONVERGENCE

Mean-square convergence refers to what happens to the expected value of the squared difference between random variable X_n and random variable X, as $n \to \infty$. The difference between $X_n(\omega)$ and $X(\omega)$ can very with ω. For some ω, $X_n(\omega)$ and $X(\omega)$ are 'very close', for other ω they are 'not so close'. But all the squared differences together must have an average that goes to zero as $n \to \infty$. This can only be verified if the limiting random variable X is known. That is usually not the case and the equivalent condition is $\lim_{n \to \infty} \mathbb{E}[(X_n - X_{n+1})^2] = 0$. Other names are L^2 convergence and convergence in quadratic mean (q.m.). Common shorthand notation is $X_n \xrightarrow{\text{msq}} X$, $X_n \xrightarrow{\text{q.m.}} X$, $X_n \xrightarrow{L^2} X$. In Chapter 3 it is used in the construction of the Itō stochastic integral.

Two other types of convergence are outlined below. These are used in the technical literature.

E.3   ALMOST SURE CONVERGENCE

This is a concept from advanced probability theory that is used in some expositions of stochastic calculus. The initial position of Brownian motion is then specified as $\mathbb{P}\{\omega; B(0, \omega) = 0\} = 1$, or $B(0) = 0$ a.s., the path continuity as $\mathbb{P}\{\omega; B(., \omega) \; is \; continuous\} = 1$, and similar statements for other properties. It can be introduced as follows.

In ordinary calculus, consider a sequence of ordinary functions f_n whose values $f_n(x)$ depend on index n. If the sequence $f_n(x)$ converges to another function $f(x)$ for all values of x, this is known as *pointwise convergence*. For example, if $f_n(x) = 1 - x^n/(1 - x)$ and $f(x) = 1/(1 - x)$ for $-\frac{1}{2} \le x \le \frac{1}{2}$, then $\lim_{n \to \infty} f_n(x) = f(x)$ for *all* x. If this same concept were applied to random variables, then X_n would be converging to X for *all* ω. It turns out that this is not possible. There are always some ω for which there cannot be convergence. But if the set of these ω has probability zero then these ω 'do not matter'. A sequence of random variables $X_1 \cdots X_n \ldots$ converges *almost surely* to a random variable X if for an arbitrarily small positive ϵ

$$\mathbb{P}[\omega : \lim_{n \to \infty} |X_n(\omega) - X(\omega)| < \epsilon] = 1$$

This is about the set of all ω for which $X_n(\omega) \to X(\omega)$ as $n \to \infty$. Convergence need not take place on a set that has probability zero of occurring, hence the qualification 'almost sure' rather than 'sure'. *Almost sure convergence* can be seen as the probabilistic version of *pointwise convergence*. $X_n(\omega)$ need not be arbitrarily close to $X(\omega)$ for *all* ω, as $n \to \infty$ but the collection of ω for which there is no closeness must have probability 0. Alternative notation is to write the event

$\lim_{n\to\infty} |X_n(\omega) - X(\omega)| < \epsilon$ as $\lim_{n\to\infty} X_n(\omega) = X(\omega)$ and its probability as $\mathbb{P}[\omega : \lim_{n\to\infty} X_n(\omega) = X(\omega)] = 1$, or as $X_n(\omega) \to X(\omega)$ and $\mathbb{P}[\omega : X_n(\omega) \to X(\omega) \text{ as } n \to \infty] = 1$. This type of convergence is also known as *convergence almost everywhere*. Common shorthand notation is $X_n \xrightarrow{\text{a.s.}} X$ or $\lim_{n\to\infty} X_n = X$ a.s. (the reference to ω is often omitted).

Almost sure convergence is a probability of a limit.

E.4   CONVERGENCE IN PROBABILITY

This is another advanced probability concept that is used in some stochastic calculus books.

For an outcome ω for which $X_n(\omega)$ is not arbitrarily close to $X(\omega)$ it holds that $|X_n(\omega) - X(\omega)| \geq \epsilon$. The set of all such outcomes ω is $\{\omega : |X_n(\omega) - X(\omega)| \geq \epsilon\}$. This set is also known as a tail event. The probability of this tail event is $\mathbb{P}[\{\omega : |X_n(\omega) - X(\omega)| \geq \epsilon\}]$ and depends on n. What happens to this probability as $n \longrightarrow \infty$ is the limit $\lim_{n\to\infty} \mathbb{P}[\{\omega : |X_n(\omega) - X(\omega)| \geq \epsilon\}]$. The tail event is often written without ω as $|X_n - X| \geq \epsilon$, and its probability as $\mathbb{P}[|X_n - X| \geq \epsilon]$. A sequence of random variables $X_1, X_2 \ldots, X_n$ converges *in probability* to a random variable X if for any $\epsilon > 0$

$$\lim_{n\to\infty} \mathbb{P}[\{\omega : |X_n(\omega) - X(\omega)| \geq \epsilon\}] = 0$$

which is equivalent to

$$\lim_{n\to\infty} \mathbb{P}[\{\omega : |X_n(\omega) - X(\omega)| < \epsilon\}] = 1$$

In other notation, it is the limit of the sequence of numbers p_n where

$$p_n \overset{\text{def}}{=} \mathbb{P}[\{\omega : |X_n(\omega) - X(\omega)| \geq \epsilon\}]$$

Common shorthand notation for this is $X_n \xrightarrow{\mathbb{P}} X$.

Convergence in probability is a limit of a sequence of probabilities.

E.5   SUMMARY

- *Almost sure convergence* requires $X_n(\omega) - X(\omega)$ to get small for almost all ω.

- *Convergence in probability* requires the probability of this difference to get small. Hence *Convergence in probability* is less demanding than *Almost sure convergence*. One says that *Convergence in probability* is the weaker of these two types of convergence. If *Almost sure convergence* is true then *Convergence in probability* is also true.

- Mean-square convergence implies *Convergence in probability* and this in turn implies *Convergence in distribution.*

CHAPTER 1

Answer [1.9.1] *Scaled Brownian Motion*

Let $X(t) \stackrel{\text{def}}{=} \sqrt{c}B(t/c)$. As $B(t/c)$ denotes the position of the Brownian motion at time t/c, its variance equals t/c. Since $X(t)$ is a positive multiple $\sqrt{c}$ of a Brownian motion, it can assume any values from $-\infty$ to ∞. The expected value of $X(t)$ and its variance can be computed without first working out the probability density of $X(t)$.

(a) The expected value of $X(t)$ is $\mathbb{E}[X(t)] = \mathbb{E}[\sqrt{c}B(t/c)] = \sqrt{c}\,\mathbb{E}[B(t/c)] = \sqrt{c}\,0 = 0$.

(b) The variance of $X(t)$ is $\text{Var}[X(t)] = \text{Var}[\sqrt{c}B(t/c)] = (\sqrt{c})^2\text{Var}[B(t/c)] = c(t/c) = t$. So the scaled Brownian motion has the same mean and variance as Brownian motion.

(c) To derive $\mathbb{P}[X(t) \leq x]$ first replace $X(t)$ by its definition, $\mathbb{P}[X(t) \leq x] = \mathbb{P}[\sqrt{c}B(t/c) \leq x]$. Next rearrange the inequality to get a random variable on the left for which the distribution is known

$$\mathbb{P}[\sqrt{c}B(\tfrac{t}{c})] \leq x] = \mathbb{P}[B(\tfrac{t}{c}) \leq \tfrac{x}{\sqrt{c}}]$$

$$\mathbb{P}[X(t) \leq x] = \int_{y=-\infty}^{x/\sqrt{c}} \frac{1}{\sqrt{t/c}\sqrt{2\pi}} \exp\left(-\frac{1}{2}\left[\frac{y}{\sqrt{t/c}}\right]^2\right) dy$$

(d) To derive the probability density of $X(t)$, differentiate the probability distribution with respect to x. That gives the expression for the integrand with y replaced by the upper integration limit $x/\sqrt{c}$, times this upper limit differentiated with respect to x. The density of $X(t)$ at $X(t) = x$ is

$$\frac{1}{\sqrt{t/c}\sqrt{2\pi}} \exp\left(-\frac{1}{2}\left[\frac{x/\sqrt{c}}{\sqrt{t/c}}\right]^2\right) \frac{d(x/\sqrt{c})}{dx}$$

$$= \frac{1}{\sqrt{t}\sqrt{2\pi}} \exp\left(-\frac{1}{2}\left[\frac{x}{\sqrt{t}}\right]^2\right)$$

which is the same as a Brownian motion density.

(e) Check if the variance over an arbitrary interval equals the length of the interval and is independent of the location of the interval.

$$\mathbb{V}\mathrm{ar}[X(t+u) - X(t)] = \mathbb{V}\mathrm{ar}[\sqrt{c}B(\tfrac{t+u}{c}) - \sqrt{c}B(\tfrac{t}{c})]$$
$$= c\mathbb{V}\mathrm{ar}[B(\tfrac{t+u}{c}) - B(\tfrac{t}{c})]$$

As $B((t+u)/c) - B(t/c)$ is the movement from time t/c to time $(t+u)/c$, its variance equals the length of this interval,

$$\mathbb{V}\mathrm{ar}[B(\tfrac{t+u}{c}) - B(\tfrac{t}{c})] = \tfrac{u}{c}$$

Thus

$$\mathbb{V}\mathrm{ar}[X(t+u) - X(t)] = \tfrac{cu}{c} = u$$

(f) This variance is independent of t so $X(t)$ is indeed a Brownian motion. As the scaled process X has the same probability distribution as Brownian motion, Brownian motion is said to be *statistically self-similar*.

Answer [1.9.2] *Seemingly Brownian Motion*

Let $X(t) \overset{\text{def}}{=} Z\sqrt{t}$ where $Z \sim N(0, 1)$.

(a) $\mathbb{E}[X(t)] = \mathbb{E}[Z\sqrt{t}] = \sqrt{t}\mathbb{E}[Z] = \sqrt{t}0 = 0$

(b) $\mathbb{V}\mathrm{ar}[X(t)] = \mathbb{V}\mathrm{ar}[Z\sqrt{t}] = (\sqrt{t})^2\mathbb{V}\mathrm{ar}[Z] = t1 = t$

Expected value and variance over time period $[0, t]$ are the same as for Brownian motion.

(c) Probability distribution of $X(t)$

$$\mathbb{P}[X(t) \le x] = \mathbb{P}[Z\sqrt{t} \le x] = \mathbb{P}[Z \le \tfrac{x}{\sqrt{t}}]$$

$$= \int_{y=-\infty}^{x/\sqrt{t}} \frac{1}{\sqrt{2\pi}} \exp(-\tfrac{1}{2}y^2)\, dy$$

(d) Probability density of $X(t)$ at $X(t) = x$

$$\frac{1}{\sqrt{2\pi}} \exp\left(-\frac{1}{2}\left[\frac{x}{\sqrt{t}}\right]^2\right) \frac{d(x/\sqrt{t})}{dx} = \frac{1}{\sqrt{2\pi}} \exp\left(-\frac{1}{2}\left[\frac{x}{\sqrt{t}}\right]^2\right)(1/\sqrt{t})$$

$$= \frac{1}{\sqrt{t}\sqrt{2\pi}} \exp\left(-\frac{1}{2}\left[\frac{x}{\sqrt{t}}\right]^2\right)$$

(e) For any two times t and $t + u$

$$\begin{aligned}
\mathbb{V}\mathrm{ar}[X(t+u) - X(t)] &= \mathbb{V}\mathrm{ar}[Z\sqrt{t+u} - Z\sqrt{t}] \\
&= \mathbb{V}\mathrm{ar}[Z(\sqrt{t+u} - \sqrt{t})] \\
&= (\sqrt{t+u} - \sqrt{t})^2 \mathbb{V}\mathrm{ar}[Z] \\
&= (\sqrt{t+u} - \sqrt{t})^2 \\
&= t + u - 2\sqrt{t+u}\sqrt{t} + t \\
&= 2t + u - \sqrt{t^2 + ut}
\end{aligned}$$

(f) For a Brownian motion $\mathbb{V}\mathrm{ar}[B(t+u) - B(t)] = u$. This is not the case here, so the process $X(t)$ is *not* a Brownian motion. It just seemed that it might be.

Increments of X

The value of Z is the same for all t, so $X(t)$ is a non-linear function of t. For example, if $Z = 0.2$ then $X(t) = 0.2\sqrt{t}$. Consider the increments of X over adjacent non-overlapping intervals $[t_1, t_2]$ and $[t_2, t_3]$

$$X(t_2) - X(t_1) = Z(\sqrt{t_2} - \sqrt{t_1}) \qquad X(t_3) - X(t_2) = Z(\sqrt{t_3} - \sqrt{t_2})$$

Their covariance is

$$\begin{aligned}
\mathbb{C}\mathrm{ov}&[Z(\sqrt{t_2} - \sqrt{t_1}), Z(\sqrt{t_3} - \sqrt{t_2})] \\
&= (\sqrt{t_2} - \sqrt{t_1})(\sqrt{t_3} - \sqrt{t_2})\mathbb{C}\mathrm{ov}[Z, Z] \\
&= (\sqrt{t_2} - \sqrt{t_1})(\sqrt{t_3} - \sqrt{t_2})1 \neq 0
\end{aligned}$$

Brownian Motion Increments

Each increment is generated by its own Z_i. These Z_i are independent.

$$B(t_2) - B(t_1) = Z_1\sqrt{t_2 - t_1} \qquad B(t_3) - B(t_2) = Z_2\sqrt{t_3 - t_2}$$

Their covariance is

$$\mathbb{Cov}[Z_1\sqrt{t_2 - t_1}, Z_2\sqrt{t_3 - t_2}] = \sqrt{t_2 - t_1}\sqrt{t_3 - t_2}]\mathbb{Cov}[Z_1, Z_2]$$
$$= \sqrt{t_2 - t_1}\sqrt{t_3 - t_2}]0 = 0$$

as $\mathbb{Cov}[Z_1, Z_2] = 0$ due to independence of Z_1 and Z_2.

Answer [1.9.3] Combination of Brownian Motions

$$\mathbb{E}[Z(t)] = \mathbb{E}\left[\alpha B(t) - \sqrt{\beta}B^*(t)\right]$$
$$= \mathbb{E}[\alpha B(t)] - \mathbb{E}\left[\sqrt{\beta}B^*(t)\right]$$
$$= \alpha\mathbb{E}[B(t)] - \sqrt{\beta}\mathbb{E}[B^*(t)]$$
$$= \alpha 0 - \sqrt{\beta}0 = 0$$

$$\mathbb{Var}[Z(t+u) - Z(t)] = \mathbb{Var}[\{\alpha B(t+u) - \sqrt{\beta}B^*(t+u)\}$$
$$-\{\alpha B(t) - \sqrt{\beta}B^*(t)\}]$$
$$= \mathbb{Var}[\alpha\{B(t+u) - B(t)\}$$
$$-\sqrt{\beta}\{B^*(t+u) - B^*(t)\}]$$
$$= \mathbb{Var}[\alpha\{B(t+u) - B(t)\}]$$
$$+\mathbb{Var}[\sqrt{\beta}\{B^*(t+u) - B^*(t)\}]$$

as B and B^* are independent, so

$$\mathbb{Var}[Z(t+u) - Z(t)] = \alpha^2 u + \beta u = (\alpha^2 + \beta)u$$

This should equal u and not depend on t.

$$\mathbb{Var}[Z(t+u) - Z(t)] = u \text{ if } \alpha^2 + \beta = 1 \text{ or } \beta = 1 - \alpha^2$$

Under that condition $Z(t)$ is Brownian motion.

Answer [1.9.4] Correlation between Brownian Motions

Correlation coefficient ρ is by definition

$$\frac{\mathbb{Cov}[B(t), B(t+u)]}{\sqrt{\mathbb{Var}[B(t)]}\sqrt{\mathbb{Var}[B(t+u)]}}$$

The numerator equals $\min(t, t+u) = t$ as derived in Chapter 1.

$$\rho = \frac{t}{\sqrt{t}\sqrt{t+u}}$$

Answer [1.9.5] *Successive Brownian Motions*

(a) At time 4, the BM position is X. Density at $X = x$ is

$$f(x) = \frac{1}{\sqrt{4}\sqrt{2\pi}} \exp\left[-\frac{1}{2}\left(\frac{x}{\sqrt{4}}\right)^2\right]$$

$\mathbb{P}[X \geq 0] = \int_{x=0}^{\infty} f(x)\,dx = 0.5$ due to symmetry of normal density.

(b) Joint density at times t_1 and t_2 is

$$f(x_1, x_2) \stackrel{\text{def}}{=} \frac{1}{\sqrt{t_1 - 0}\sqrt{2\pi}} \exp\left[-\frac{1}{2}\left(\frac{x_1 - 0}{\sqrt{t_1 - 0}}\right)^2\right]$$

$$\times \frac{1}{\sqrt{t_2 - t_1}\sqrt{2\pi}} \exp\left[-\frac{1}{2}\left(\frac{x_2 - x_1}{\sqrt{t_2 - t_1}}\right)^2\right]$$

Gate specification:

gate 1	time t_1	$x_{1\text{Low}}$	$x_{1\text{High}}$
gate 2	time t_2	$x_{2\text{Low}}$	$x_{2\text{High}}$

Probability of BM paths passing through gate 1 and gate 2

$$p_{12} \stackrel{\text{def}}{=} \int_{x_2=x_{2\text{Low}}}^{x_{2\text{High}}} \int_{x_1=x_{1\text{Low}}}^{x_{1\text{High}}} f(x_1, x_2)\,dx_1\,dx_2$$

Probability of both path positions being positive

$$p_{12} = \int_{x_2=0}^{\infty} \int_{x_1=0}^{\infty} f(x_1, x_2)\,dx_1\,dx_2$$

(c) $\dfrac{1}{p_{12}} \displaystyle\int_{x_2=0}^{\infty} x_2\, f(x_2)\,dx_2$

The results can all be computed by numerical integration and verified by simulation.

Answer [1.9.8] *Brownian Bridge*

(a) $X(1) = B(1) - 1B(1) = 0;$ $X(0) = B(0) - 0B(0) = 0;$ so
$X(1) = X(0)$

(b) $\text{Cov}[X(t), X(t+u)] = \text{Cov}[B(t) - tB(1), B(t+u)$
$$-(t+u)B(1)]$$
$$= \text{Cov}[B(t), B(t+u)]$$
$$+ \text{Cov}[-tB(1), B(t+u)]$$
$$+ \text{Cov}[B(t), -(t+u)B(1)]$$
$$+ \text{Cov}[-tB(1), -(t+u)B(1)]$$
$$= \min(t, t+u) - t\text{Cov}[B(1), B(t+u)]$$
$$- (t+u)\text{Cov}[B(t), B(1)]$$
$$+ t(t+u)\text{Cov}[B(1), B(1)]$$
$$= \min(t, t+u) - t\min(1, t+u)$$
$$- (t+u)\min(t, 1) + t(t+u)\text{Var}[B(1)]$$

The time interval is $0 \leq t \leq 1$. For $t < 1$ and $t + u < 1$ the above equals

$$t - t(t+u) - (t+u)t + t(t+u)1 = t - t(t+u)$$

Alternative Derivation

Use the definition of covariance

$$\text{Cov}[X(t), X(t+u)] = \mathbb{E}[\{X(t) - \mathbb{E}[X(t)]\}\{X(t+u)$$
$$- \mathbb{E}[X(t+u)]\}]$$

Substituting $\mathbb{E}[X(t)] = 0$ and $\mathbb{E}[X(t+u)] = 0$ leaves

$$\text{Cov}[X(t), X(t+u)] = \mathbb{E}[X(t)X(t+u)]$$

Substituting the expressions for $X(t)$ and $X(t+u)$ and multiplying terms gives

$$\mathbb{E}[B(t)B(t+u) - (t+u)B(t)B(1) - tB(1)B(t+u)$$
$$+ t(t+u)B(1)^2] = \mathbb{E}[B(t)B(t+u)] - (t+u)\mathbb{E}[B(t)B(1)]$$
$$- t\mathbb{E}[B(1)B(t+u)] + t(t+u)\mathbb{E}[B(1)^2]$$

The remainder is as above.

Answer [1.9.9] *Brownian Motion through Gates*

Let the position of the Brownian motion path at time t be denoted x. The density at time t_1 is

$$f(x_1) = \frac{1}{\sqrt{t_1}\sqrt{2\pi}} \exp\left[-\frac{1}{2}\left(\frac{x_1 - 0}{\sqrt{t_1 - 0}}\right)^2\right]$$

The joint density at times t_1 and t_2 is

$$f(x_1, x_2) \overset{\text{def}}{=} \frac{1}{\sqrt{t_1}\sqrt{2\pi}} \exp\left[-\frac{1}{2}\left(\frac{x_1 - 0}{\sqrt{t_1 - 0}}\right)^2\right]$$

$$\times \exp\left[-\frac{1}{2}\left(\frac{x_2 - x_1}{\sqrt{t_2 - t_1}}\right)^2\right]$$

(a) $p_1 \overset{\text{def}}{=} \mathbb{P}[B(t_1) \geq 0] = \int_{x_1=0}^{\infty} f(x_1)dx_1 = 0.5$

$$p_{12} \overset{\text{def}}{=} \mathbb{P}[B(t_1) \geq 0 \text{ and } B(t_2) \geq 0]$$

$$= \int_{x_2=0}^{\infty} \int_{x_1=0}^{\infty} f(x_1, x_2)\, dx_1\, dx_2$$

Then the 'average' position at time t_1 of all paths that have a positive position at time t_1 is

$$\mathbb{E}[B(t_1) \text{ conditional upon } B(t_1) \geq 0] = \frac{1}{p_1} \int_{x_1=0}^{\infty} x_1 f(x_1)dx_1$$

Note the scaling by p_1

Specifying t_1 permits numerical calculation. For example when $t_1 = 1$

$$f(x_1) = \frac{1}{\sqrt{2\pi}} \exp\left[-\tfrac{1}{2}x_1^2\right]$$

and

$$\mathbb{E}[B(1) \text{ conditional upon } B(1) \geq 0]$$

$$= \frac{1}{0.5} \int_{x_1=0}^{\infty} x_1 \frac{1}{\sqrt{2\pi}} \exp\left[-\tfrac{1}{2}x_1^2\right] dx_1$$

$$= \frac{-2}{\sqrt{2\pi}} \int_{x_1=0}^{\infty} \exp\left[-\tfrac{1}{2}x_1^2\right] d\left(-\tfrac{1}{2}x_1^2\right)$$

$$= \frac{-2}{\sqrt{2\pi}} \exp\left[-\tfrac{1}{2}x_1^2\right]\Big|_{x_1=0}^{\infty} = \frac{-2}{\sqrt{2\pi}}[0-1]$$

$$= 0.7979$$

(b) The average position at t_2 of all paths that have a positive position at time t_1 and time t_2 is

$$\mathbb{E}[B(t_2) \text{ conditional upon } B(t_1)$$

$$\geq 0 \text{ and } B(t_2) \geq 0] = \frac{1}{p_{12}} \int_{x_2=0}^{\infty} \int_{x_1=0}^{\infty} x_2 f(x_1, x_2) \, dx_1 \, dx_2$$

Note the scaling by p_{12}

(c) Expected value of increment over $[t_1, t_2]$ of paths passing through gate 1 and gate 2

$$\frac{1}{p_{12}} \int_{x_2=x_{2\text{Low}}}^{x_{2\text{High}}} \int_{x_1=x_{1\text{Low}}}^{x_{1\text{High}}} (x_2 - x_1) f(x_1, x_2) \, dx_1 \, dx_2$$

The results can all be computed by numerical integration and verified by simulation.

CHAPTER 2

Answer [2.8.1]

This is the same as the process discussed in Section 2.6.1.

Answer [2.8.2]

For process $S_n^* \overset{\text{def}}{=} S_n/(pu + qd)^n$ to be a martingale, it must hold that $\mathbb{E}[S_{n+1}^*|\Im_n] = S_n^*$ where $S_{n+1}^* = S_{n+1}/(pu + qd)^{n+1}$. In the up-state, $S_{n+1} = uS_n$ so $S_{n+1}^* = uS_n/(pu + qd)^{n+1}$. Similarly in the down-state,

$S_{n+1}^* = dS_n/(pu+qd)^{n+1}$. Thus

$$\mathbb{E}[S_{n+1}^*|\Im_n] = p\left\{\frac{uS_n}{(pu+qd)^{n+1}}\right\} + q\left\{\frac{dS_n}{(pu+qd)^{n+1}}\right\}$$
$$= \frac{(pu+qd)S_n}{(pu+qd)^{n+1}}$$
$$= \frac{S_n}{(pu+qd)^n}$$

Thus $\mathbb{E}[S_{n+1}^*|\Im_n] = S_n^*$ and S_n^* *is a martingale.*

Answer [2.8.3]

Let $s < t$. Substituting $B(t) = B(s) + \{B(t) - B(s)\}$ gives

$$\mathbb{E}[B(t) + 4t|\Im(s)]$$
$$= \mathbb{E}[B(s) + \{B(t) - B(s)\} + 4t|\Im(s)]$$
$$= \mathbb{E}[B(s)|\Im(s)] + \mathbb{E}[B(t) - B(s)|\Im(s)] + \mathbb{E}[4t|\Im(s)]$$
$$= B(s) + 0 + 4t$$

So $\mathbb{E}[B(t) + 4t|\Im(s)] \neq B(s) + 4s$. Thus $B(t) + 4t$ is *not* a martingale. There was no need to decompose t.

Answer [2.8.4]

(a) This is the continuous counterpart of the discrete process S_n^2 discussed in the text. To evaluate $\mathbb{E}[B(t)^2|\Im(s)]$, write $B(t) = B(s) + \{B(t) - B(s)\}$. That gives

$$\mathbb{E}[[B(s) + \{B(t) - B(s)\}]^2|\Im(s)]$$
$$= \mathbb{E}[B(s)^2 + 2B(s)\{B(t) - B(s)\} + \{B(t) - B(s)\}^2|\Im(s)]$$
$$= \mathbb{E}[B(s)^2|\Im(s)] + \mathbb{E}[2B(s)\{B(t) - B(s)\}|\Im(s)]$$
$$+ \mathbb{E}[\{B(t) - B(s)\}^2|\Im(s)]$$

In the first term, $B(s)$ is known, so $\mathbb{E}[B(s)^2|\Im(s)] = B(s)^2$. In the second term, $B(t) - B(s)$ is independent of $B(s)$, so $\mathbb{E}[B(s)\{B(t) - B(s)\}|\Im(s)]$ can be written as the product $\mathbb{E}[B(s)|\Im(s)]\mathbb{E}[B(t) - B(s)|\Im(s)] = B(s)0 = 0$. The third term, $\mathbb{E}[\{B(t) - B(s)\}^2|\Im(s)] = (t - s)$. Thus $\mathbb{E}[B(t)^2|\Im(s)] = B(s)^2 + (t - s) \neq B(s)^2$ and the random pocess $B(t)^2$ is *not* a martingale. The difference is the variance over the period s to t.

(b) Let $Z \overset{\text{def}}{=} B(t)^2$. To find the probability density of Z, first derive its probability distribution, $\mathbb{P}[Z \leq z]$, which is a function of z, denoted $F(z)$.

$$\mathbb{P}[Z \leq z] = \mathbb{P}[B(t)^2 \leq z] = \mathbb{P}[-\sqrt{z} \leq B(t) \leq \sqrt{z}]$$
$$= \mathbb{P}[B(t) \leq \sqrt{z}] - \mathbb{P}[B(t) \leq -\sqrt{z}]$$

The last term $\mathbb{P}[B(t) \leq -\sqrt{z}] = \mathbb{P}[B(t) \geq \sqrt{z}] = 1 - \mathbb{P}[B(t) \leq \sqrt{z}]$ as the left tail equals the right tail since the probability density of $B(t)$ is symmetric. So

$$\mathbb{P}[Z \leq z] = \mathbb{P}[B(t) \leq \sqrt{z}] - (1 - \mathbb{P}[B(t) \leq \sqrt{z}])$$
$$= 2\mathbb{P}[B(t) \leq \sqrt{z}] - 1$$

$$F(z) = 2 \int_{x=-\infty}^{\sqrt{z}} \frac{1}{\sqrt{t}\sqrt{2\pi}} \exp\left[-\frac{1}{2}\left(\frac{x}{\sqrt{t}}\right)^2\right] dx - 1$$

Probability density at $Z = z$ is

$$\frac{d}{dz}F(z) = 2\frac{1}{\sqrt{t}\sqrt{2\pi}} \exp\left[-\frac{1}{2}\left(\frac{\sqrt{z}}{\sqrt{t}}\right)^2\right] \frac{d}{dz}(\sqrt{z})$$

where the last term $d/dz(\sqrt{z}) = \frac{1}{2}1/(\sqrt{z})$. The probability density, for $z>0$, is

$$2\frac{1}{\sqrt{t}\sqrt{2\pi}} \exp\left[-\frac{1}{2}\left(\frac{\sqrt{z}}{\sqrt{t}}\right)^2\right]\frac{1}{2}\frac{1}{\sqrt{z}} = \frac{1}{\sqrt{z}}\frac{1}{\sqrt{t}\sqrt{2\pi}} \exp\left[-\frac{1}{2}\frac{z}{t}\right]$$

For any t this has the shape of a negative exponential.

Answer [2.8.5]

(a) This is the continuous counterpart of the discrete process $S_n^2 - n$ discussed in the text. Decompose t into $s + (t - s)$. Repeating the steps above shows that $B(t)^2 - t$ *is* a martingale. The term that was subtracted to transform $B(t)^2$ into a martingale is known as a *compensator*.

(b) Let $Z \overset{\text{def}}{=} B(t)^2 - t$. To find the probability density of Z, first derive its probability distribution, $\mathbb{P}[Z \leq z]$, which is a function of z, denoted $F(z)$.

$$\mathbb{P}[Z \le z] = \mathbb{P}[B(t)^2 - t \le z] = \mathbb{P}[B(t)^2 \le z + t]$$
$$= \mathbb{P}[-\sqrt{z + t} \le B(t) \le \sqrt{z + t}]$$

Let $z + t$ be denoted z^*. Then the rest of the derivation is the same as in [2.8.4(b)] with z replaced by z^*. For $z^* > 0$, that is, for $z > -t$

$$F(z) = \mathbb{P}[Z \le z] = 2 \int_{x=-\infty}^{\sqrt{z+t}} \frac{1}{\sqrt{t}\sqrt{2\pi}} \exp\left[-\frac{1}{2}\left(\frac{x}{\sqrt{t}}\right)^2\right] dx - 1$$

Probability density is

$$\frac{1}{\sqrt{z+t}} \frac{1}{\sqrt{t}\sqrt{2\pi}} \exp\left[-\frac{1}{2}\frac{z+t}{t}\right]$$

Answer [2.8.6]

It is to be verified whether

$$\mathbb{E}\left[\exp\left[-\varphi B(t) - \tfrac{1}{2}\varphi^2 t\right] | \Im(s)\right] = \exp\left[-\varphi B(s) - \tfrac{1}{2}\varphi^2 s\right]$$

Write $B(t)$ as the known value $B(s)$ plus the random variable $[B(t) - B(s)]$, and t as $\{s + [t - s]\}$. The left-hand side is then

$$\mathbb{E}[\exp[-\varphi\{B(s) + [B(t) - B(s)]\} - \tfrac{1}{2}\varphi^2\{s + [t - s]\}]|\Im(s)]$$

Collecting all values that are known at time s, and taking them outside the $\mathbb{E}$ operator gives

$$\exp\left[-\varphi B(s) - \tfrac{1}{2}\varphi^2 s\right] \mathbb{E}\left[\exp\left\{-\varphi[B(t) - B(s)] - \tfrac{1}{2}\varphi^2[t - s]\right\} | \Im(s)\right]$$

Consider the exponent of the second term and call it Y, so

$$Y \stackrel{\text{def}}{=} -\varphi[B(t) - B(s)] - \tfrac{1}{2}\varphi^2[t - s]\}$$

Then

$$\mathbb{E}\left\{\exp\left[-\varphi[B(t) - B(s)] - \tfrac{1}{2}\varphi^2[t - s]\right] | \Im(s)\right\}$$
$$= \mathbb{E}\{\exp[Y]\} = \exp\left\{\mathbb{E}[Y] + \tfrac{1}{2}\mathbb{V}\mathrm{ar}[Y]\right\}$$

As $\mathbb{E}[Y] = -\tfrac{1}{2}\varphi^2[t - s]$ and $\mathbb{V}\mathrm{ar}[Y] = (-\varphi)^2(t - s)$,

$$\exp\left\{\mathbb{E}[Y] + \tfrac{1}{2}\mathbb{V}\mathrm{ar}[Y]\right\} = \exp\left\{-\tfrac{1}{2}\varphi^2[t - s] + \tfrac{1}{2}\varphi^2(t - s)\right\}$$
$$= \exp\{0\} = 1$$

Thus $\mathbb{E}\left[\exp\left[-\varphi B(t) - \frac{1}{2}\varphi^2 t\right] | \Im(s)\right] = \exp\left[-\varphi B(s) - \frac{1}{2}\varphi^2 s\right]$, and the martingale property has been shown. The origin of this question is in verifying whether $\exp[-\varphi B(t)]$ is a martingale, and, if not, whether it can be modified to a martingale by a suitable choice of a term in the exponent which compensates the movement of $-\varphi B(t)$ in expectation.

$$\mathbb{E}[\exp[-\varphi B(t)]|\Im(s)] = \mathbb{E}[\exp[-\varphi\{B(s) + [B(t) - B(s)]\}|\Im(s)]]$$
$$= \exp[-\varphi B(s)]\mathbb{E}\{\exp[-\varphi[B(t) - B(s)]|\Im(s)]\}$$

The exponent of the second term, $-\varphi[B(t) - B(s)]$, has mean 0 and variance $\varphi^2(t - s)$, thus

$$\mathbb{E}\{\exp[-\varphi[B(t) - B(s)]|\Im(s)]\} = \exp\left[0 + \frac{1}{2}\varphi^2(t - s)\right]$$

so the result is

$$\mathbb{E}[\exp[-\varphi B(t)]|\Im(s)] = \exp[-\varphi B(s)]\exp\left[\frac{1}{2}\varphi^2(t - s)\right]$$

This does *not* equal $\exp[-\varphi B(s)]$, so $\exp[-\varphi B(t)]$ is *not* a martingale. To transform it into a martingale, move $\exp\left[\frac{1}{2}\varphi^2 t\right]$ to the left-hand. That gives

$$\mathbb{E}\left[\exp\left[-\varphi B(t) - \frac{1}{2}\varphi^2 t\right]|\Im(s)\right] = \exp\left[-\varphi B(s) - \frac{1}{2}\varphi^2 s\right]$$

which *is* the expression for a martingale.

Introduce the notation $Z \stackrel{\text{def}}{=} \exp\left[-\varphi B(t) - \frac{1}{2}\varphi^2 t\right]$. As $\ln[Z] = -\varphi B(t) - \frac{1}{2}\varphi^2 t$ is a normal random variable, Z has a lognormal distribution with parameters

$$m \stackrel{\text{def}}{=} \mathbb{E}\{\ln[Z]\} = -\frac{1}{2}\varphi^2 t \qquad s \stackrel{\text{def}}{=} \mathcal{S}tdev\{\ln[Z]\} = \varphi\sqrt{t}$$

The density of Z at $Z = z$ is

$$\frac{1}{zs\sqrt{2\pi}}\exp\left[-\frac{1}{2}\left(\frac{\ln(z) - m}{s}\right)^2\right]$$

Answer [2.8.7]

$$\mathbb{E}[\{M(u) - M(s)\}^2|\Im(s)] = \mathbb{E}[M(u)^2|\Im(s)]$$
$$- 2\mathbb{E}[M(u)M(s)|\Im(s)] + \mathbb{E}[M(s)^2|\Im(s)]$$

When $\Im(s)$ is given, $M(s)$ is known, so

$$\mathbb{E}[M(u)M(s)|\Im(s)] = M(s)\mathbb{E}[M(u)|\Im(s)]$$

and

$$\mathbb{E}[M(s)^2|\Im(s)] = M(s)^2$$

The martingale property of M implies that

$$\mathbb{E}[M(u)|\Im(s)] = M(s)$$

Substituting these gives

$$\mathbb{E}[M(u)^2 - M(s)^2|\Im(s)]$$

or

$$\mathbb{E}[M(u)^2|\Im(s)] - M(s)^2$$

CHAPTER 3

Answer [3.9.1]

(a) $I_n \overset{\text{def}}{=} \displaystyle\sum_{k=0}^{n-1} f(t_k)[B(t_{k+1}) - B(t_k)]$

Why could I_n be a martingale? The fact that $[B(t_{k+1}) - B(t_k)]$ is a martingale, and that I_n is just a combination of several of these, with weightings $f(t_k)$ that are known before the movements are generated. For this reason I_n is also called a *martingale transform* (a transformation of martingales). It is to be verified whether $\mathbb{E}[I_{n+1}|\Im_n] = I_n$. Since $I_n = \mathbb{E}[I_n|\Im_n]$ this is the same as verifying whether $\mathbb{E}[I_{n+1} - I_n|\Im_n] = 0$.

$$I_{n+1} = \sum_{k=0}^{n-1} f(t_k)[B(t_{k+1}) - B(t_k)] + f(t_n)[B(t_{n+1}) - B(t_n)]$$

$$I_{n+1} - I_n = f(t_n)[B(t_{n+1}) - B(t_n)]$$

$$\begin{aligned}\mathbb{E}[I_{n+1} - I_n|\Im_n] &= \mathbb{E}[f(t_n)[B(t_{n+1}) - B(t_n)]|\Im_n] \\ &= f(t_n)\mathbb{E}[B(t_{n+1}) - B(t_n)|\Im_n]\end{aligned}$$

since $f(t_n)$ is known when $\Im_n$ is given. As $\mathbb{E}[B(t_{n+1}) - B(t_n)|\Im_n] = 0$ the result is $\mathbb{E}[I_{n+1} - I_n|\Im_n] = 0$. Thus I_n is a *discrete*-time martingale.

If f is a Brownian motion then

$$I(n) = \sum_{k=0}^{n-1} B(t_k)[B(t_{k+1}) - B(t_k)]$$

is a discrete-time martingale.

(b) Expression for variance derived in Section 3.3 is $\mathbb{V}\mathrm{ar}[I(n)] = \sum_{k=0}^{n-1} \mathbb{E}[f_k^2]\Delta t$ where $\Delta t = T/n$. Substituting $f_k = B(t_k)$ gives $\mathbb{V}\mathrm{ar}[I] = \sum_{k=0}^{n-1} \mathbb{E}[B(t_k)^2](T/n)$. Using $\mathbb{E}[B(t_k)^2] = t_k = k(T/n)$ gives

$$\mathbb{V}\mathrm{ar}[I] = \sum_{k=0}^{n-1} k\left(\tfrac{T}{n}\right)\left(\tfrac{T}{n}\right) = \left(\tfrac{T}{n}\right)^2 \sum_{k=0}^{n-1} k$$

$$= \left(\tfrac{T}{n}\right)^2 \tfrac{1}{2}(n-1)n = \tfrac{1}{2}T^2\left(1 - \tfrac{1}{n}\right)$$

(c) As $n \to \infty$, $\mathbb{V}\mathrm{ar}[I] = \tfrac{1}{2}T^2(1 - 1/n) \to \tfrac{1}{2}T^2$

Answer [3.9.2]

(a) $f^{(n+1)} = B(t_{\frac{1}{2}})\,1_{[t_{1/2},t_1)} + \sum_{k=1}^{n-1} B(t_k)\,1_{[t_k,t_{k+1})}$

$$= B(t_{\frac{1}{2}})\,1_{[t_{1/2},t_1)} + f^{(n)}$$

(b) Integrands $f^{(n+1)}$ and $f^{(n)}$ differ by $B(t_{\frac{1}{2}})\,1_{[t_{1/2},t_1)}$. The magnitude of that difference is measured by the norm of f. The squared norm is

$$\|f^{(n)}(t,\omega) - f^{(n+1)}(t,\omega)\|^2 = \int_{t=0}^{T} \mathbb{E}[\{f^{(n)}(t,\omega) - f^{(n+1)}(t,\omega)\}^2]\,dt$$

As $[f^{(n)}(t,\omega) - f^{(n+1)}(t,\omega)]^2 = [B(t_{\frac{1}{2}})\,1_{[t_{1/2},t_1)}]^2$, the integrand is zero outside $[t_{\frac{1}{2}}, t_1)$ so the only remaining integral term is $\int_{t=t_{1/2}}^{t_1}$

$\mathbb{E}\left[B(t_{\frac{1}{2}})\right]\,dt$ and

$$\int_{t=t_{1/2}}^{t_1} \tfrac{1}{2}(T/n)\,dt = \tfrac{1}{2}\left(\tfrac{T}{n}\right)\int_{t=t_{1/2}}^{t_1} dt = \tfrac{1}{2}\left(\tfrac{T}{n}\right)(t_1 - t_{\frac{1}{2}})$$

$$= \tfrac{1}{2}\left(\tfrac{T}{n}\right)\tfrac{1}{2}\left(\tfrac{T}{n}\right)$$

So the norm of the difference between successive approximations $f^{(n)}$ and $f^{(n+1)}$ is $\| f^{(n)} - f^{(n+1)} \| = \frac{1}{2}(T/n)$.

(c) $\frac{1}{2}(T/n) \to 0$ as $n \to \infty$, so it has been shown that $f^{(n)}$ converges to f in the norm of f.

(d) $\displaystyle I^{(n)} = \sum_{k=0}^{n-1} B(t_k, \omega)[B(t_{k+1}) - B(t_k)]$

$I^{(n+1)} = B(t_{\frac{1}{2}})[B(t_1) - B(t_{\frac{1}{2}})] + I^{(n)}$

(e) $I^{(n+1)}$ differs from $I^{(n)}$ by the term $B(t_{\frac{1}{2}})[B(t_1) - B(t_{\frac{1}{2}})]$. So

$[I^{(n)} - I^{(n+1)}]^2 = B(t_{\frac{1}{2}})^2[B(t_1) - B(t_{\frac{1}{2}})]^2$

$\mathbb{E}[\{B(t_{\frac{1}{2}})^2[B(t_1) - B(t_{\frac{1}{2}})]\}^2] = \mathbb{E}[\mathbb{E}[B(t_{\frac{1}{2}})^2[B(t_1) - B(t_{\frac{1}{2}})]^2 | \Im_{\frac{1}{2}}]]$

$\mathbb{E}[\{B(t_{\frac{1}{2}})^2[\mathbb{E}[\{B(t_1) - B(t_{\frac{1}{2}})\}^2] | \Im_{\frac{1}{2}}] = \mathbb{E}[B(t_{\frac{1}{2}})^2 \frac{1}{2}\left(\frac{T}{n}\right)]$
$$= \frac{1}{2}\left(\frac{T}{n}\right)\frac{1}{2}\left(\frac{T}{n}\right)$$

So

$\| I^{(n)} - I^{(n+1)} \|^2 = \frac{1}{2}\left(\frac{T}{n}\right)\frac{1}{2}\left(\frac{T}{n}\right)$

The norm is $\| I^{(n)} - I^{(n+1)} \| = \frac{1}{2}\left(\frac{T}{n}\right)$.

(f) $\frac{1}{2}(T/n) \to 0$ as $n \to \infty$.

(g) It has been shown that the sequence of discrete stochastic Itō integrals $I^{(n)}$, which correspond to successively refined partitions, converges in mean-square. The limit is called the stochastic integral. This only proves existence, it does not result in an expression for the stochastic integral. Note that the above confirms that the numerical values of the norm of the integrand and norm of the integral are equal,

$\| f^{(n)} - f^{(n+1)} \| = \| I^{(n)} - I^{(n+1)} \|$

Answer [3.9.3]

(a) $\mathbb{E}\{\int_{t=0}^{T} B(t)\, dB(t)\}^2 = \int_{t=0}^{T} \mathbb{E}[B(t)^2]\, dt$ is an ordinary Riemann integral. Substituting $\mathbb{E}[B(t)^2] = t$ gives

$$\int_{t=0}^{T} t\, dt = \frac{1}{2}t^2 \Big|_{t=0}^{T} = \frac{1}{2}T^2$$

(b) Use the result $\int_{t=0}^{T} B(t)dB(t) = \frac{1}{2}B(T)^2 - \frac{1}{2}T$. Then

$$\left\{ \int_{t=0}^{T} B(t)\,dB(t) \right\}^2 = \frac{1}{4}[B(T)^2 - T]^2$$

$$= \frac{1}{4}[B(T)^4 - 2B(T)^2T + T^2]$$

Taking the expectation of this expression gives

$$\mathbb{E}\left\{ \int_{t=0}^{T} B(t)\,dB(t) \right\}^2 = \frac{1}{4}\mathbb{E}\{[B(T)^4 - 2B(T)^2T + T^2]\}$$

$$= \frac{1}{4}\{\mathbb{E}[B(T)^4] - 2T\mathbb{E}[B(T)^2] + T^2\}$$

$$= \frac{1}{4}\{3T^2 - 2TT + T^2\} = \frac{1}{2}T^2$$

which agrees with the result of (a).

(c) $\mathrm{Var}[I(T)] = \mathbb{E}[\{I(T)\}^2] - \{\mathbb{E}[I(T)]\}^2$

As $I(T)$ is an Itō stochastic integral, $\mathbb{E}[I(T)] = 0$, the answer is (a).

Answer [3.9.4]

The inner integral $\int_{y=t_{i-1}}^{s} dB(y)$ can be written as $B(s) - B(t_{i-1})$. Substituting this gives

$$\int_{s=t_{i-1}}^{t_i} \left[\int_{y=t_{i-1}}^{s} dB(y) \right] dB(s)$$

$$= \int_{s=t_{i-1}}^{t_i} [B(s) - B(t_{i-1})]\,dB(s)$$

$$= \int_{s=t_{i-1}}^{t_i} B(s)\,dB(s) - \int_{s=t_{i-1}}^{t_i} B(t_{i-1})\,dB(s)$$

The first integral is

$$\int_{s=t_{i-1}}^{t_i} B(s)\,dB(s) = \frac{1}{2}\{B(t_i)^2 - B(t_{i-1})^2\} - \frac{1}{2}(t_i - t_{i-1})$$

and the second integral is

$$\int_{s=t_{i-1}}^{t_i} B(t_{i-1})\,dB(s) = B(t_{i-1})[B(t_i) - B(t_{i-1})]$$

Substituting these gives

$$\tfrac{1}{2}[B(t_i)^2 - B(t_{i-1})^2] - \tfrac{1}{2}(t_i - t_{i-1}) - B(t_{i-1})[B(t_i) - B(t_{i-1})]$$
$$= \tfrac{1}{2}B(t_i)^2 - \tfrac{1}{2}B(t_{i-1})^2 - B(t_i)B(t_{i-1}) + B(t_{i-1})^2 - \tfrac{1}{2}(t_i - t_{i-1})$$
$$= \tfrac{1}{2}B(t_i)^2 - B(t_i)B(t_{i-1}) + \tfrac{1}{2}B(t_{i-1})^2 - \tfrac{1}{2}(t_i - t_{i-1})$$
$$= \tfrac{1}{2}[B(t_i) - B(t_{i-1})]^2 - \tfrac{1}{2}(t_i - t_{i-1})$$

Answer [3.9.5]

The discrete stochastic integral which converges to $\int_{t=0}^{T} B(t)\,dB(t)$ is

$$I_n \overset{\text{def}}{=} \sum_{k=0}^{n-1} B(t_k)[B(t_{k+1}) - B(t_k)]$$

Use $ab = \tfrac{1}{2}(a+b)^2 - \tfrac{1}{2}a^2 - \tfrac{1}{2}b^2$ where $a \overset{\text{def}}{=} B(t_k)$ and $b \overset{\text{def}}{=} B(t_{k+1}) - B(t_k)$. Then $a + b = B(t_{k+1})$ and

$$I_n = \tfrac{1}{2}\sum_{k=0}^{n-1} B(t_{k+1})^2 - \tfrac{1}{2}\sum_{k=0}^{n-1} B(t_k)^2 - \tfrac{1}{2}\sum_{k=0}^{n-1}[B(t_{k+1}) - B(t_k)]^2$$

$$= \tfrac{1}{2}\Big[B(t_n)^2 - B(t_0)^2\Big] - \tfrac{1}{2}\sum_{k=0}^{n-1}[B(t_{k+1}) - B(t_k)]^2$$

where $B(t_0) = 0$

$$= \tfrac{1}{2}B(T)^2 - \tfrac{1}{2}\sum_{k=0}^{n-1}[B(t_{k+1}) - B(t_k)]^2$$

As $n \to \infty$, the sum converges in mean square to T, as shown in Annex C. Thus I_n converges in mean square to $\tfrac{1}{2}B(T)^2 - \tfrac{1}{2}T$, so

$$\int_{t=0}^{T} B(t)\,dB(t) = \tfrac{1}{2}B(T)^2 - \tfrac{1}{2}T$$

If the time period is $[T_1, T_2]$ instead of $[0, T]$, this becomes

$$\int_{t=T_1}^{T_2} B(t)\,dB(t) = \tfrac{1}{2}[B(T_2)^2 - B(T_1)^2] - \tfrac{1}{2}(T_2 - T_1)$$

Answer [3.9.6]

The variance of $TB(T) - \int_{t=0}^{T} B(t)\,dt$ is derived in two ways. Introduce the notation

$$I(T) \stackrel{\text{def}}{=} \int_{t=0}^{T} t\,dB(t)$$

$$J(T) \stackrel{\text{def}}{=} TB(T) - \int_{t=0}^{T} B(t)\,dt$$

so

$$I(T) = J(T)$$

Method 1

$$\mathbb{V}\mathrm{ar}[I(T)] = \mathbb{E}[I(T)^2] = \mathbb{E}\left[\left\{\int_{t=0}^{T} t\,dB(t)\right\}^2\right]$$

as I is an Itō stochastic integral. Due to the Itō isometry, this can be written as the ordinary integral $\int_{t=0}^{T} \mathbb{E}[t^2]\,dt$. As t^2 is non-random, the integral equals $\int_{t=0}^{T} t^2\,dt = \frac{1}{3}T^3$.

Method 2

The same variance is now derived using the expression $J(T) = TB(T) - \int_{t=0}^{T} B(t)\,dt$

$$\mathbb{V}\mathrm{ar}[J(T)] = \mathbb{E}[J(T)^2] - \{\mathbb{E}[J(T)]\}^2$$

In the last term

$$\mathbb{E}[J(T)] = \mathbb{E}[TB(T)] - \mathbb{E}\left[\int_{t=0}^{T} B(t)\,dt\right]$$

Using the fact that $\int_{t=0}^{T} B(t)\,dt$ is normal with mean 0 (as derived in Section 3.7)

$$\mathbb{E}[J(T)]\} = T\mathbb{E}[B(T)] - 0 = 0$$

The first term in the variance can be written as

$$\mathbb{E}[J(T)^2] = \mathbb{E}\left[\left\{TB(T) - \int_{t=0}^{T} B(t)\,dt\right\}^2\right]$$

$$= \mathbb{E}\left[T^2 B(T)^2 + \left\{\int_{t=0}^{T} B(t)\,dt\right\}^2 - 2TB(T)\int_{t=0}^{T} B(t)\,dt\right]$$

The first term of the latter expression can be directly evaluated as

$$\mathbb{E}[T^2 B(T)^2] = T^2 \mathbb{E}[B(T)^2] = T^2 T = T^3$$

In the second term, $\int_{t=0}^{T} B(t)\,dt$ is a normally distributed random variable, say $X(T)$, with mean 0 and variance $\frac{1}{3}T^3$, as shown in Section 3.7. So

$$\mathbb{E}\left[\left\{\int_{t=0}^{T} B(t)\,dt\right\}^2\right] = \mathbb{E}[X(T)^2] = \mathbb{V}\text{ar}[X(T)] = \frac{1}{3}T^3$$

The third term

$$\mathbb{E}\left[-2T\,B(T)\left\{\int_{t=0}^{T} B(t)\,dt\right\}\right] = -2T\,\mathbb{E}[B(T)X(T)]$$

involves the product of two random variables, $B(T)$ and $X(T)$, which are both dependent on T, and therefore have a covariance. By the definition of covariance

$$\mathbb{C}\text{ov}[B(T), X(T)] = \mathbb{E}[B(T)X(T)] - \mathbb{E}[B(T)]\mathbb{E}[X(T)]$$

as in the last term both expected values are 0

$$\mathbb{E}[B(T)X(T)] = \mathbb{C}\text{ov}[B(T), X(T)]$$

The book by *Epps* derives on page 485 that $\mathbb{C}\text{ov}[B(T), X(T)] = \frac{1}{2}T^2$. Substituting this gives the final result as $T^3 + T^3/3 - 2T\frac{1}{2}T^2 = T^3/3$, the same as in the first method.

Answer [3.9.7]

$$I(T) \stackrel{\text{def}}{=} \int_{t=0}^{T} \sqrt{|B(t)|}\,dB(t)$$

As $I(T)$ is a Itō stochastic integral and $\mathbb{E}[I(T)] = 0$, $\mathbb{V}\text{ar}[I(T)] = \mathbb{E}[I(T)^2]$.

$$\mathbb{E}[I(T)^2] = \mathbb{E}\left[\left\{\int_{t=0}^{T} \sqrt{|B(t)|}\,dB(t)\right\}^2\right]$$

$$= \int_{t=0}^{T} \mathbb{E}\left[\left(\sqrt{|B(t)|}\right)^2\right]\,dt$$

$$= \int_{t=0}^{T} \mathbb{E}[|B(t)|]\,dt$$

The integrand, $\mathbb{E}[|B(t)|]$, is worked out in Annex A, as equal to $\sqrt{2/\pi}\sqrt{t}$. Thus

$$\mathbb{V}\mathrm{ar}[I(T)] = \int_{t=0}^{T} \sqrt{\tfrac{2}{\pi}}\sqrt{t}\,dt = \sqrt{\tfrac{2}{\pi}\tfrac{2}{3}}T^{3/2}$$

Answer [3.9.8]

$$X \stackrel{\mathrm{def}}{=} \int_{t=0}^{T} [B(t)+t]^2 \, dB(t)$$

As X is an Itō stochastic integral, $\mathbb{E}[X] = 0$.

$$\mathbb{V}\mathrm{ar}[X] = \mathbb{E}[X^2]$$
$$= \mathbb{E}\left[\left\{\int_{t=0}^{T} [B(t)+t]^2 \, dB(t)\right\}^2\right]$$
$$= \int_{t=0}^{T} \mathbb{E}[\{B(t)+t\}^4]\,dt$$

The integrand expands into

$$\mathbb{E}[B(t)^4 + 4B(t)^3 t + 6B(t)^2 t^2 + 4B(t)t^3 + t^4]$$
$$= \mathbb{E}[B(t)^4] + 4t\mathbb{E}[B(t)^3] + 6t^2\mathbb{E}[B(t)^2] + 4t^3\mathbb{E}[B(t)] + t^4$$
$$= 3t^2 + 4t0 + 6t^2 t + 4t^3 0 + t^4$$
$$= 3t^2 + 6t^3 + t^4$$

Integrating this from $t = 0$ to $t = T$ gives the answer

$$T^3 + \frac{6}{4}T^4 + \frac{1}{5}T^4$$

CHAPTER 4

For greater readability, subscript t is omitted in the derivation steps but shown in the final answer.

Answer [4.10.1]

$f \stackrel{\mathrm{def}}{=} (1/3)B(t)^3$ is a function of single variable $B(t)$ so the d notation for ordinary derivatives can be used instead of the ∂ notation for partial derivatives.

$$df = \frac{df}{dB} dB + \frac{1}{2} \frac{d^2 f}{dB^2} (dB)^2$$

$$(dB)^2 = dt \quad \frac{df}{dB} = B^2 \quad \frac{d^2 f}{dB^2} = 2B$$

$$dF = B^2 dB + \frac{1}{2} 2B \, dt$$

$$df(t) = B(t) \, dt + B(t)^2 \, dB(t)$$

Answer [4.10.2]

$f \stackrel{\text{def}}{=} B(t)^2 - t$ is a function of two variables, t and $B(t)$.

$$df = \frac{\partial f}{\partial t} dt + \frac{\partial f}{\partial B} dB + \frac{1}{2} \frac{\partial^2 f}{\partial t^2} (dt)^2 + \frac{1}{2} \frac{\partial^2 f}{\partial B^2} (dB)^2$$
$$+ \frac{\partial^2 f}{\partial t \partial B} dt \, dB$$

$$(dt)^2 = 0 \quad (dB)^2 = dt \quad dt \, dB = 0$$

$$\frac{\partial f}{\partial t} = -1 \quad \frac{\partial f}{\partial B} = 2B \quad \frac{\partial^2 f}{\partial B^2} = 2$$

Since $(dt)^2 = 0$, $\partial^2 f / \partial t^2$ is not needed. Since $dt \, dB = 0$, $\partial^2 f / \partial t \partial B$ is not needed.

$$df = -1 \, dt + 2B \, dB + \frac{1}{2} 2 \, dt$$
$$df(t) = 2B(t) \, dB(t)$$

Answer [4.10.3]

$f \stackrel{\text{def}}{=} \exp[B(t)]$ is a function of single variable B.

$$df = \frac{df}{dB} dB + \frac{1}{2} \frac{d^2 f}{dB^2} (dB)^2$$

$$(dB)^2 = dt \quad \frac{df}{dB} = \exp(B) = f \quad \frac{d^2 f}{dB^2} = \frac{df}{dB} = f$$

$$df = f \, dB + \frac{1}{2} F \, dt$$

$$df(t) = \frac{1}{2} f(t) \, dt + f(t) \, dB(t) \quad \text{dynamics of } f(t)$$

$$\frac{df(t)}{f(t)} = \frac{1}{2} dt + dB(t) \quad \text{dynamics of proportional change in } f(t)$$

Answer [4.10.4]

$f \overset{\text{def}}{=} \exp[B(t) - \frac{1}{2}t]$ is a function of two variables, t and $B(t)$.

$$df = \frac{\partial f}{\partial t} dt + \frac{\partial f}{\partial B} dB + \frac{1}{2} \frac{\partial^2 f}{\partial t^2} (dt)^2 + \frac{1}{2} \frac{\partial^2 f}{\partial B^2} (dB)^2$$

$$+ \frac{\partial^2 f}{\partial t \partial B} dt\, dB$$

$(dt)^2 = 0 \quad (dB)^2 = dt \quad dt\, dB = 0$

$\dfrac{\partial f}{\partial t} = \exp\left[B - \frac{1}{2}t\right](-\frac{1}{2}) = -\frac{1}{2}f \quad \frac{\partial^2 f}{\partial t^2}$

is not needed as $(dt)^2 = 0$

$\dfrac{\partial f}{\partial B} = \exp\left[B - \frac{1}{2}t\right]1 = f \quad \dfrac{\partial^2 f}{\partial B^2} = \dfrac{\partial f}{\partial B} = f \quad \dfrac{\partial^2 f}{\partial t \partial B}$

is not needed as $dt\, dB = 0$

$$df = -\frac{1}{2}f\, dt + f\, dB + \frac{1}{2}Z\, dt$$

$df(t) = f(t)\, dB(t) \quad$ dynamics of $f(t)$

$\dfrac{df(t)}{f(t)} = dB(t) \quad$ dynamics of proportional change in $f(t)$

Answer [4.10.5]

$S \overset{\text{def}}{=} \exp\left[(\mu - \frac{1}{2}\sigma^2)t + \sigma B(t)\right]$ is a function of two variables, t and $B(t)$.

$$dS = \frac{\partial S}{\partial t} dt + \frac{\partial S}{\partial B} dB + \frac{1}{2} \frac{\partial^2 S}{\partial t^2} (dt)^2$$

$$+ \frac{1}{2} \frac{\partial^2 S}{\partial B^2} (dB)^2 + \frac{\partial^2 S}{\partial t \partial B} dt\, dB$$

$(dt)^2 = 0 \quad (dB)^2 = dt \quad dt\, dB = 0$

$\dfrac{\partial S}{\partial t} = \exp\left[(\mu - \frac{1}{2}\sigma^2)t + \sigma B(t)\right](\mu - \frac{1}{2}\sigma^2) = S\left(\mu - \frac{1}{2}\sigma^2\right)$

$\dfrac{\partial S}{\partial B} = \exp\left[(\mu - \frac{1}{2}\sigma^2)t + \sigma B(t)\right]\sigma = S\sigma$

$\dfrac{\partial^2 S}{\partial B^2} = \sigma \dfrac{\partial S}{\partial B} = \sigma S\sigma = S\sigma^2$

Since $(dt)^2 = 0$, $\frac{\partial^2 S}{\partial t^2}$ is not needed. Since $dt\, dB = 0$, $\partial^2 S/\partial t \partial B$ is not needed.

$$dS = S\left(\mu - \tfrac{1}{2}\sigma^2\right) dt + S\sigma\ dB + \tfrac{1}{2}S\sigma^2\, dt$$
$$dS(t) = S(t)[\mu\ dt + \sigma\ dB(t)]$$
$$\frac{dS(t)}{S(t)} = \mu\ dt + \sigma\, dB(t)$$

Answer [4.10.6]

$\ln[S]$ is a function of single variable $S(t)$.

$$d \ln[S] = \frac{d \ln[S]}{dS} dS + \frac{1}{2}\frac{d^2 \ln[S]}{dS^2}(dS)^2$$
$$dS = S[\mu\, dt + \sigma\, dB]$$
$$(dS)^2 = S^2[\mu\, dt + \sigma\, dB]^2 = S^2[\mu\, dt + \sigma\, dB]^2$$
$$= S^2[\mu^2\, (dt)^2 + 2\mu\sigma\, dt\, dB + \sigma^2\, (dB)^2]$$
$$(dt)^2 = 0 \quad dt\, dB = 0 \quad (dB)^2 = dt$$
$$(dS)^2 = S^2[\mu\, dt + \sigma\, dB]^2 = S^2[\mu\, dt + \sigma\, dB]^2$$
$$= S^2[\mu^2\, (dt)^2 + 2\mu\sigma\, dt\, dB + \sigma^2\, (dB)^2]$$
$$= S^2\sigma^2\, dt$$
$$\frac{d \ln[S]}{dS} = \frac{1}{S} \quad \frac{d^2 \ln[S]}{dS^2} = \frac{-1}{S^2}$$
$$d \ln[S] = \frac{1}{S}S[\mu\, dt + \sigma\, dB] + \frac{1}{2}\frac{-1}{S^2}S^2\sigma^2\, dt$$
$$d \ln[S(t)] = \left(\mu - \tfrac{1}{2}\sigma^2\right) dt + \sigma\, dB(t)$$

The random term in the dynamics of $\ln[S(t)]$ is the same as in the dynamics of the proportional change $dS(t)/S(t)$, but the dt term is smaller by $\tfrac{1}{2}\sigma^2$.

Answer [4.10.7]

$f \overset{\text{def}}{=} 1/S(t)$ is a function of single variable S.

$$df = \frac{df}{dS} dS + \frac{1}{2}\frac{d^2 f}{dS^2}(dS)^2$$
$$dS = \mu S\, dt + \sigma S\, dB$$

$$(dS)^2 = (\mu S \, dt + \sigma S \, dB)^2$$
$$= \mu^2 S^2 (dt)^2 + 2\mu S \sigma S \, dt \, dB + \sigma^2 S^2 (dB)^2$$
$$(dt)^2 = 0 \quad dt \, dB = 0 \quad (dB)^2 = dt$$
$$\frac{\partial f}{\partial S} = \frac{-1}{S^2} \frac{\partial^2 f}{\partial S^2} = \frac{2}{S^3} \quad (dS)^2 = \sigma^2 S^2 \, dt$$
$$df = \frac{-1}{S^2}[\mu S \, dt + \sigma S \, dB] + \tfrac{1}{2}\frac{2}{S^3}\sigma^2 S^2 \, dt$$
$$df = \frac{1}{S}[(\sigma^2 - \mu) \, dt - \sigma \, dB]$$
$$df(t) = f(t)[(\sigma^2 - \mu) \, dt - \sigma \, dB(t)]$$
$$\frac{df(t)}{f(t)} = (\sigma^2 - \mu) \, dt - \sigma \, dB(t)$$

Answer [4.10.8]

$R = 1/Q$ is a function of single variable Q.

$$dR = \frac{dR}{dQ} \, dQ + \frac{1}{2}\frac{d^2 R}{dQ^2} (dQ)^2$$

Substituting

$$dQ = Q[\mu_Q \, dt + \sigma_Q \, dB]$$
$$(dQ)^2 = Q^2 \sigma_Q^2 \, dt$$
$$\frac{dR}{dQ} = \frac{-1}{Q^2} \quad \frac{d^2 R}{dQ^2} = \frac{2}{Q^3}$$

gives

$$dR = \frac{-1}{Q^2} Q[\mu_Q \, dt + \sigma_Q \, dB] + \frac{1}{2}\frac{2}{Q^3} Q^2 \sigma_Q^2 \, dt$$
$$= \frac{-1}{Q}[\mu_Q \, dt + \sigma_Q \, dB] + \frac{1}{Q}\sigma_Q^2 \, dt$$
$$= -R[\mu_Q \, dt + \sigma_Q \, dB] + R\sigma_Q^2 \, dt$$
$$= R(-\mu_Q + \sigma_Q^2) \, dt - R\sigma_Q \, dB(t)$$

Dividing by $R(t) \neq 0$ gives the dynamics of the relative change

$$\frac{dR(t)}{R(t)} = (-\mu_Q + \sigma_Q^2) \, dt - \sigma_Q \, dB(t)$$

Answer [4.10.9]

P is a function of two variables, r and t

$$dP = \frac{\partial P}{\partial t} dt + \frac{\partial P}{\partial r} dr + \frac{1}{2} \frac{\partial^2 P}{\partial r^2} (dr)^2 + \frac{\partial^2 P}{\partial t \partial r} dt\, dr$$

Substituting

$$dr = \mu\, dt + \sigma\, dB$$
$$(dr)^2 = \sigma^2\, dt$$
$$dt\, dr = dt\,(\mu\, dt + \sigma\, dB) = \mu\,(dt)^2 + \sigma\, dt\, dB = 0$$

gives

$$dP = \frac{\partial P}{\partial t} dt + \frac{\partial P}{\partial r}(\mu\, dt + \sigma\, dB) + \frac{1}{2} \frac{\partial^2 P}{\partial r^2}\sigma^2\, dt$$

$$= \left[\frac{\partial P}{\partial t} + \mu \frac{\partial P}{\partial r} + \frac{1}{2}\sigma^2 \frac{\partial^2 P}{\partial r^2} \right] dt + \sigma \frac{\partial P}{\partial r}\, dB(t)$$

Answer [4.10.10]

As X is a function of two variables, t and $M(t)$

$$dX = \frac{\partial X}{\partial t} dt + \frac{\partial X}{\partial M} dM + \frac{1}{2} \frac{\partial^2 X}{\partial M^2} (dM)^2$$

Substituting

$$\frac{\partial X}{\partial t} = -1 \qquad \frac{\partial X}{\partial M} = 2M \qquad \frac{\partial^2 X}{\partial M^2} = 2 \quad (dM)^2 = dt$$

gives

$$dX = -1\, dt + 2M\, dM + \tfrac{1}{2}2\, dt$$
$$= 2M(t)\, dM(t)$$

In integral form over $[s \le t \le u]$

$$X(u) - X(s) = 2 \int_{t=s}^{u} M(t)\, dM(t)$$

$$\mathbb{E}[X(u)|\Im(s)] = X(s) + 2\mathbb{E}[\int_{t=s}^{u} M(t)\, dM(t)|\Im(s)]$$

X is a martingale if the rhs expectation equals zero. It is shown in *Kuo* that this is indeed the case.

Answer [4.10.11]

The discrete stochastic integral with respect to continuous martingale M is

$$J_n \stackrel{\text{def}}{=} \sum_{k=0}^{n-1} f(t_k)[M(t_{k+1}) - M(t_k)]$$

One would expect J_n to be a martingale because $[M(t_{k+1}) - M(t_k)]$ is a martingale, and J_n is just a combination of several of these, with weightings $f(t_k)$ that are known before the movements of M are generated. The question is: $\mathbb{E}[J_{n+1}|\mathfrak{I}_n] \stackrel{?}{=} J_n$. Writing J_n as $[J_n|\mathfrak{I}_n]$, the question is rephrased as: $\mathbb{E}[J_{n+1} - J_n|\mathfrak{I}_n] \stackrel{?}{=} 0$.

$$J_{n+1} = \sum_{k=0}^{n-1} f(t_k)[M(t_{k+1}) - M(t_k)]$$
$$+ f(t_n)[M(t_{n+1}) - M(t_n)]$$
$$J_{n+1} - J_n = f(t_n)[M(t_{n+1}) - M(t_n)]$$
$$\mathbb{E}[J_{n+1} - J_n|\mathfrak{I}_n] = \mathbb{E}[f(t_n)[M(t_{n+1}) - M(t_n)]|\mathfrak{I}_n]$$
$$= f(t_n)\mathbb{E}[M(t_{n+1}) - M(t_n)|\mathfrak{I}_n]$$

since $f(t_n)$ is known when $\mathfrak{I}_n$ is given. As M is a martinale, $\mathbb{E}[M(t_{n+1}) - M(t_n)|\mathfrak{I}_n] = 0$, so $\mathbb{E}[J_{n+1} - J_n|\mathfrak{I}_n] = 0$. Thus J_n is a *discrete*-time martingale.

CHAPTER 5

Answer [5.12.1]

$$X(t) = \exp(-\lambda t)\left[X(0) + \sigma \int_{s=0}^{t} \exp(\lambda s)\, dB(s)\right]$$

Beware: Treating $X(t)$ as a function of Brownian motion is conceptually seriously wrong !!!

Introduce $Z(t) \stackrel{\text{def}}{=} \int_{s=0}^{t} \exp(\lambda s)\, dB(s)$. Then $X(t) = \exp(-\lambda t)[X(0) + \sigma Z(t)]$ is a function of t and Z.

$$dX = \frac{\partial X}{\partial t}\, dt + \frac{\partial X}{\partial Z}\, dZ + \frac{1}{2}\frac{\partial^2 X}{\partial t^2}\, (dt)^2$$
$$+ \frac{1}{2}\frac{\partial^2 X}{\partial Z^2}\, (dZ)^2 + \frac{\partial^2 X}{\partial t \partial Z}\, dt\, dZ$$

The various terms on the right hand side are:

- keeping Z constant, $\frac{\partial X}{\partial t} = -\lambda X(t)$
- keeping t constant, $\frac{\partial X}{\partial Z} = \exp(-\lambda t)\sigma$
- $\frac{\partial^2 X}{\partial Z^2} = \frac{\partial}{\partial Z}(\frac{\partial X}{\partial Z}) = \frac{\partial}{\partial Z}\sigma \exp(-\lambda t) = 0$
- $dZ = \exp(\lambda t)\,dB(t)$
- $dt\,dZ = \exp(\lambda t)\,dt\,dB(t) = 0$
- $(dt)^2 = 0$

Substituting these gives

$$dX = -\lambda X(t) + \sigma \exp(-\lambda t)\exp(\lambda t)\,dB(t)$$
$$= -\lambda X(t)\,dt + \sigma\,dB(t)$$

as was to be shown.

Alternative Method

$X(t)$ is the product of a non-random term $\exp(-\lambda t)$ and a random term [...]. To determine $dX(t)$, use the product rule

$$d[Y(t)Z(t)] = Y(t)\,dZ(t) + Z(t)\,dY(t) + dY(t)\,dZ(t)$$

Applying this to the product of $\exp(-\lambda t)$ and $[X(0) + \sigma \int_{s=0}^{t} \exp(\lambda s)\,dB(s)]$ gives

$$dX(t) = [...]\,d[\exp(-\lambda t)] + \exp(-\lambda t)\,d[...] + d[\exp(-\lambda t)]\,d[...]$$

The expression for $d[...]$ is $\sigma \exp(\lambda t)\,dB(t)$. The expression for $d[\exp(-\lambda t)]$ is $-\lambda \exp(-\lambda t)\,dt$. Multiplying these expressions produces the cross product $dt\,dB(t)$ which is zero.

$$dX(t) = [...](-\lambda)\exp(-\lambda t)\,dt + \exp(-\lambda t)\sigma \exp(\lambda t)\,dB(t)$$
$$= -\lambda\{\exp(-\lambda t)[X(0) + \sigma \int_{s=0}^{t} \exp(\lambda s)\,dB(s)]\} + \sigma\,dB(t)$$

As the expression in $\{...\}$ is $X(t)$

$$dX(t) = -\lambda X(t)\,dt + \sigma\,dB(t)$$

as was to be shown.

Answer [5.12.2]

The random process specified in these questions is known as a Brownian bridge.

(a) Random process X is specified for $0 \le t < T$ (strictly) as

$$X(t) \stackrel{\text{def}}{=} (T - t) \int_{s=0}^{t} \frac{1}{T - s}\, dB(s) \quad X(0) = 0$$

Let the stochastic integral be denoted $Z(t)$, $Z(t) \stackrel{\text{def}}{=} \int_{s=0}^{t} [1/(T - s)]\, dB(s)$, which in shorthand is $dZ(t) = [1/(T - t)]\, dB(t)$. Then $X(t) \stackrel{\text{def}}{=} (T - t)Z(t)$ is a function of t and $Z(t)$. Itō's formula gives

$$
\begin{aligned}
dX(t) &= -Z(t)\, dt + (T - t)\, dZ(t) \\
&= -Z(t)\, dt + (T - t)\frac{1}{T - t}\, dB(t) \\
&= -Z(t)\, dt + dB(t)
\end{aligned}
$$

To get this in terms of X, substitute $Z(t) = X(t)/(T - t)$, giving

$$dX(t) = -\frac{X(t)}{T - t}\, dt + dB(t)$$

(b) Random process Y is specified for $0 \le t < T$ (strictly) as

$$Y(t) \stackrel{\text{def}}{=} a\left(1 - \frac{t}{T}\right) + b\frac{t}{T} + (T - t)\int_{s=0}^{t} \frac{1}{T - s}\, dB(s) \quad Y(0) = a$$

Let the stochastic integral be denoted $Z(t)$, $Z(t) \stackrel{\text{def}}{=} \int_{s=0}^{t} [1/(T - s)]\, dB(s)$, which in shorthand is $dZ(t) = [1/(T - t)\, dB(t)$. Then

$$Y(t) \stackrel{\text{def}}{=} a\left(1 - \frac{t}{T}\right) + b\frac{t}{T} + (T - t)Z(t)$$

is a function of t and $Z(t)$. Itō's formula gives

$$
\begin{aligned}
dY(t) &= \left[-a\frac{1}{T} + b\frac{1}{T} - Z(t)\right]\, dt + (T - t)\frac{1}{T - t}\, dB(t) \\
&= \left[-a\frac{1}{T} + b\frac{1}{T} - Z(t)\right]\, dt + dB(t)
\end{aligned}
$$

Write $Z(t)$ in terms of $Y(t)$. First rearrange $Y(t)$ as

$$Y(t) = a\frac{T-t}{T} + b\frac{t}{T} + (T-t)Z(t)$$

So

$$Z(t) = \frac{Y(t)}{T-t} - a\frac{1}{T} - b\frac{t}{T}\frac{1}{T-t}$$

Substituting gives

$$dY(t) = \left[-a\frac{1}{T} + b\frac{1}{T} - \frac{Y(t)}{T-t} + a\frac{1}{T} + b\frac{t}{T}\frac{1}{T-t}\right] dt + dB(t)$$

$$= \frac{b - Y(t)}{T-t} dt + dB(t)$$

For $a = 0$ and $b = 0$, the case of question (a) is recovered.

Answer [5.12.3]

$$dr(t) = [b(t) - ar(t)] dt + \sigma\, dB(t)$$

Simplify the drift by introducing the new random process $X(t) \overset{\text{def}}{=} \exp(at)r(t)$ which is a function of the two variables t and r. Itō's formula give its dynamics as

$$dX = \frac{\partial X}{\partial t} dt + \frac{\partial X}{\partial r} dr + \frac{\partial^2 X}{\partial r^2} (dr)^2$$

$$\frac{\partial X}{\partial t} = aX \quad \frac{\partial X}{\partial r} = \exp(at) \quad \frac{\partial^2 X}{\partial r^2} = 0 \quad (dr)^2 = \sigma^2\, dt$$

$$dX(t) = aX(t)\, dt + \exp(at)\{[b(t) - ar(t)]\, dt + \sigma\, dB(t)\}$$
$$= aX(t)\, dt + \exp(at)b(t)\, dt - a\exp(at)r(t)\, dt + \sigma\exp(at)\, dB(t)$$
$$= aX(t)\, dt + \exp(at)b(t)\, dt - aX(t)\, dt + \sigma\exp(at)\, dB(t)$$

Term $aX(t)\, dt$ cancels, leaving the simplified SDE

$$dX(t) = \exp(at)b(t)\, dt + \sigma\exp(at)\, dB(t)$$

Using s for running time

$$dX(s) = \exp(as)b(s)\, ds + \sigma\exp(as)\, dB(s)$$

The corresponding integral expression is

$$X(t) = X(0) + \int_{s=0}^{t} \exp(as)b(s)\,ds + \int_{s=0}^{t} \sigma \exp(as)\,dB(s)$$

Converting back to r gives

$$\exp(at)r(t) = \exp(a0)r(0) + \int_{s=0}^{t} \exp(as)b(s)\,ds$$

$$+ \int_{s=0}^{t} \sigma \exp(as)\,dB(s)$$

$$r(t) = \exp(-at)\left\{ r(0) + \int_{s=0}^{t} \exp(as)b(s)\,ds + \sigma \int_{s=0}^{t} \exp(as)\,dB(s) \right\}$$

where $r(0)$ is a known non-random value.

Answer [5.12.4]

As g is a function of the single variable r

$$dg = \frac{dg}{dr}\,dr + \frac{1}{2}\frac{d^2g}{dr^2}(dr)^2$$

Substitute

$$dr = -\lambda[r - \bar{r}]\,dt + \sigma\sqrt{r}\,dB(t)$$

$$(dr)^2 = \sigma^2 r\,dt$$

Then

$$dg = \frac{dg}{dr}\{-\lambda[r - \bar{r}]\,dt + \sigma\sqrt{r}\,dB(t)\} + \tfrac{1}{2}\frac{d^2g}{dr^2}\sigma^2 r\,dt$$

$$= \left\{ -\lambda[r - \bar{r}]\frac{dg}{dr} + \tfrac{1}{2}\sigma^2 r\frac{d^2g}{dr^2} \right\}dt + \sigma\sqrt{r}\,\frac{dg}{dr}\,dB(t)$$

The diffusion coefficient is $\sigma\sqrt{r}\,(dg/dr)$. Setting this equal to 1 implies that g must satisfy

$$\frac{dg}{dr} = \frac{1}{\sigma\sqrt{r}}$$

Ordinary integration of the function g of r, with respect to r, gives

$$g = \tfrac{2}{\sigma}\sqrt{r} \quad \text{and} \quad r = \tfrac{1}{4}\sigma^2 g^2$$

In the drift coefficient

$$\frac{d^2g}{dr^2} = \frac{d}{dr}\left(\frac{dg}{dr}\right) = \frac{-1}{2\sigma r\sqrt{r}}$$

The SDE for g in terms of r is

$$dg = \left\{-\lambda[r-\bar{r}]\frac{1}{\sigma\sqrt{r}} + \frac{1}{2}\sigma^2 r\,\frac{-1}{2\sigma r\sqrt{r}}\right\}dt + dB(t)$$

Expressing r in terms of g then gives the final expression for dg in terms of g.

The diffusion coefficient (volatility) has become constant but the drift coefficient $\{...\}$ is now much more complicated.

Answer [5.12.5]

$$d\ln[r(t)] = \theta(t)\,dt + \sigma\,dB(t)$$

From geometric Brownian motion it is known that by going from dr/r to $d[\ln r]$ the drift of dr/r is reduced by $\frac{1}{2}$volatility^2. Here it is the other way around. Thus

$$\frac{dr(t)}{r(t)} = \left[\theta(t) + \frac{1}{2}\sigma^2\right]dt + \sigma\,dB(t)$$

This can also be seen as follows. Let $X \stackrel{\text{def}}{=} \ln(r)$. Then $r = \exp(X)$, which is a function of single variable X.

$$dr = \frac{dr}{dX}dX + \frac{1}{2}\frac{d^2r}{dX^2}(dX)^2 = r\,dX + \frac{1}{2}r\,(dX)^2$$

The stochastic differential of X is

$$dX = d\ln r = \theta(t)\,dt + \sigma\,dB(t)$$

and

$$(dX)^2 = \sigma^2\,dt$$

Substituting these gives

$$dr = r[\theta(t)\,dt + \sigma\,dB(t)] + \frac{1}{2}r\sigma^2\,dt$$

and

$$\frac{dr(t)}{r(t)} = [\theta(t) + \frac{1}{2}\sigma^2]dt + \sigma\,dB(t)$$

Answer [5.12.6]

SDE

$$dX(t) = X(t)\,dt + dB(t) \qquad X(0) \text{ known}$$

is of the general linear form with

$$\mu_{1X} = 0 \qquad \mu_{2X} = 1$$
$$\sigma_{1X} = 1 \qquad \sigma_{2X} = 0$$

SDE for Y

$$\mu_Y = \mu_{2X} = 1$$
$$\sigma_Y = \sigma_{2X} = 0$$
$$dY(t)/Y(t) = 1\,dt + 0\,dB(t) = dt$$
$$\ln[Y(t)] = t + c$$
$$\ln[Y(0)] = \ln[1] = 0 = 0 + c$$
$$c = 0$$
$$Y(t) = \exp(t)$$

SDE for Z

$$\mu_Z = \frac{\mu_{1X} - \sigma_Y \sigma_{1X}}{Y} = 0$$

$$\sigma_Z = \frac{\sigma_{1X}}{Y} = \frac{1}{Y} = e^{-t}$$
$$dZ(t) = 0\,dt + e^{-t}\,dB(t) = e^{-t}\,dB(t) \qquad Z(0) = X(0)$$
$$Z(t) = Z(0) + \int_{s=0}^{t} e^{-s}\,dB(s) = X(0) + \int_{s=0}^{t} e^{-s}\,dB(s)$$

Solution

$$X(t) = Y(t)Z(t)$$
$$= \exp(t)[X(0) + \int_{s=0}^{t} \exp(-s)\,dB(s)]$$

Verifying the Solution
Write the solution as $X(t) = Y(t)Z(t)$ where $Y(t) = e^t$ and $Z(t) = X(0) + \int_{s=0}^{t} e^{-s}\,dB(s)$.

$$dX = Y\,dZ + Z\,dY + dY\,dZ$$

Substituting $dY = e^t\, dt$ and $dZ(t) = e^{-t}\, dB(t)$ gives

$$dX(t) = e^t e^{-t}\, dB(t) + \left[X(0) + \int_{s=0}^{t} e^{-s}\, dB(s) \right] e^t\, dt$$
$$+ e^t\, dt\, e^{-t}\, dB(t)$$

The last term is 0, and the coefficient of dt is $X(t)$, leaving the original SDE

$$dX(t) = dB(t) + X(t)\, dt$$

Answer [5.12.7]

SDE

$$dX(t) = -X(t)\, dt + e^{-t}\, dB(t) \qquad X(0) \text{ known}$$

is of the general linear form with

$$\mu_{1X} = 0 \qquad \mu_{2X} = -1$$
$$\sigma_{1X} = e^{-t} \qquad \sigma_{2X} = 0$$

SDE for Y

$$\mu_Y = \mu_{2X} = -1$$
$$\sigma_Y = \sigma_{2X} = 0$$
$$dY(t)/Y(t) = -1\, dt + 0\, dB(t) = -dt$$
$$\ln[Y(t)] = -t + c$$
$$\ln[Y(0)] = \ln[1] = 0 = 0 + c$$
$$c = 0$$
$$Y(t) = e^{-t}$$

SDE for Z

$$\mu_Z = \frac{\mu_{1X} - \sigma_Y \sigma_{1X}}{Y} = 0$$
$$\sigma_Z = \frac{\sigma_{1X}}{Y} = \frac{e^{-t}}{e^{-t}} = 1$$
$$dZ(t) = 0\, dt + dB(t) = dB(t) \qquad Z(0) = X(0)$$
$$Z(t) = Z(0) + B(t) = X(0) + B(t)$$

Solution

$$X(t) = Y(t)Z(t)$$
$$= e^{-t}[X(0) + B(t)]$$

Verifying the Solution
Write the solution as $X(t) = Y(t)Z(t)$ where $Y(t) = e^{-t}$ and $Z(t) = [X(0) + B(t)]$.

$$dX = Y\,dZ + Z\,dY + dY\,dZ$$

Substituting $dY = -e^{-t}\,dt$ and $dZ(t) = dB(t)$ gives

$$dX(t) = e^{-t}\,dB(t) - [X(0) + B(t)]e^{-t}\,dt - e^{-t}\,dt\,dB(t)$$

The last term is 0, and the coefficient of dt is $-X(t)$, leaving the original SDE

$$dX(t) = e^{-t}\,dB(t) - X(t)\,dt$$

Answer [5.12.8]

As SDE

$$dX(t) = m\,dt + \sigma X(t)\,dB(t) \quad X(0) \text{ known}$$

is linear in the unknown process X, use the general solution method for linear SDEs. In general form

$$dX(t) = [\mu_{1X}(t) + \mu_{2X}(t)X(t)]\,dt + [\sigma_{1X}(t) + \sigma_{2X}(t)X(t)]\,dB(t)$$

Let $X(t) = Y(t)Z(t)$ where

$$\frac{dY(t)}{Y(t)} = \mu_Y(t)\,dt + \sigma_Y(t)\,dB(t) \text{ with } Y(0) = 1$$
$$dZ(t) = \mu_Z(t)\,dt + \sigma_Z(t)\,dB(t) \text{ with } Z(0) = X(0)$$

For the SDE of X

$$\mu_{1X}(t) = m \qquad \mu_{2X}(t) = 0$$
$$\sigma_{1X}(t) = 0 \qquad \sigma_{2X}(t) = \sigma$$

Then for Y

$$\mu_Y(t) = \mu_{2X}(t) = 0$$
$$\sigma_Y(t) = \sigma_{2X}(t) = \sigma$$
$$\frac{dY(t)}{Y(t)} = \sigma \, dB(t) \text{ with } Y(0) = 1$$

The solution to the SDE for Y is

$$Y(t) = \exp[-\tfrac{1}{2}\sigma^2 t + \sigma B(t)]$$

For Z

$$\mu_Z(t) = \frac{\mu_{1X}(t) - \sigma_Y(t)\sigma_{1X}(t)}{Y(t)} = \frac{m}{Y(t)}$$
$$= m \exp[\tfrac{1}{2}\sigma^2 t - \sigma B(t)]$$
$$\sigma_Z(t) = \frac{\sigma_{1X}(t)}{Y(t)} = 0$$
$$dZ(t) = m \exp\left[\tfrac{1}{2}\sigma^2 t - \sigma B(t)\right] dt \text{ with } Z(0) = X(0) \text{ known}$$

The solution to the SDE for Z is

$$Z(t) = Z(0) + m \int_{s=0}^{t} \exp\left[\tfrac{1}{2}\sigma^2 s - \sigma B(s)\right] ds$$

As $Z(0) = X(0)$

$$Z(t) = X(0) + m \int_{s=0}^{t} \exp\left[\tfrac{1}{2}\sigma^2 s - \sigma B(s)\right] ds$$

Multiplying $Y(t)$ and $Z(t)$ gives

$$X(t) = Y(t)Z(t) = \exp\left[-\tfrac{1}{2}\sigma^2 t + \sigma B(t)\right]$$
$$\times \left\{ X(0) + m \int_{s=0}^{t} \exp\left[\tfrac{1}{2}\sigma^2 s - \sigma B(s)\right] ds \right\}$$

Verifying the Solution
Write the solution as $X(t) = Y(t)Z(t)$ where

$$Y(t) = \exp[-\tfrac{1}{2}\sigma^2 t + \sigma B(t)]$$
$$Z(t) = m \int_{s=0}^{t} \exp[\tfrac{1}{2}\sigma^2 s - \sigma B(s)] \, dt + X(0)$$

Then use $dX = Y \, dZ + Z \, dY + dY \, dZ$. Itō's formula applied to Y gives $dY = \sigma Y \, dB(t)$. And the SDE notation for Z is

$$dZ(t) = m \exp\left[\tfrac{1}{2}\sigma^2 t - \sigma B(t)\right] dt$$

Substition gives

$$dX(t) = \exp\left[-\tfrac{1}{2}\sigma^2 t + \sigma B(t)\right] m \exp\left[\tfrac{1}{2}\sigma^2 t - \sigma B(t)\right] dt$$
$$+ \left\{ X(0) + m \int_{s=0}^{t} \exp\left[\tfrac{1}{2}\sigma^2 s - \sigma B(s)\right] dt \right\}$$
$$\times \left\{ \sigma \exp\left[-\tfrac{1}{2}\sigma^2 t + \sigma B(t)\right] dB(t) \right\}$$
$$= m \, dt + \sigma X(t) \, dB(t)$$

as $dY \, dZ = 0$. The original SDE has been recovered.

CHAPTER 7

Answer [7.10.1]

x, a straight line

Answer [7.10.2]

Random variable X has probability distribution $N(0, 1)$ under probability $\mathbb{P}$. Its density under $\mathbb{P}$ is $1/\sqrt{2\pi} \exp\left[-\tfrac{1}{2}x^2\right]$ at $X = x$. Random variable Y is defined as $Y \stackrel{\text{def}}{=} X + \mu$. Density of Y under $\mathbb{P}$ at $X = x$ is $1/\sqrt{2\pi} \exp\left[-\tfrac{1}{2}(x - \mu)^2\right]$. Density of Y under $\widehat{\mathbb{P}}$ at $Y = y$ must be $1/\sqrt{2\pi} \exp\left[-\tfrac{1}{2}y^2\right]$.

$$(\text{density of } Y \text{ under } \widehat{\mathbb{P}}) = \frac{d\widehat{\mathbb{P}}}{d\mathbb{P}} \times (\text{density of } X \text{ under } \mathbb{P})$$

$$\frac{1}{\sqrt{2\pi}} \exp\left[-\tfrac{1}{2}y^2\right] = \frac{d\widehat{\mathbb{P}}}{d\mathbb{P}} \frac{1}{\sqrt{2\pi}} \exp\left[-\tfrac{1}{2}x^2\right]$$

For a particular realization $X = x$ and $Y = y$

$$\frac{d\widehat{\mathbb{P}}}{d\mathbb{P}} = \frac{1}{\sqrt{2\pi}} \exp\left[-\tfrac{1}{2}y^2\right] / \frac{1}{\sqrt{2\pi}} \exp\left[-\frac{1}{2}x^2\right]$$
$$= \exp\left[-\tfrac{1}{2}\{(x + \mu)^2 - x^2\}\right]$$
$$= \exp\left[-\tfrac{1}{2}(x^2 + 2x\mu + \mu^2 - x^2)\right]$$
$$= \exp\left[-\mu x - \tfrac{1}{2}\mu^2\right]$$

$d\widehat{\mathbb{P}}/d\mathbb{P}$ is the *positive* random variable $\exp\left[-\mu X - \frac{1}{2}\mu^2\right]$, a *function of random variable* X; it is the Radon–Nikodym derivative with respect to random variable X. Under $\mathbb{P}$, random variable Y has mean μ, but under $\widehat{\mathbb{P}}$ it has mean 0. The mean has been changed by a change in probability.

Verification
It is useful to verify whether the expected value of Y under $\widehat{\mathbb{P}}$ is indeed 0, when using the $d\widehat{\mathbb{P}}/d\mathbb{P}$ that has just been derived.

$$\mathbb{E}_{\widehat{\mathbb{P}}}[Y] = \mathbb{E}_{\mathbb{P}}\left[\frac{d\widehat{\mathbb{P}}}{d\mathbb{P}}Y\right] = \mathbb{E}_{\mathbb{P}}\left[\frac{d\widehat{\mathbb{P}}}{d\mathbb{P}}(X+\mu)\right]$$

where the right-hand side uses $X \sim N(0, 1)$.

$$\mathbb{E}_{\widehat{\mathbb{P}}}[Y] = \mathbb{E}_{\mathbb{P}}\left\{\exp\left[-\mu X - \frac{1}{2}\mu^2\right](X+\mu)\right\}$$
$$= \int_{x=-\infty}^{\infty} \exp\left[-\mu x - \frac{1}{2}\mu^2\right](x+\mu)\frac{1}{\sqrt{2\pi}}\exp\left[-\frac{1}{2}x^2\right] dx$$

where lower case x is a value of X. The exponent is

$$\left[-\mu x - \frac{1}{2}\mu^2 - \frac{1}{2}x^2\right] = -\frac{1}{2}[x^2 + 2x\mu + \mu^2] = -\frac{1}{2}(x+\mu)^2$$

so

$$\mathbb{E}_{\widehat{\mathbb{P}}}[Y] = \int_{x=-\infty}^{\infty} (x+\mu)\frac{1}{\sqrt{2\pi}}\exp\left[-\frac{1}{2}(x+\mu)^2\right] dx$$

Change variable to $z \stackrel{\text{def}}{=} x + \mu$; this has the same integration limits as x and $dz = dx$. Thus

$$\mathbb{E}_{\widehat{\mathbb{P}}}[Y] = \int_{x=-\infty}^{\infty} z\frac{1}{\sqrt{2\pi}}\exp\left[-\frac{1}{2}z^2\right] dz$$

which is the expected value of a standard normal and equals 0 as was to be shown.

It is also useful to verify that the density of Y under $\widehat{\mathbb{P}}$ is indeed standard normal.

density of Y under $\widehat{\mathbb{P}}$ equals $(\frac{d\widehat{\mathbb{P}}}{d\mathbb{P}}|_{X=x}) \times$ (density of X under $\mathbb{P}$)

$$= \exp\left[-\mu x - \frac{1}{2}\mu^2\right]\frac{1}{\sqrt{2\pi}}\exp\left[-\frac{1}{2}x^2\right]$$

The exponent is

$$\left[-\mu x - \tfrac{1}{2}\mu^2 - \tfrac{1}{2}x^2\right] = -\tfrac{1}{2}[x^2 + 2x\mu + \mu^2]$$
$$= -\tfrac{1}{2}[x + \mu]^2 = -\tfrac{1}{2}y^2$$

Density of Y under $\widehat{\mathbb{P}}$ is $(1/\sqrt{2\pi})\exp[-\tfrac{1}{2}y^2]$, as was to be shown.

Answer [7.10.3]

To go from $\widehat{\mathbb{P}}$ to $\mathbb{P}$ requires

$$\frac{d\mathbb{P}}{d\widehat{\mathbb{P}}} = (\text{density of } X \text{ under } \mathbb{P})/(\text{density of } Y \text{ under } \widehat{\mathbb{P}})$$

$$= \frac{1}{\sqrt{2\pi}}\exp\left[-\tfrac{1}{2}x^2\right]\Big/\tfrac{1}{\sqrt{2\pi}}\exp\left[-\tfrac{1}{2}y^2\right]$$

If $Y = y$ then $x = y - \mu$. Substituting x gives

$$\exp\left[-\tfrac{1}{2}(y - \mu)^2\right]\Big/\exp\left[-\tfrac{1}{2}y^2\right]$$

and after rearranging, $d\mathbb{P}/d\widehat{\mathbb{P}} = \exp\left[\mu y - \tfrac{1}{2}\mu^2\right]$ when $Y = y$. In general,

$$\frac{d\mathbb{P}}{d\widehat{\mathbb{P}}} = \exp\left[\mu Y - \tfrac{1}{2}\mu^2\right]$$

a *positive function of random variable* Y. $d\mathbb{P}/d\widehat{\mathbb{P}}$ is the Radon–Nikodym derivative with respect to random variable Y.

Verification of Solution
Verify that $\mathbb{E}_{\mathbb{P}}[X] = 0$.

$$\mathbb{E}_{\mathbb{P}}[X] = \mathbb{E}_{\widehat{\mathbb{P}}}\left[\frac{d\mathbb{P}}{d\widehat{\mathbb{P}}}X\right] = \mathbb{E}_{\widehat{\mathbb{P}}}\left[\frac{d\mathbb{P}}{d\widehat{\mathbb{P}}}(Y - \mu)\right]$$

where

$$\frac{d\mathbb{P}}{d\widehat{\mathbb{P}}} = \exp\left[\mu Y - \tfrac{1}{2}\mu^2\right] \text{ and } Y \sim N(0, 1) \text{ under } \widehat{\mathbb{P}}$$

$$= \int_{y=-\infty}^{\infty} \exp\left[\mu y - \tfrac{1}{2}\mu^2\right](y - \mu)\frac{1}{\sqrt{2\pi}}\exp\left[-\tfrac{1}{2}y^2\right]dy$$

The exponent equals

$$\left[\mu y - \tfrac{1}{2}\mu^2 - \tfrac{1}{2}y^2\right] = -\tfrac{1}{2}[y^2 - 2y\mu + \mu^2] = -\tfrac{1}{2}(y - \mu)^2$$

The integral equals $\int_{y=-\infty}^{\infty}(y-\mu)\exp\left[-\frac{1}{2}(y-\mu)^2\right]dy$. Change to variable $w \stackrel{\text{def}}{=} y-\mu$. This has the same integration limits as y, and $dw = dy$. The integral is then

$$\int_{w=-\infty}^{\infty} w \frac{1}{\sqrt{2\pi}} \exp\left[-\frac{1}{2}w^2\right] dw$$

which is the expected value of a standard normal and equals 0, so $\mathbb{E}_\mathbb{P}[X] = 0$, as was to be shown. Verification that the density of X under $\mathbb{P}$ is $N(0, 1)$ is done in the same way as in Exercise [7.10.2].

Answer [7.10.4]

Over the time period $[0, T]$, the random terminal stock price $S(T)$ is related to the given initial stock price $S(0) = 1$ by

$$S(T) = \exp\left[\left(\mu - \frac{1}{2}\sigma^2\right) T + \sigma B(T)\right]$$

under original probability $\mathbb{P}$, where μ is growth rate of stock price. Under risk-neutral probability $\widehat{\mathbb{P}}$ it is

$$\widehat{S}(T) = \exp\left[\left(r - \frac{1}{2}\sigma^2\right) T + \sigma \widehat{B}(T)\right]$$

where r is risk-free interest rate.

(a) Expected value of $S(T)$ under $\mathbb{P}$. Under $\mathbb{P}$, $S(T) = \exp[(\mu - \frac{1}{2}\sigma^2)T + \sigma B(T)]$. It is convenient to introduce separate notation for the exponent

$$X(t) \stackrel{\text{def}}{=} \left(\mu - \frac{1}{2}\sigma^2\right) T + \sigma B(T)$$

$X(t)$ is normal because $B(T)$ is normal, and rescaling by σ and adding constant $\left(\mu - \frac{1}{2}\sigma^2\right) T$ maintains normality. Thus

$$\mathbb{E}[S(t)] = \mathbb{E}[\exp[X(t)]] = \exp\left\{\mathbb{E}[X(t)] + \frac{1}{2}\mathbb{V}\text{ar}[X(t)]\right\}$$

where $\mathbb{E}[X(t)] = \left(\mu - \frac{1}{2}\sigma^2\right) T$ and $\mathbb{V}\text{ar}[X(t)] = \sigma^2 T$.

$$\mathbb{E}[S(t)] = \exp\left[\left(\mu - \frac{1}{2}\sigma^2\right) T + \frac{1}{2}\sigma^2 T\right] = \exp(\mu T)$$

(b) Radon–Nikodym derivative $d\widehat{\mathbb{P}}/d\mathbb{P}$ transforms the density of $B(t)$ to the density of $\widehat{B}(T)$. First step is to derive dynamics of S under both probabilities. $S(T)$ is a function of T and $B(T)$. Using Itō's formula

$$\frac{dS(T)}{S(T)} = m\,dT + \sigma\,dB(T)$$

where $m = \mu$ under $\mathbb{P}$ and $m = r$ under $\widehat{\mathbb{P}}$; details of derivation omitted here. Write the drift of $dS(T)/S(T)$ under $\mathbb{P}$ as

$$[r + (\mu - r)]\,dT = r\,dT + (\mu - r)\,dT$$

Apply the transformation $\sigma\,\widehat{B}(T) \stackrel{\text{def}}{=} (\mu - r)\,T + \sigma\,B(T)$, divide by σ, let $\varphi \stackrel{\text{def}}{=} (\mu - r)/\sigma$, then

$$\widehat{B}(T) = \varphi T + \sigma B(T) \quad d\widehat{B}(T) = \varphi\,dT + \sigma\,dB(T)$$

According to Girsanov, $\widehat{B}$ is a standard Brownian motion under probability $\widehat{\mathbb{P}}$. Let x denote a value of random variable $B(T)$, then $x + \varphi T$ is the corresponding value of $\widehat{B}(T)$. Under $\mathbb{P}$, at $B(T) = x$, $d\widehat{\mathbb{P}}/d\mathbb{P} = (\text{density of } \widehat{B})/(\text{density of } B)$

$$\frac{1}{\sqrt{T}\sqrt{2\pi}} \exp\left[-\frac{1}{2}\left(\frac{x + \varphi T}{\sqrt{T}}\right)^2\right] \Big/ \frac{1}{\sqrt{T}\sqrt{2\pi}} \exp\left[-\frac{1}{2}\left(\frac{x}{\sqrt{T}}\right)^2\right]$$

$$= \exp\left[-\frac{1}{2}\left(\frac{x + \varphi T}{\sqrt{T}}\right)^2 + \frac{1}{2}\left(\frac{x}{\sqrt{T}}\right)^2\right]$$

$$= \exp\left[-\varphi x - \tfrac{1}{2}\varphi^2 T\right]$$

$d\widehat{\mathbb{P}}/d\mathbb{P} = \exp\left[-\varphi B(T) - \tfrac{1}{2}\varphi^2 T\right]$ in terms of $B(T)$

(c) From the expected value of $\widehat{S}(T)$ under $\widehat{\mathbb{P}}$ to the expected value of $S(T)$ under $\mathbb{P}$:

$$\widehat{\mathbb{E}}[\widehat{S}(T)] = \exp(rT) \text{ by the same method as used in (a).}$$

Transformation from $\widehat{\mathbb{P}}$ to $\mathbb{P}$ requires the Radon–Nikodym derivative $d\mathbb{P}/d\widehat{\mathbb{P}}$.

$$\frac{d\mathbb{P}}{d\widehat{\mathbb{P}}} = 1 \Big/ \frac{d\widehat{\mathbb{P}}}{d\mathbb{P}} = \exp\left[\varphi B(T) + \tfrac{1}{2}\varphi^2 T\right]$$

in terms of $B(T)$. To use this under $\widehat{\mathbb{P}}$, it has to be in terms $\widehat{B}(T)$, so substitute $B(T) = \widehat{B}(T) - \varphi T$, giving

$$\frac{d\mathbb{P}}{d\widehat{\mathbb{P}}} = \exp\left[\varphi\{\widehat{B}(T) - \varphi T\} + \tfrac{1}{2}\varphi^2 T\right]$$

$$= \exp\left[\varphi\widehat{B}(T) - \tfrac{1}{2}\varphi^2 T\right]$$

Note that the second term in the exponential has changed sign. $\widehat{\mathbb{E}}[\widehat{S}(T)\frac{d\mathbb{P}}{d\widehat{\mathbb{P}}}]$ produces $\mathbb{E}[S(T)]$.

$$\widehat{S}(T)\frac{d\mathbb{P}}{d\widehat{\mathbb{P}}} = \exp\left[\left(r - \tfrac{1}{2}\sigma^2\right)T + \sigma\widehat{B}(T)\right]\exp\left[\varphi\widehat{B}(T) - \tfrac{1}{2}\varphi^2 T\right]$$

Let the exponent

$$Y(T) \stackrel{\text{def}}{=} \left(r - \tfrac{1}{2}\sigma^2\right)T + \sigma\widehat{B}(T) + \varphi\widehat{B}(T) - \tfrac{1}{2}\varphi^2 T$$

$$= rT + (\sigma + \varphi)\widehat{B}(T) - \tfrac{1}{2}(\sigma^2 + \varphi^2)T$$

$Y(T)$ is normal

$$\widehat{\mathbb{E}}[Y(T)] = rT - \tfrac{1}{2}(\sigma^2 + \varphi^2)T \qquad \widehat{\text{Var}}[Y(T)] = (\sigma + \varphi)^2 T$$

$$\widehat{\mathbb{E}}\left[\widehat{S}(T)\frac{d\mathbb{P}}{d\widehat{\mathbb{P}}}\right] = \widehat{\mathbb{E}}\{\exp[Y(T)]\} = \exp\left\{\widehat{\mathbb{E}}[Y(T)] + \tfrac{1}{2}\widehat{\text{Var}}[Y(T)]\right\}$$

$$= \exp\left\{rT - \tfrac{1}{2}(\sigma^2 + \varphi^2)T + \tfrac{1}{2}(\sigma + \varphi)^2 T\right\}$$

$$= \exp(rT + \sigma\varphi T)$$

Using $\varphi \stackrel{\text{def}}{=} (\mu - r)/\sigma$ gives

$$\exp\left(rT + \sigma\frac{\mu - r}{\sigma}T\right) = \exp(rT + \mu T - rT) = \exp(\mu T)$$

$$\mathbb{E}[S(T)] = \widehat{\mathbb{E}}\left[\widehat{S}(T)\frac{d\mathbb{P}}{d\widehat{\mathbb{P}}}\right] = \exp(\mu T)$$

as was to be shown.

Answer [7.10.5]

The random process W defined by the Girsanov tranformation $W(t) \stackrel{\text{def}}{=} B(t) + \varphi t$ is a Brownian motion under the new probability distribution

$\widehat{\mathbb{P}}$ that is created from the orginal probability distribution $\mathbb{P}$ of $B(t)$ by the Radon–Nikodym derivative

$$Z(t) = \exp\left[-\varphi B(t) - \tfrac{1}{2}\varphi^2 t\right] = \frac{d\widehat{\mathbb{P}}}{d\mathbb{P}}$$

with $\mathbb{E}[Z(t)] = 1$. Itō's formula applied to Z as a function of variables t and $B(t)$ readily gives the geometric Brownian motion expression

$$\frac{dZ(t)}{Z(t)} = -\varphi\, dB(t)$$

As this SDE has no drift, Z is a martingale. As $W(t)$ is a Brownian motion under $\widehat{\mathbb{P}}$, it should have $\widehat{\mathbb{E}}[W(t)] = 0$. This is now verified via $\widehat{\mathbb{E}}[W(t)] = \mathbb{E}[W(t)Z(t)]$. Note that $W(t)$ and $Z(t)$ are both functions of the same random variable $B(t)$ of which the probability density under $\mathbb{P}$ is known. At $B(t) = x$, $W(t)Z(t)$ has the value $(x + \varphi t)\exp\left[-\varphi x - \tfrac{1}{2}\varphi^2 t\right]$ and the probability density is

$$\frac{1}{\sqrt{t}\sqrt{2\pi}}\exp\left[-\frac{1}{2}\left(\frac{x}{\sqrt{t}}\right)^2\right]$$

Thus

$$\mathbb{E}[W(t)Z(t)] = \int_{x=-\infty}^{\infty}(x + \varphi t)\exp\left[-\varphi x - \tfrac{1}{2}\varphi^2 t\right]$$
$$\times \underbrace{\frac{1}{\sqrt{t}\sqrt{2\pi}}\exp\left[-\frac{1}{2}\left(\frac{x}{\sqrt{t}}\right)^2\right]}_{\text{density of } B(t)} dx$$

Changing the density to standard normal by $u = x/\sqrt{t}$ gives

$$\int_{x=-\infty}^{\infty}(u\sqrt{t} + \varphi t)\exp\left[-\varphi u\sqrt{t} - \tfrac{1}{2}\varphi^2 t\right]\frac{1}{\sqrt{2\pi}}\exp\left[-\tfrac{1}{2}u^2\right]du$$

The exponent can be combined to $-\tfrac{1}{2}(u + \varphi\sqrt{t})^2$ so the above can be written as

$$= \int_{x=-\infty}^{\infty}(u\sqrt{t} + \varphi t)\frac{1}{\sqrt{2\pi}}\exp\left[-\tfrac{1}{2}(u + \varphi\sqrt{t})^2\right]du$$
$$= \sqrt{t}\int_{x=-\infty}^{\infty}u\frac{1}{\sqrt{2\pi}}\exp\left[-\tfrac{1}{2}(u + \varphi\sqrt{t})^2\right]du$$
$$+\varphi t\int_{x=-\infty}^{\infty}\frac{1}{\sqrt{2\pi}}\exp\left[-\tfrac{1}{2}(w + \varphi\sqrt{t})^2\right]dw$$

As the first integral is the expected value of a normal random variable with mean $-\varphi\sqrt{t}$, it equals $-\varphi\sqrt{t}$. As the second integral is the area under the probability density of that random variable, it equals 1. Thus

$$\widehat{\mathbb{E}}[W(t)] = \sqrt{t}(-\varphi)\sqrt{t} + \varphi t 1 = 0$$

as was to be shown.

Answer [7.10.6]

(a) Under original probability $\mathbb{P}$

$$S^*(T) = S(0)\exp\left[(\mu - r - \tfrac{1}{2}\sigma^2)T + \sigma B(T)\right]$$
$$= S(0)\exp\left[(\mu - r - \tfrac{1}{2}\sigma^2)T\right]\exp\left[\sigma B(T)\right]$$

and

$$Z(T) = \exp\left[-\varphi B(T) - \tfrac{1}{2}\varphi^2 T\right]$$

So

$$Z(T)S^*(T) = S(0)\exp\left[(\mu - r - \tfrac{1}{2}\sigma^2 - \tfrac{1}{2}\varphi^2)T\right]$$
$$\times\exp\left[(\sigma - \varphi)B(T)\right]$$

Then

$$\mathbb{E}_{\mathbb{P}}[Z(T)S^*(T)]$$
$$= S(0)\exp\left[(\mu - r - \tfrac{1}{2}\sigma^2 - \tfrac{1}{2}\varphi^2)T\right]\mathbb{E}_{\mathbb{P}}\left[\exp\left[(\sigma - \varphi)B(T)\right]\right]$$

With $\mathbb{E}[\exp[(\sigma - \varphi)B(T)]] = \exp[\tfrac{1}{2}(\sigma - \varphi)^2 T] = \exp[\tfrac{1}{2}\sigma^2 T - \sigma\varphi T + \tfrac{1}{2}\varphi^2 T]$. As $\sigma\varphi = \sigma\frac{\mu-r}{\sigma} = \mu - r$, $\mathbb{E}[\exp[(\sigma - \varphi)B(T)]] = \exp[\tfrac{1}{2}\sigma^2 T - (\mu - r)T + \tfrac{1}{2}\varphi^2 T]$ and $\mathbb{E}[Z(T)S^*(T)] = S(0)$, due to cancellations in the exponents.

(b) Under original probability $\mathbb{P}$

$$V^*(T) = \exp(-rT)V(T)$$
$$= \exp(-rT)\left[\alpha(T)S(T) + \exp(-rT)\right]$$
$$= \alpha(T)\exp(-rT)S(T) + 1$$
$$= \alpha(T)S^*(T) + 1$$

So

$$Z(T)V^*(T) = [\alpha(t)Z(T)S^*(t) + Z(T)]$$

and

$$\begin{aligned}
\mathbb{E}[Z(T)V^*(T)] &= \mathbb{E}[\alpha(t)Z(T)S^*(t)] + \mathbb{E}[Z(T)] \\
&= \alpha(t)\mathbb{E}[Z(T)S^*(t)] + 1 \\
&= \alpha(t)S(0) + 1 \\
&= V(0)
\end{aligned}$$

(c) $$\begin{aligned}
\mathbb{E}_{\mathbb{P}}[Z(T)V^*(T)] &= \int_{x=-\infty}^{\infty} V^*(T)\underbrace{\exp\left[-\varphi x - \tfrac{1}{2}\varphi^2 T\right] d\mathbb{P}}_{d\widehat{\mathbb{P}}} \\
&= \int_{x=-\infty}^{\infty} V^*(T)d\widehat{\mathbb{P}} \\
&= \widehat{\mathbb{E}}[V^*(T)]
\end{aligned}$$

since $Z(T) = d\widehat{\mathbb{P}}(T)/d\mathbb{P}(T)$.

Answer [7.10.7]

Original probability mass

$$\mathbb{P}[X = k] = \frac{\lambda_1^k}{k!}\exp(-\lambda_1)$$

New probability mass

$$\widehat{\mathbb{P}}[X = k] = \frac{\lambda_2^k}{k!}\exp(-\lambda_2)$$

Radon–Nikodym derivative

$$\frac{d\widehat{\mathbb{P}}}{d\mathbb{P}} = \frac{\lambda_2^k}{k!}\exp(-\lambda_2)\Big/\frac{\lambda_1^k}{k!}\exp(-\lambda_1) = \left(\frac{\lambda_2}{\lambda_1}\right)^k \exp[-(\lambda_2 - \lambda_1)]$$

Answer [7.10.8]

At time 0, the barrier starts at level L. At time t the vertical position of the barrier is $L + \mu t$. The first passage time is the earliest time at which $B(t) = L + \mu t$. Then $B(t) - \mu t = L$, which is the same as the process $B(t) - \mu t$ first hitting *horizontal* barrier L. Introduce the new Brownian motion $\widehat{B}(t) \overset{\text{def}}{=} B(t) - \mu t$ under the new probability $\widehat{\mathbb{P}}$. Then $\widehat{B}(t)$ hitting the horizontal barrier is same as $B(t)$ hitting the sloped barrier

$$\widehat{B}(t) = L \Leftrightarrow B(t) - \mu t = L$$

Let T_{hor} denote the random time of first hitting a horizontal barrier at level L. Its probability density at $T_{\text{hor}} = t$ is known, say $f(t)$; the expression is derived in Annex A, Section A.4, but is not needed to answer this question. Note that $\widehat{B}(t)$ is here the 'original' Brownian motion for which the density of first passage of a horizontal barrier is known. So the Girsanov transformation is from $\widehat{B}(t)$ to $B(t)$

$$B(t) = \widehat{B}(t) + \mu t$$

As $\widehat{B}(t)$ is under $\widehat{\mathbb{P}}$, density f is under $\widehat{\mathbb{P}}$. The Radon–Nikodym derivative for going from $\widehat{B}(t)$ to $B(t)$ is

$$\frac{d\mathbb{P}}{d\widehat{\mathbb{P}}} = \exp\left[-\tfrac{1}{2}\mu^2 t - \mu\widehat{B}(t)\right]$$

This is in terms of the 'original' Brownian motion, $\widehat{B}(t)$, and both coefficients in the exponent have a minus sign because the term μt in the Girsanov transformation has a plus sign. At the time of hit, $\widehat{B}(t) = L$, and the probability density is at that time, so $d\mathbb{P}/d\widehat{\mathbb{P}}$ at $\widehat{B}(t) = L$ is $\exp\left(-\tfrac{1}{2}\mu^2 t - \mu L\right)$. The probability density of the time of first hit of the sloping barrier is therefore

$$f(t) \exp\left(-\tfrac{1}{2}\mu^2 t - \mu L\right)$$

which can be verified by simulation.

References

Probability Theory

(listing in this section in order of increasing technical level and scope)

Haigh, J. *Probability Models*. ISBN 1-85233-431-2. Springer 'Undergraduate Mathematics Series' 2002. Elementary. Fully worked out solutions to exercises.

Ross, S. *A First Course in Probability* 7/ed. ISBN 0132018179. Prentice Hall 2005. Elementary text on standard computational probability theory. Highly readable, widely used.

Foata, D. and Fuchs, A. *Calcul des probabilités* 2e éd. ISBN 2-10-004104-5. Dunod 1998. Excellent French textbook on computational probability with accessible coverage of basic advanced probability. In France aimed at students in *Écoles d'ingénieurs*.

Jacod, J. and Protter, P. *Probability Essentials* 2/ed. ISBN 3-540-66419-X. Springer 'Universitext' 2004. Compact and comprehensive textbook on conceptual probability 'at the beginning graduate level'. The material is laid out in 28 short topics. Recommended for the mathematically inclined reader.

Stochastic Calculus

Bass, R. *The Basics of Financial Mathematics*. Lecture notes available on the author's website *www.math.uconn.edu/~bass*, section 'Lecture notes', entry 'Financial mathematics'. Permission by Professor Bass to include this reference is gratefully acknowledged.

Brzeźniak, Z. and Zastawniak, T. *Basic Stochastic Processes*. ISBN 3-540-76175-6. Springer 'Undergraduate Mathematics Series' 7th printing 2005. Much coverage of conditional expectations. No applications to option valuation. Fully worked out solutions to exercises.

Capasso, V. and Bakstein, D. *An Introduction to Continuous-Time Stochastic Processes; Theory, Models, and Applications to Finance, Biology, and Medicine*. ISBN 0-8176-3234-4. Birkhäuser 2005. Concise exposition for the reader who prefers rigorous mathematics and is comfortable with the real analysis and advanced probability theory this entails.

Klebaner, F.C. *Introduction to Stochastic Calculus with Applications* 2/ed. ISBN 1-86094-566-X. World Scientific Publishing 2005. Includes overview of prerequisite real analysis and probability. Uses convergence in probability. Also covers semi-martingales. For the mathematically inclined reader. Two chapters on option valuation and interest rate modelling. Answers to most exercises.

Kuo, H.-H. *Introduction to Stochastic Integration*. ISBN 0-387-28720-5. Springer 'Universitext' 2006. Well motivated exposition 'for anyone who wants to learn Itô calculus in a short period of time'. Includes chapter on option valuation. Highly recommended mathematics.

Lin, X.S. *Introductory Stochastic Analysis for Finance and Insurance*. ISBN 0-471-71642-1. Wiley Series in Probability and Statistics 2006. Nice elementary introduction.

Mikosch, T. *Elementary Stochastic Calculus with Finance in view*. ISBN 981-02-3543-7. World Scientific Publishing 1998. Earliest elementary (mostly) text with applications in finance.[1]

Shreve, S. *Stochastic Calculus for Finance II, Continuous-Time Models*. ISBN 0-387-40101-6. Springer 'Finance' 2004. Introduces all probability concepts and stochastic calculus in a finance context with intuitive explanation, and numerous examples. Also covers calculus of jump processes and its use in option valuation.

Option Valuation – Binomial

Roman, S. *Introduction to the Mathematics of Finance; From Risk Management to Options Pricing*. ISBN 0-387-21364-3. Springer 'Undergraduate Texts in Mathematics' 2004. Mostly models in discrete-time framework. Very readable coverage of probability concepts, both discrete and continuous, that are of direct relevance to financial mathematics. 'The mathematics is not watered down but is appropriate for the intended audience. No measure theory is used and only a small amount of linear algebra is required. All necessary probability theory is developed in several chapters throughout the book, on an as-needed basis.'

Shreve, S. *Stochastic Calculus for Finance I, The Binomial Asset Pricing Model*. ISBN 0-387-40100-8. Springer 'Finance' 2004. Stepping stone for continuous-time modelling.

Option Valuation – General

with coverage of Stochastic Calculus

Benth, F.E. *Option Theory with Stochastic Analysis*. ISBN 3-540-40502-X. Springer 'Universitext' 2004. Very readable and engaging elementary yet comprehensive

---

[1] Earliest text that focused on applications is *Stochastic Differential Equations: Theory and Applications* by *L. Arnold*, 1973. Applications from engineering. Now a classic, out of print, but in many libraries.

introduction to mathematical finance which makes ideal companion reading for option valuation methodology in *Brownian Motion Calculus*. Also covers basics of simulation techniques (in Excel/VBA), enough for various simulation exercises in *Brownian Motion Calculus*. Fully worked out answers to all exercises. Highly recommended.

Björk,T. *Arbitrage Theory in Continuous Time* 2/ed. ISBN 0-19-927126-7. Oxford University Press 2004. Well-known textbook on option theory and interest rate modelling. Introductory chapters on stochastic calculus. Technical appendix on conceptual probability theory.

Dana, R.-A. and Jeanblanc, M. *Financial Markets in Continuous Time.* ISBN 3-540-43403-8. Springer 'Finance' series. Mainly on discrete time and continuous time valuation of equity based options, and interest rate modelling. Also coverage of general equilibrium theory for financial markets. Nice informal coverage of stochastic calculus in appendices. Brief exposition of numerical methods for PDEs and simulation concepts.

Epps, T.W. *Pricing Derivative Securities*, 2/ed. ISBN 981-02-4298-0. World Scientific Publishing 2007. Major broad ranging textbook on options theory and interest rate modelling. Initial chapters review probability concepts and stochastic calculus. Also numerical methods and programs.

Jiang, L. *Mathematical Modeling and Methods of Option Pricing.* ISBN 981-256-369-5. World Scientific Publishing 2005. Comprehensive book on PDE approach to option valuation.

Korn, R. and Korn, E. *Option Pricing and Portfolio Optimization: Modern Methods of Financial Mathematics.* ISBN 0-8218-2123-7. American Mathematical Society 2001. Detailed expositions of stochastic calculus concepts with full proofs in a sequence of 'excursions'. Attractive reference on rigorous mathematical aspects for the persistent reader.

Lamberton, L. and Lapeyre. B. *Introduction to Stochastic Calculus applied to Finance*, 2/ed. ISBN 978-1-584-88626-6. Chapman & Hall 'Financial Mathematics Series' 2007. Concise coverage of basic stochastic calculus, standard equity option valuation, interest rate modelling, option valuation under a jump process, and numerical methods.

Wilmott, P., Howison, S. and Dewynne, J. *The Mathematics of Financial Derivatives – A Student Introduction.* ISBN 0-521-49789-2. Cambridge University Press 1995. Very readable introduction to PDE approach for equity options and some interest rate modelling. Excellent coverage of numerical methods for American options.

Interest Rate Modelling

Cairns, A.J.G. *Interest Rate Models, An Introduction.* ISBN 0-691-11894-9. Princeton University Press 2004. Comprehensive introduction to main models in textbook format. Appendix recaps prerequisite stochastic calculus.

Zagst, R. *Interest Rate Management.* ISBN 3-540-67594-9. Springer 'Finance' 2002. Fine precise review of stochastic calculus in two initial chapters. Will appeal to mathematically oriented readers.

Simulation, etc.

Asmussen, S. and Glynn, P. W. *Stochastic Simulation: Algorithms and Analysis*. ISBN 0-387-30679X. Springer 'Stochastic Modelling and Applied Probability' 2007.

Glasserman, P. *Monte-Carlo Methods for Financial Engineering*. ISBN 0-387-00451-3. Springer 'Stochastic Modelling and Applied Probability' 2003. Authoritative text on simulation methods for finance.

McLeish, D.L. *Monte Carlo Simulation & Finance*. ISBN 0-471-67778-7. Wiley Finance 2005. Entertaining, easily readable exposition with many examples. Supporting appendices on author's University of Waterloo website.

Stojanovic, S. *Computational Financial Mathematics using MATHEMATICA*. ISBN 0-8176-4197-1. Birkhäuser 2003.

Real Analysis

Abbott, S. *Understanding Analysis*. ISBN 978-0-387-95060-0. Springer 'Undergraduate Texts in Mathematics' 2002. Title speaks for itself. Explains wonderfully what those well-known concepts of real analysis really mean.

Brannan, D.A. *A First Course in Mathematical Analysis*. ISBN 0-521-68424-2. Cambridge University Press 2006. Very well presented. Ideal for brushing up on basic concepts of convergence and integration. Even discusses non-differentiable functions!

Index

Index compiled by Terry Halliday

Printed and bound by CPI Group (UK) Ltd, Croydon, CR0 4YY

16/04/2025

14658502-0002